T0297783

Cellular Internet of Things

Cellular Internet of Things
Technologies, Standards, and Performance

Olof Liberg
Ericsson Business Unit Networks
Mårten Sundberg
Ericsson Business Unit Networks
Y.-P. Eric Wang
Ericsson Research
Johan Bergman
Ericsson Business Unit Networks
Joachim Sachs
Ericsson Research

ACADEMIC PRESS
An imprint of Elsevier

Academic Press is an imprint of Elsevier
125 London Wall, London EC2Y 5AS, United Kingdom
525 B Street, Suite 1800, San Diego, CA 92101-4495, United States
50 Hampshire Street, 5th Floor, Cambridge, MA 02139, United States
The Boulevard, Langford Lane, Kidlington, Oxford OX5 1GB, United Kingdom

Library of Congress Cataloging-in-Publication Data
A catalog record for this book is available from the Library of Congress

British Library Cataloguing-in-Publication Data
A catalogue record for this book is available from the British Library

ISBN: 978-0-12-812458-1

For information on all Academic Press publications visit our
website at https://www.elsevier.com/books-and-journals

Working together
to grow libraries in
developing countries

www.elsevier.com • www.bookaid.org

Publisher: Mara E. Conner
Acquisition Editor: Tim Pitts
Editorial Project Manager: Charlotte Kent
Production Project Manager: Mohana Priyan Rajendran
Designer: Greg Harris

Typeset by TNQ Books and Journals

Contents

About the Authors

Olof Liberg is a Master Researcher at Ericsson Business Unit Networks. After studies in Sweden, USA, Germany, and Switzerland, he received a bachelor's degree in Business and Economics and a master's degree in Engineering Physics, both from Uppsala University, Sweden. He joined Ericsson in 2008 and has in recent years specialized in the design and standardization of cellular systems for machine-type communications and Internet of Things. He has, over the years, actively contributed to the work in several standardization bodies such as 3GPP, ETSI, and the MulteFire Alliance. He was the chairman of 3GPP TSG GERAN and its Working Group 1, during the 3GPP prestudy on new radio access technologies for Internet of Things leading up to the specification of EC-GSM-IoT and NB-IoT.

Mårten Sundberg is a Senior Specialist in GSM radio access technology at Ericsson Business Unit Networks. After studies at Uppsala University in Sweden, receiving a master's degree in Engineering Physics, he joined Ericsson in 2005 and has continued his work at Ericsson with focus on physical layer and RF-related standardization as a 3GPP delegate since 2006, and later also as a delegate in the ETSI Technical Committees Mobile Standards Group. As Rapporteur of the 3GPP TSG GERAN Work Item on EC-GSM-IoT, he led the technical work to standardize the new GSM-based features dedicated for Internet of Things.

Johan Bergman is a Master Researcher at Ericsson Business Unit Networks. He received his master's degree in Engineering Physics from Chalmers University of Technology in Sweden. He joined Ericsson in 1997, initially working with baseband receiver algorithm design for 3G cellular systems. Since 2005, he has been working with 3G/4G physical layer standardization in 3GPP TSG RAN Working Group 1. He has contributed to more than 50 US patents. As Rapporteur of the 3GPP TSG RAN Work Items on LTE for machine-type communications (MTC) in Releases 13, 14, and 15, he has led the technical work to standardize the new LTE-based features dedicated for Internet of Things.

Y.-P. Eric Wang is a Principal Researcher at Ericsson Research. He holds a PhD degree in electrical engineering from the University of Michigan, Ann Arbor. In 2001 and 2002, he was a member of the executive committee of the IEEE Vehicular Technology Society and served as the society's Secretary. Dr. Wang was an Associate Editor of the IEEE Transactions on Vehicular Technology from 2003 to 2007. He is a technical leader in Ericsson Research in the area of Internet of Things (IoT) connectivity. Dr. Wang was a corecipient of Ericsson's Inventors of the Year award in 2006. He has contributed to more than 100 US patents and more than 50 IEEE articles.

Joachim Sachs is a Principal Researcher at Ericsson Research. After studies in electrical engineering in Germany, France, Norway, and Scotland, he received a diploma degree from RWTH Aachen University, Germany, and a PhD from Technical University of Berlin. He is with Ericsson since 1997 and spent a sabbatical as visiting scholar with Stanford University in 2009. Dr. Sachs has worked on various topics in the area of wireless communication systems, and in recent years with a focus on 4G and 5G design for machine-type communications and Internet of Things. Since 1995, he has been active in the IEEE and the German VDE Information Technology Society, where he is currently cochair of the technical committee on communication networks and systems. He serves on the editorial board of the *IEEE Internet of Things Journal*. In 2006, he received the Ericsson Inventor of the Year award and in 2010, the research award of the Vodafone foundation for scientific research.

Preface

The Internet of Things is the latest rising star in the information and communications technology industry. It embodies the vision of connecting virtually anything with everything and builds on a global growth of the overall number of connected devices. To support, and perhaps further boost, this growth the *Third Generation Partnership Project* (3GPP) standards development organization has in its Release 13 developed three new technologies known as *Extended Coverage GSM Internet of Things* (EC-GSM-IoT), *LTE for Machine-Type Communications* (LTE-M), and *Narrowband Internet of Things* (NB-IoT) dedicated to providing cellular services to a massive number of IoT devices.

This book sets out to introduce, characterize, and, when relevant, in detail describe these three new technologies, defining a new concept known as the *Cellular Internet of Things*. To start with, Chapter 1 introduces the overall content of the book. Chapter 2 provides an overview of the accomplishments of 3GPP in the field of Machine-Type Communications until Release 13. Chapters 3, 5, and 7 describe the design objectives as well as the technical foundations of EC-GSM-IoT, LTE-M, and NB-IoT. Both physical layer details and idle and connected mode procedures are presented.

Chapters 4, 6, and 8 go through the performance objectives of the three technologies and present extensive evaluation results in terms of coverage, throughput, latency, battery lifetime, system capacity, and device complexity. The details behind each of these performance objectives including the methodologies for evaluating them are presented.

Chapter 9 summarizes the findings of the earlier chapters and describes the merits of the three technologies and provides a comparison between them. The chapter also introduces a set of solutions for short- and long-range communications operating in unlicensed frequency bands. The advantages of licensed operation over unlicensed operation are highlighted.

Finally, Chapter 10 gives a glance of how the Cellular Internet of Things fits into the vision of the fifth generation (5G) wireless communication system.

Acknowledgments

We would like to thank all our colleagues in Ericsson who have contributed to the development of wireless IoT technologies including EC-GSM-IoT, LTE-M, and NB-IoT. Without their efforts this book would never have been written. In particular, we would like to express our gratitude to our friends John Diachina, Björn Hofström, Zhipeng Lin, Ulf Händel, Nicklas Johansson, Xingqin Lin, Uesaka Kazuyoshi, Sofia Ek, Andreas Höglund, Björn Nordström, Emre Yavuz, Håkan Palm, Mattias Frenne, Oskar Mauritz, Tuomas Tirronen, Santhan Thangarasa, Bela Rathonyi, Anna Larmo, Johan Torsner, Erika Tejedor, Ansuman Adhikary, Yutao Sui, Jonas Kronander, Gerardo Agni Medina Costa, and Chenguang Lu for their help in reviewing and improving the content of the book.

We would also like to express our sincere appreciation to Johan Sköld. Without his advice and encouragement in the early phases of our work, this project would never have taken off.

Further we would like to thank our colleagues in 3GPP for contributing to the successful standardization of EC-GSM-IoT, LTE-M, and NB-IoT. In particular, we would like to thank Alberto Rico-Alvarino, Debdeep Chatterjee, and Matthew Webb for their help in reviewing the technical details of the book.

Finally, we would like to express our admiration for our families: Ellen, Hugo, and Flora; Katharina, Benno, Antonia, and Josefine; Wan-Ling, David, Brian, Kuo-Hsiung and Ching-Chih. It is needless to say how important their inspiration and support have been during the long process of writing this book.

THE CELLULAR INTERNET OF THINGS

1

CHAPTER OUTLINE

Abstract

This chapter introduces the overall content of the book. It contains a brief introduction to the massive Machine-Type Communications (mMTC) category of use cases, spanning a wide range of applications such as smart metering and wearables. When discussing these applications, special attention is given to the service requirements associated with mMTC, for example, in terms of reachability, throughput, and latency. The chapter continues and introduces the concept of the Cellular Internet of Things (CIoT) and the three technologies *Extended Coverage Global System for Mobile Communications Internet of Things* (EC-GSM-IoT), *Narrowband Internet of Things* (NB-IoT), and *Long-Term Evolution for Machine-Type Communications* (LTE-M) that can be said to define this concept. While EC-GSM-IoT and LTE-M are backward compatible solutions based on GSM and LTE, respectively, NB-IoT is a brand new radio access technology.

The final part of the chapter looks beyond the set of cellular access technologies and introduces the Low Power Wide Area Network (LPWAN) range of solutions that already have secured a significant footprint in the mMTC market. Unlike the cellular systems, these LPWANs have been designed to operate in licensed exempt spectrum. An initial discussion around the pros and cons of licensed exempt operation is presented to prepare the reader for the final chapters of the book, where a closer look is taken at operation in the unlicensed frequency domain.

Cellular Internet of Things. http://dx.doi.org/10.1016/B978-0-12-812458-1.00001-0

1.1 INTRODUCTION

The Internet of Things, commonly referred to as IoT, is the latest rising star in the *information and communications technology* (ICT) industry and embodies the vision of connecting virtually anything with everything. Cisco estimates that 12 billion devices will be connected by 2020 [1]. Ericsson goes even further in its vision of 18 billion connected devices in 2022 [2]. Regardless of which of these two leading ICT network providers has made the best estimate, the anticipated number of devices is nothing short of dazzling. As a comparison, the total number of mobile cellular subscriptions currently amounts to 7.3 billion [2].

So what is it that will become connected, that is not already connected? Traditional use cases such as connecting utility meters to support, e.g., distribution and billing, will likely increase in numbers. A recent example is the Great Britain Smart Metering Implementation Programme, where the British government has decided to replace 53 million meters in roughly 30 million premises with advanced electricity and gas meters intended to support customers with "near real-time information on their energy consumption" [3]. Similar projects are either in progress or in the planning stages in a majority of the European Union member states [4]. At the same time, new use cases, for example, in the category of *Wearables*, are gaining momentum with increasing market traction being realized.

When viewed in totality, the overall number of connected devices is already undergoing an exponential growth where connectivity over cellular networks serves as a key enabler for this growth. Between 2015 and 2021, it has been approximated that the volume of devices connected to the Internet via cellular technologies alone will experience a compounded annual growth rate of roughly 25% [1].

This accelerated growth in cellular devices is supported, or perhaps driven, by the recent work performed by the *Third Generation Partnership Project* (3GPP) standards development organization in Release 13 of its featured technologies. 3GPP has already been responsible for standardization of the GSM (2G), *UMTS* (3G), and LTE (4G) radio access technologies. The early development and evolution of these technologies has mainly been driven by traditional service requirements defined by voice and mobile broadband services. In the past couple of years, new requirements for machine connectivity have emerged in support of the IoT and transformed the process of standards evolution into what can be seen as a fast-moving revolution. Accordingly, the designs of GSM and LTE have been rethought, and a new radio access technology dedicated to supporting the IoT has been developed, resulting in three new CIoT technologies becoming globally available within a very short time frame. The mission of this book is to introduce, characterize, and, when relevant, in detail describe these three new technologies known as EC-GSM-IoT, NB-IoT, and LTE-M.

EC-GSM-IoT is a fully backward compatible solution that can be installed onto existing GSM deployments, which by far represent the world's largest and most widespread cellular technology. EC-GSM-IoT has been designed to provide connectivity to IoT type of devices under the most challenging radio coverage conditions in frequency deployments as tight as 600 kHz.

NB-IoT is a brand new radio access technology that in some aspects reuses technical components from LTE to facilitate operation within an LTE carrier. The technology also supports stand-alone operation. As the name reveals, the entire system is operable in a narrow spectrum, starting from only 200 kHz, providing unprecedented deployment flexibility because of the minimal spectrum requirements. The 200 kHz is divided into channels as narrow as 3.75 kHz to support a combination of extreme coverage and high uplink capacity, considering the narrow spectrum deployment.

LTE-M is based on LTE, which is by far the fastest growing cellular technology. LTE-M provides just as EC-GSM-IoT and NB-IoT ubiquitous coverage and highly power efficient operation. Using a flexible system bandwidth of 1.4 MHz or more, the technology is capable of serving higher-end applications with more stringent requirements on throughput and latency than what can be supported by EC-GSM-IoT and NB-IoT.

The fact that they can provide reliable communication under extreme coverage conditions while supporting battery life of many years and also realizing securely encrypted communication is said to be common to all of these technologies. All three technologies have also been designed to enable ultra-low device complexity and cost, which are important factors considering the objective of providing connectivity to billions of devices supporting diverse services and applications.

3GPP is, however, not the only organization supporting the accelerated growth in a number of connected devices. This book therefore goes beyond EC-GSM-IoT, NB-IoT, and LTE-M to provide an overview of the competitive IoT landscape and introduces a few of the most promising technologies operating in unlicensed spectrum. Both short-range solutions and long-range solutions established in the so-called LPWAN market segment are presented. As these technologies are targeting unlicensed frequency bands, they are following different design principles compared to the 3GPP technologies, which are developed for licensed frequency bands. To be able to distinguish between technologies for licensed and unlicensed operation, the book sets out to introduce and analyze the regulations associated with device operation in the most recognized license exempt frequency bands.

The book will finally provide a glimpse of the future and describe how the CIoT technologies are expected to evolve to support new use case and meet the latest market requirements. Also, the work on a fifth generation (5G) communication technology ongoing within the *International Telecom Union* (ITU) and 3GPP will be presented. In this context 5G features such as mMTC and *Ultra Reliable and Low Latency Communication* (URLLC) will be introduced.

1.2 NEW APPLICATIONS AND REQUIREMENTS
1.2.1 LEADING UP TO THE CELLULAR INTERNET OF THINGS

The CIoT may be the latest hype in the mobile industry, but cellular communication for mobile telephony has actually been well established for more than three decades. Over time, it has evolved from being a niche product in developed countries to become an everyday commodity reaching almost every corner of the world. It has, since its introduction in the early 1980s, had a considerable impact on many aspects of the way we live our lives. A significant milestone in the short but intense history of mobile telephony was the introduction of the smartphone. With it came a framework for developing and distributing smart and compact applications that pushed mobile phone usage far beyond the traditional use cases of making voice calls and sending text messages. Today a smartphone is as advanced as any computer and allows the user to manage everything from personal banking services to social interaction with friends and family while listening to the latest music from the hit charts. The obvious advantage of the smartphone over other connected devices is its convenient form factor and its seamless wireless connectivity that allows the user to access the Internet whenever and wherever.

The introduction of the smartphone not only changed the usage of the phone but also had a profound impact on the requirements on the underlying communication systems providing the wireless connectivity. For a classic mobile phone supporting voice and text messages, the most important requirements are acceptable voice quality and reliable delivery of messages. Since the introduction of the smartphone, expectations go far beyond this and requirements in terms of, e.g., high data rates and low latency have become more prominent. The requirements set by the ITU on the world's first 4G systems were also to a significant extent shaped by this and focused on the ability of the candidate solutions to provide mobile broadband services. For base station to device transmission, a peak downlink data rate of 1.5 Gbit/100 MHz/s was one of the ITU targets. For the uplink direction, i.e., transmission from the device to the base station, the target was slightly relaxed to 675 Mbit/100 MHz/s. A latency of less than 10 ms was another important requirement intended to secure an acceptable mobile broadband experience [5].

With the introduction of *Machine-to-Machine* (M2M), MTC, and IoT applications and services, the expectations as well as the set of requirements placed on the mobile communication technologies changed again. A not too advanced smartphone user may, for example, stream several gigabits of data every month with high requirements on quality of experience while it may be sufficient for an average utility meter to access the network once per day to send the latest billing information of a few bytes to a centralized billing system. The requirements associated with M2M and IoT services are not, however, only being relaxed, they are becoming more diverse also, as new applications are emerging. Although the mentioned utility meter displays relaxed requirements in terms of latency and throughput, other requirements may be far more stringent than what is expected from the smartphone use case. It is not rare for utility meters located deep indoors, for example, in basements, to place high requirements on the wireless coverage provided by the supporting communications systems. It is furthermore expected that the number of M2M and IoT devices will, by far, eventually outnumber the smartphone population and thereby effectively set new requirements on system capacity and network availability.

1.2.2 MASSIVE MACHINE-TYPE COMMUNICATIONS AND ULTRA RELIABLE AND LOW LATENCY COMMUNICATIONS

From a service, applications, and requirements point of view, the IoT market is often said to be divided into at least two categories: mMTC and URLLC. The Next Generation Mobile Networks Alliance is in its 5G white paper [6] describing these two categories in terms of typical use cases and their associated requirements. Smart Wearables and Sensor Networks are two industrial vertical features mentioned as belonging to the mMTC market category. Smart wearables comprise not only, e.g., smart watches but also sensors integrated in clothing. A main use case is sensing health-related metrics such as body temperature and heartbeat. It is clear that if this trend gains traction, the number of devices per person will go far beyond what we see today, which will put new requirements on the capacity that must be supported by cellular networks providing IoT services. It can furthermore be expected that in order for clothing manufacturers to find wearables an appealing concept, the devices must be extremely compact to support seamless integration in the clothing. The devices must also be of ultra-low cost to attract clothing manufacturers as well as consumers.

Sensor networks is a family name for various utility meters such as gas, water, and electricity meters. Potentially, every home is equipped with a multitude of sensors that will put high requirement on the capacity of the communication system providing them with connectivity. As utility meters are associated with stringent requirements on coverage, which is radio resource consuming, the task to provide sufficient capacity for these becomes even more challenging. Meters may in addition entirely rely on battery power, which will put high requirements on device energy efficiency to facilitate operation for years on small and low-cost batteries.

URLLC can, on the other hand, be exemplified by high-end applications such as automated driving, industrial automation, and eHealth. The news is filled with articles about traditional car manufacturers and giants from the ICT industry competing in the development of autonomous vehicles. If such applications are to be supported by cellular communication networks, the network needs to offer close to perfect reliability combined with support for extreme latency requirements. It is also not far-fetched to imagine remote steering as an attractive alternative to, or perhaps first step toward, fully autonomous vehicles. In this case also requirements on support for high data rates to support high-resolution video may come into the picture. Similar requirements can obviously also be mapped to the industrial automation and eHealth verticals.

Figure 1.1 summarizes the just-made observations with a high level illustration of expected requirements for the mMTC category and the URLLC category in terms of coverage, number of supported connections, latency, throughput, mobility, device complexity, and device battery life. For comparison also typical mobile broadband requirements discussed in Section 2.1 are depicted. The center of the radar chart corresponds to relaxed requirements while the outskirts of the chart map to stringent requirements.

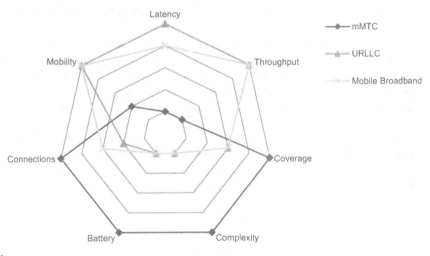

FIGURE 1.1

mMTC, URLLC, and mobile broadband requirements.

1.2.3 INTRODUCING EC-GSM-IOT, NB-IOT, AND LTE-M

The three technologies EC-GSM-IoT, NB-IoT, and LTE-M, described in Chapters 3–8 in this book, were to a large extent designed to serve use cases belonging to the category of mMTC. The work on

LTE-M started in September 2011 with the 3GPP feasibility study named *Study on Provision of Low-Cost MTC UEs Based on LTE* [7], referred to as the *LTE-M study item* in the following. The main justification for this study item was to extend the LTE device capabilities in the low-end MTC domain, provide an alternative to *General Packet Radio Service* (GPRS) and *Enhanced General Packet Radio Service* (EGPRS) devices, and facilitate a migration of GSM networks toward LTE. As the aim was to replace GPRS/EGPRS as a bearer for MTC services, the study naturally used GPRS/EGPRS performance as the benchmark when setting its objectives. It was, e.g., required that data rates, spectrum efficiency, and power consumption should be at least as good as EGPRS. The main focus of the study was, however, to provide a solution with device complexity and cost on par with GPRS. In September 2012, it was decided to add a new objective on study of the feasibility of a coverage improvement of 20 dB beyond normal LTE coverage. LTE devices have typically been considered far more expensive than GPRS/EGPRS devices, mainly because of the improved capabilities provided by LTE. Therefore for LTE to become competitive in the low-end mMTC market, a reduction in device cost and complexity was considered crucial. The coverage enhancement was considered to be needed to facilitate deep indoor coverage in locations where, for example, utility meters are expected to be located.

The work on EC-GSM-IoT and NB-IoT started one release later in the 3GPP feasibility study named *Cellular System Support for Ultra-low Complexity and Low Throughput Internet of Things* [8], referred to as the *Cellular IoT study item* in the following. Many aspects of this study can be recognized from the LTE-M study item. The overall objective did, however, change from targeting a solution comparable to GPRS/EGPRS to finding a solution competitive in the Low Power Wide Area segment, which at the time to a large extent was defined by technologies for unlicensed operation designed for ultra-low complexity, extreme coverage range, and long device battery life as discussed in more detail in Section 1.3.

As for the work on LTE-M, an important part of the Cellular IoT study was to meet an objective of extending coverage with 20 dB. While the LTE-M study item used LTE coverage as reference, the Cellular IoT study item defined the improvement in relation to GPRS coverage. The GPRS technology is often seen as one of the most capable technologies in terms of coverage, and today it is still the main technology providing cellular IoT connectivity. But just as for LTE, its capability was believed to be insufficient to cater for deep indoor coverage. Part of the reasoning behind this objective was that modern building materials besides their ability to provide excellent thermal insulation often are characterized by high attenuation of radio waves. A further aspect that was taken into consideration is that many IoT devices have a compact form factor where efficient antenna design is not prioritized. This limits the achievable antenna gain and further increases the need for coverage and increased *Maximum Coupling Loss* (MCL).

MCL is defined as the maximum loss in the conducted power level that a system can tolerate and still be operable (defined by a minimum acceptable received power level). Hence, it can be calculated as the difference between the conducted power levels measured at the transmitting and receiving antenna ports, and it is an attractive metric to define coverage because of its simplicity. Because it is defined using the antenna connector as the reference point, the directional gain of the antenna is not considered when calculating MCL. Coverage can also be expressed by the *Maximum Path Loss* (MPL) a technology can support. The path loss is defined by the loss in the signal path (e.g., distant-dependent propagation loss, building penetration loss, etc.) of the radiated power. Hence, it can be calculated by the difference in radiated power levels at the transmitting and receiving antennas. To determine the MPL also the antenna gain at the transmitter and receiver need to be considered. The difference between MCL and MPL is illustrated in Figure 1.2. MCL has been chosen by 3GPP as the metric to evaluate coverage enhancements. In the remainder of the book, the focus will therefore be on MCL with a few exceptions when describing the capabilities of technologies operating in unlicensed spectrum. The concept of MCL is described in Section 3.2.8.1 in detail.

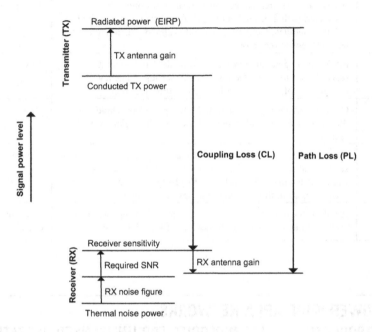

FIGURE 1.2

Illustration of coupling loss and path loss.

In addition to coverage enhancement, it was required in the Cellular IoT study to show that device battery life of at least 10 years were feasible to support, for example, utility meters with no access to a mains power source. To minimize the maintenance cost of applications and equipment relying on limited battery capacity, low energy consumption leading to long battery life is of high importance. A network supporting a large-scale deployment of millions of devices requiring a fresh battery every year would be difficult to maintain on commercial terms.

Furthermore, the candidate technologies were required to present a network capacity sufficient to support IoT type of services for central London under an assumption of 40 connected devices per household, resulting in over 60,000 devices/km^2. Last but not least, it was requested to achieve all this with stringent requirements on keeping the device complexity ultra-low to secure as low device cost as possible. Table 1.1 summarizes these performance objectives required to be fulfilled by the candidate technologies developed within the Cellular IoT study item, including EC-GSM-IoT and NB-IoT.

Table 1.1 Overview of the performance objectives in the CIoT study item [8]

Objective	Description	Requirement
Coverage	The MCL of the system was required to surpass GPRS coverage by 20 dB. The GPRS coverage reference was estimated to be equal to 144 dB.	164 dB
Throughput	A coverage requirement becomes meaningful only when associated with a throughput target. It was therefore required that the candidate technologies should support a data rate of at least 160 bps at the coverage limit.	160 bps
Latency	mMTC is, in general, expected to provide delay tolerant services. But to guarantee high-alert reporting, a latency requirement of 10 s was set for high-priority reports.	10 s
Capacity	Capacity supporting 40 connected devices per household in downtown London with roughly 1500 households/km^2 was required.	60,000 devices/km^2
Power efficiency	To secure operation regardless of the access to a mains power source, and with minimal requirements on battery capacity, it was expected that a 5-watt-hour (Wh) battery would last for at least 10 years in the most extreme coverage situations.	10 years/5 Wh
Complexity	An ultra-low device complexity was required to support mass production and deployment of devices of ultra-low cost.	Ultra-low device complexity

1.3 LOW POWER WIDE AREA NETWORKS
1.3.1 AN INTRODUCTION TO TECHNOLOGIES FOR UNLICENSED OPERATION

As mentioned in Section 1.1, the 3GPP cellular technologies are not the only solutions competing for IoT traffic. Also well-known technologies such as *Bluetooth* and *Wi-Fi* can serve as bearers for MTC traffic. A distinction between the group of cellular technologies and Bluetooth and Wi-Fi is that the former is intended for operation in licensed spectrum while the latter two belong to the group of systems operating in unlicensed spectrum, in so-called license exempt frequency bands.

Licensed spectrum corresponds to a part of the public frequency space that has been licensed by national or regional authorities to a private company, typically a mobile network operator, under the condition of providing a certain service to the public such as cellular connectivity. At its best, a licensed frequency band is globally available, which is of considerable importance for technologies aiming for worldwide presence. The huge success of GSM is, for example, to a significant extent built around the availability of the GSM 900 MHz band in large parts of the world. Licensed spectrum is, however, commonly associated with high costs, and the media frequently give reports of spectrum auctions bringing in significant incomes to national authorities all across the world.

Unlicensed spectrum, on the other hand, corresponds to portions of the public frequency space that can be said to remain public and therefore free of licensing costs. Equipment manufacturers using this public spectrum must, however, meet a set of national or regional technical regulations for technologies deployed within that spectrum. Among of the most popular license exempt frequency bands are the so-called *industrial, scientific and medical* (ISM) bands identified in article 5.150 of the ITU Radio Regulations [9]. Regional variations for some of these bands exist, for example, in the frequency range around 900 MHz while other bands such as the range around 2.4 GHz can be said to be globally available. In general, the regulations associated with license exempt bands aim at limiting harmful interference to other technologies operating within as well as outside of the unlicensed band.

As the name implies, the ISM bands were originally intended for ISM applications. Later, it was made available for devices providing more general types of services such as short-range devices benefiting from operation in a set of harmonized frequency bands [10]. Bluetooth and Wi-Fi, and thereto related technologies such as *Bluetooth Low Energy*, *ZigBee*, and *Wi-Fi Halow*, commonly use the ISM bands to provide relatively short-range communication, at least in relation to the cellular technologies. Bluetooth can be said to be part of a *Wireless Personal Area Network* while Wi-Fi provides connectivity in a *Wireless Local Area Network* (WLAN). In recent years, a new set of technologies have emerged in the category of LPWANs. These are designed to meet the regulatory requirements associated with the ISM bands, but in contrast to WPAN and WLAN technologies they provide long-range connectivity, which is an enabler for supporting wireless devices in locations where WPAN and WLAN systems cannot provide sufficient coverage.

1.3.2 LICENSED AND LICENSE EXEMPT BAND REGULATIONS

To understand the potential of LPWAN solutions operating in license exempt bands in relation to those designed for licensed bands, it is important to understand the regulations setting the ultimate boundary for the design of the systems. Systems deployed within both licensed and unlicensed bands operate in accordance with regulations that are determined on a national or regional basis. For the cellular systems these regulations as a general rule follow the requirements set by the 3GPP technical specifications. For license exempt bands, the regulations are not coordinated across regions to the same extent, and the local variations are higher. For example, in the United States it is the *Federal Communications Commission* (FCC) that publishes the *Electronic Code of Federal Regulations*, which defines the regulations for operation in the license exempt bands 902–928 MHz and 2400–2483.5 MHz [11], while in Europe it is the *European Telecom Standards Institute* (ETSI) that publishes the *Harmonized standards* for the license exempt bands 863–870 MHz [12] and 2400–2483.5 MHz [13].

The 3GPP requirements are, on one hand, defined to secure coexistence between 3GPP systems operating adjacent to each other within the same band as well as between systems operating in different bands. On the other hand, they define in-band requirements that guarantee a minimum level of performance of the system as a whole as well as the performance on a per link basis. In case the regulatory bodies add or modify requirements compared with the ones set by 3GPP, the impact is typically limited to emission levels outside of the 3GPP bands, which have limited impact on the fundamental aspects of the system design.

The requirements for license exempt bands are defined to support coexistence toward other systems outside the band as well as inside the band. Inside the band the different systems are not as in licensed operation separated in the frequency domain but are overlapping in the sense that they may use the same frequency resource at any point in time. To limit the interference between the unlicensed systems, regulations are commonly defined to limit the used output power. In addition, requirements on the *duty cycle* as well as the *dwell time* on a specific frequency resource may be defined. The duty cycle defines the ratio by which a transmitting device may use a radio resource. The dwell time, on the other hand, sets the maximum contiguous time by which a transmitting device may use a radio resource. These requirements establish strict design boundaries that make it challenging for any systems operating in the ISM bands to provide high and robust coverage for a multitude of devices while meeting service requirements on, e.g., latency and throughput.

Chapter 9 will further review the regulations for unlicensed operation including the herein introduced requirements.

1.3.3 LOW POWER VERSUS WIDE AREA

A LPWAN system is characterized by its ability to provide high coverage while using low device output power. The cellular IoT systems EC-GSM-IoT, NB-IoT, and LTE-M can, for example, all be said to belong to the category of LPWAN systems. Coverage of a wireless device is, in general, limited by the device-intrinsic thermal noise. The power of the noise is linearly increasing with increasing temperature and system bandwidth. The steepness of the linear increase is determined by the device's *Noise Factor*, which resembles the difference in noise power measured in an ideal device and a real device implementation.

As the simplest approach to combat thermal noise power and improve the *Signal to Noise power Ratio* (SNR) is to increase a system's useful signal power, it may seem like a contradiction to associate low power with wide area coverage as done for LPWANs. The use of low power is, however, highly motivated because it facilitates low device cost, enables compact device design, and supports flexible usage of a device for diverse applications. As explained in Section 1.3.2, low output power is also a typical requirement in systems operating in licensed exempt bands to limit the harmful interference within and between the systems using these bands.

Fortunately, there are means besides increased signal power to cope with challenging coverage conditions. Commonly, it is the choice of system bandwidth that becomes the decisive factor impacting the absolute noise level measured in a wireless device. This implies that a system designed to use a low signal bandwidth enjoys the benefit of operating at a lower absolute noise level, which for a given useful power level improves the SNR.

Figure 1.3 illustrates the impact on SNR when increasing the useful signal bandwidth from 1 kHz to 1 MHz for a system operating at a transmission power of 20 dBm, i.e., 100 mW, and at a constant ambient temperature of 20°C. The SNR is calculated for a coupling loss between the transmitter and receiver of 164 dB and assumes a 3dB receiver *Noise Figure* (NF), which is the logarithmic representation of the Noise Factor. 164 dB coupling loss corresponds to very challenging coverage conditions as elaborated in Section 1.2 while an NF of 3 dB is a value that can be considered realistic for a base station implementation.

In their regulations for the ISM bands the FCC and ETSI explicitly mention *spread spectrum technologies* as a means of conveying information. Spread spectrum is a family of technologies where a transmitter transforms, or spreads, a low bit rate information signal to a high bit rate and wideband carrier of low power spectral density. The intended recipient of the carrier is able to regenerate the useful signal to extract the payload information therein, while other users ideally experience the wideband carrier as a weak interferer with *Gaussian noise*—like characteristics. One of the most popular spread spectrum techniques is the *Direct Sequence Spread Spectrum* (DSSS) modulation

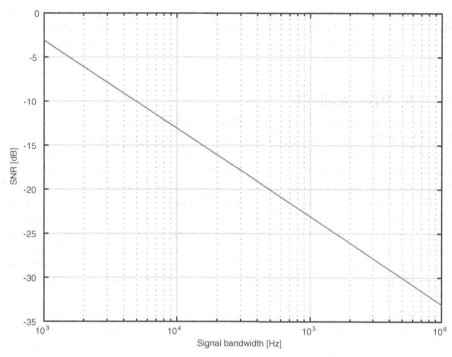

FIGURE 1.3

SNR versus system bandwidth at 164 dB MCL, a NF of 3 dB, and a signal power of 20 dBm.

where the transmitter spreading operation is defined by the multiplication of the useful signal with a known pseudo random spreading code of high bit rate. The ratio between the bit rate of the spreading signal, i.e., the *chip rate*, and the useful signal bit rate defines the processing gain by which the receiver can improve the SNR of the useful signal compared with the SNR over the received wideband carrier. In fact, with the spread spectrum technique, the achieved processing gain compensates for the lower SNR as a result of using a wider bandwidth signal shown in Figure 1.3, and eventually the SNR penalty of using a wider bandwidth signal is fully compensated for.

When reading about the 3GPP systems in Chapters 3–8, it will become apparent that narrowband and spread spectrum modulations are merely two alternatives among a multitude of available methods that can be adopted to improve the coverage of a system. Improving the *code rate* of a technology by trading useful throughput for increased redundancy is one of the most established methods that is commonly used to increase the robustness of a wireless system. Another frequently used method is to introduce repetition-based transmissions to increase the overall transmission time of a radio block, which allows a system to improve the receiver processing gain. Just as in the case of the spread spectrum technique, through the achieved receiver processing gain, the repetition-based approach will eventually equalize the SNR penalty that can be associated with a wideband transmission, compared with transmission of relatively smaller bandwidth not using repetitions.

As explained in later chapters, these and other methods are combined in the design of EC-GSM-IoT, NB-IoT, and LTE-M to make sure that the systems using these technologies meet the ambitious MCL that 3GPP considers to be sufficient to provide the services required by a LPWAN solution.

1.3.4 A COMPETITIVE MARKET

The linear relationship between bandwidth and SNR depicted in Figure 1.3 is a fact utilized, e.g., by French LPWAN vendor Sigfox with their *Ultra Narrowband Modulation*. It uses a narrow bandwidth carrier to support a claimed MPL of 162 dB at the European 868 MHz and US 902 MHz ISM frequencies [14]. Sigfox is among the most successful LPWAN actors and supports coverage in considerable parts of Europe including nationwide coverage in Portugal, Spain, and France [15].

LoRa Alliance [16] is a first example of a successful player in the LWPAN market using spread spectrum technology to meet the ISM band regulations. They are using *Chirp Spread Spectrum* modulation [17], which is a technique using frequency modulation to spread the signal. A radio bearer is modulated with up and down chirps, where an up chirp corresponds to a pulse of finite length with increasing frequency, while a down chirp is a pulse of decreasing frequency. The LoRa Alliance claims to provide a MCL of 155 dB in the European 867−869 MHz band, and 154 dB in the US 902−928 MHz band [17].

A second example of an LPWAN vendor using spread spectrum is Ingenu with their *Random Phase Multiple Access* (RPMA) technology. RPMA is a DSSS-based modulation complemented by a pseudo random time of arrival that helps distinguishing users multiplexed on the same radio resource. Ingenu claims to achieve an MPL of 172 dB in the United States and 168 dB in Europe [18]. While Sigfox and LoRa Alliance are using the US 902−928 MHz and European 867−869 MHz ISM bands to gain the coverage advantage associated with low-frequency bands, Ingenu is focusing on the 2.4 GHz license exempt band. They claim that the higher coupling loss associated with 2.4 GHz is compensated by the benefits of this band in terms of higher allowed output power and as an enabler for compact antenna design facilitating receive diversity without compromising the device form factor. They furthermore bring forward that the 2.4 GHz band is globally available and provides a bandwidth of up to 80 MHz [19].

These three and other LPWAN vendors have attracted considerable market interest and media attention because of their operation in license exempt bands in combination with claims of support for high link budgets, long device battery life, low device complexity, and high system capacity. As explained in Section 1.2, these are all important capabilities for a technology aiming to serve applications in the mMTC market segment. This did ultimately bring the attention to the shortcomings of the traditional cellular technologies regarding their inability to provide full support for IoT type of services without further evolution. As a consequence, 3GPP triggered a massive effort in its Release 13 to start the development of the three Cellular IoT technologies described in Chapters 3−8. This is to empower traditional mobile network operators with a path to remain competitive with Sigfox, LoRa, and Ingenu and their likes while avoiding the deployment of a parallel wireless infrastructure targeting support for IoT services.

Chapter 9 will, as mentioned earlier, review the license exempt regulations. It will in addition discuss the pros and cons of the licensed LPWAN versus unlicensed LPWAN operation. During this discussion further details on the technologies developed by Sigfox, LoRa, and Ingenu will be presented. Chapter 9 will finally introduce a few of the most prominent technologies for short-range communication in unlicensed spectrum.

REFERENCES

[1] Cisco, The Zettabyte Era: Trends and Analysis, Cisco White Paper, 2016.
[2] Ericsson, Ericsson Mobility Report, November 2016.
[3] Department of Energy & Climate Change, Smart Metering Implementation Programme, Third Annual Report on the Rollout of Smart Meters, Report, Crown, London, 2014.
[4] European Commission, Benchmarking Smart Metering Deployment in the EU-27 with a Focus on Electricity, COM(2014) 356 Final, Report, Brussels, June 17, 2014.
[5] ITU-R, Report ITU-R M.2134, Requirements Related to Technical Performance for IMT-Advanced Radio Interface(s), 2008.
[6] Next Generation Mobile Networks Ltd, A Deliverable by the NGMN Alliance, NGMN 5G White Paper, 2015.
[7] Third Generation Partnership Project, Technical Report 36.888 v12.0.0, Study on Provision of Low-cost Machine-Type Communications (MTC) User Equipments (UEs) Based on LTE, 2013.
[8] Third Generation Partnership Project, Technical Report 45.820 v13.0.0, Cellular System Support for Ultra-low Complexity and Low Throughput Internet of Things, 2016.
[9] ITU, Radio Regulations, Articles, 2012.
[10] ITU-R, Recommendation ITU-R SM.1896, Frequency Ranges for Global or Regional Harmonization of Short-range Devices, November 2011.
[11] U.S. Government Publishing Office, Electronic Code of Federal Regulations, Article 15.247, July 7, 2016.
[12] European Telecom Standards Institute, EN 300 220-1 V2.4.1 Electromagnetic Compatibility and Radio Spectrum Matters; Short Range Devices; Radio Equipment to Be Used in the 25 MHz to 1 000 MHz Frequency Range with Power Levels Ranging up to 500 MW; Part 1: Technical Characteristics and Test Methods, January 2012.
[13] European Telecom Standards Institute, EN 300 328 V2.0.20 Wideband Transmission Systems; Data Transmission Equipment Operating in the 2,4 GHz ISM Band and Using Wide Band Modulation Techniques; Harmonised Standard Covering the Essential Requirements of Article 3.2 of the Directive 2014/53/EU, May 2016.
[14] Sigfox, About Sigfox, 2016. Available: http://makers.sigfox.com/.
[15] Sigfox, Sigfox Coverage, 2016. Available: http://www.sigfox.com/coverage.
[16] LoRa Alliance, LoRa Alliance Wide Area Networks for IoT, 2016. Available: https://www.lora-alliance.org/.
[17] LoRa Alliance Technical Marketing Workgroup 1.0, LoRaWAN™, What Is it?, A Technical Overview of LoRa® and LoRaWAN™, White Paper, November 2015.
[18] Ingenu, RPMA Technology, For the Internet of Things, Connecting Like Never Before, White Paper.
[19] Ingenu, 2.4 GHz & 900 MHz, Unlicensed Spectrum Comparison, White Paper.

WORLD-CLASS STANDARDS

2

CHAPTER OUTLINE

Abstract

This chapter presents the Third Generation Partnership Project (3GPP), including its ways of working, organization, and linkage to the world's largest regional Standardization Development Organizations (SDOs). 3GPP's work on Machine-Type Communication (MTC) from Release 8 to Release 13 is examined in detail. First, the main achievements made in Release 8–11 are described, and a focus is placed on the aspect of handling the large numbers of MTC devices expected to be served by cellular technologies. The chapter continues with a look at the work completed in Release 12 wherein the suitability of Long-Term Evolution (LTE) to support a large number of devices making infrequent small data transmissions, which is assumed to characterize many Internet of Things (IoT) applications, is reviewed. Also the features of *Power Saving Mode* (PSM) and *extended Discontinuous Reception* (eDRX) specified in Release 12 and 13 to facilitate device battery life of several years are presented. Finally, a glance at the Release 12 and 13 feasibility studies leading up to the normative work on *Extended Coverage Global System for Mobile Communications Internet of Things* (EC-GSM-IoT), *Narrowband Internet of Things* (NB-IoT), and *Long-Term Evolution for Machine-Type Communications* (LTE-M) is given.

2.1 THIRD GENERATION PARTNERSHIP PROJECT

3GPP is the global standardization forum behind the evolution and maintenance of Global System for Mobile Communications (GSM), Universal Mobile Telecommunications System (UMTS), and LTE. The project is coordinated by seven regional SDOs representing Europe, the United States, China, Korea, Japan, and India. 3GPP has since its start in 1998 organized its work in release cycles and has now in 2017 reached Release 15. 3GPP has already established plans for Release 15 to secure the first delivery of a fifth generation (5G) system by 2018.

Cellular Internet of Things. http://dx.doi.org/10.1016/B978-0-12-812458-1.00002-2

A release contains a set of work items where each typically delivers a feature that is made available to the cellular industry at the end of the release cycle through a set of *Technical Specifications* (TSs). A feature is specified in four stages where Stage 1 contains the service requirements, Stage 2 a high-level feature description, and Stage 3 the detailed description that is needed to implement the feature. The fourth and final stage contains the development of the performance requirements and conformance testing procedures for ensuring proper implementation of the feature. Each feature is implemented in a distinct version of the 3GPP TSs that maps to the release within which the feature is developed. At the end of a release cycle the version of the specifications used for feature development is frozen and published. In the next release a new version of each TS is opened as needed for new features associated with that release. Each release contains a wide range of features providing functionality spanning across GSM, UMTS, and LTE as well as providing interworking between the three. In each release it is further ensured that GSM, UMTS, and LTE can coexist in the same geographical area. That is, the introduction of, for example, LTE into a frequency band should not have a negative impact on GSM and UMTS operation.

The technical work is distributed over a number of *Technical Specifications Groups* (TSGs), each supported by a set of *Working Groups* (WGs) with technical expertise representing different companies in the industry. When EC-GSM-IoT, NB-IoT and LTE-M was specified in Release 13 the 3GPP organizational structure was built around four TSGs:

- *TSG Service and system Aspects* (SA),
- *TSG Core network and Terminals* (CT),
- *TSG GSM/EDGE Radio Access Network* (GERAN), and
- *TSG Radio Access Network* (RAN).

TSG SA is responsible for the system aspects and service requirements, i.e., the *Stage 1* requirements, and TSG CT for core network aspects and specifications. Both TSG SA and TSG CT cover all RANs, whereas TSG GERAN was responsible for GSM development and maintenance, and TSG RAN is responsible for UMTS, LTE, and their evolution.

The overall project management is handled by the *Project Coordination Group* (PCG) that, for example, holds the final right to appoint TSG Chairmen, to adopt new work items and approve correspondence with external bodies of high importance, such as the International Telecommunications Union. Above the PCG are the seven SDOs: *ARIB* (Japan), *CCSA* (China), ETSI (Europe), *ATIS* (US), *TTA* (Korea), *TTC* (Japan), and *TSDSI* (India). Within 3GPP these SDOs are known as the *Organizational Partners* (OPs) that hold the ultimate authority to create or terminate TSGs and are responsible for the overall scope of 3GPP.

The Release 13 specification work for EC-GSM-IoT, NB-IoT, and LTE-M was led by TSG GERAN and TSG RAN. TSG GERAN initiated its work through a feasibility study resulting in *Technical Report* (TR) 45.820 Cellular System Support for Ultra-Low Complexity and Low Throughput Internet of Things [1]. It is common that 3GPP before going to normative specification work for a new feature performs a study of the feasibility of that feature and records the outcome of the work in a TR. In this specific case the TR recommended to continue with normative work items on EC-GSM-IoT and NB-IoT. While GERAN took on the responsibility of the EC-GSM-IoT work item, the work item on NB-IoT was transferred to TSG RAN. TSG RAN also took responsibility for the work item associated with LTE-M, which just as NB-IoT is part of the LTE series of specifications.

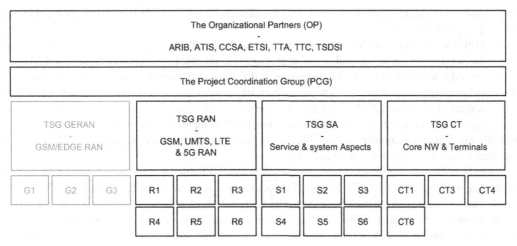

FIGURE 2.1

Organizational structure of 3GPP.

After 3GPP Release 13, i.e., after completion of the EC-GSM-IoT specification work, TSG GERAN and its WGs GERAN1, GERAN2, and GERAN3 were closed and their responsibilities were transferred to TSG RAN and its WGs RAN5 and RAN6. As a consequence, TSG RAN is now responsible for NB-IoT and GSM, including EC-GSM-IoT, in addition to being responsible for UMTS and LTE and the development of a new 5G RAN.

Figure 2.1 gives an overview of the 3GPP organizational structure at the end of Release 13, indicating the four levels: OP including its regional SDOs, PCG, the three active TSGs, and the WGs of each TSG. The TSG GERAN branch is shown in gray to highlight that it was transferred to TSG RAN after Release 13 and is therefore no longer active.

2.2 FROM MACHINE-TYPE COMMUNICATIONS TO THE CELLULAR INTERNET OF THINGS

2.2.1 ACCESS CLASS AND OVERLOAD CONTROL

Although Chapter 1 may have given the impression that 3GPP did not address the Internet of Things (IoT) before Release 13, the work on MTC started far earlier. This section presents the work done by 3GPP for GSM and LTE in the area of MTC from the very start of 3GPP in Release 99 until Release 13. Although UMTS is not within the scope of this overview, the interested reader should note that most of the features presented for LTE are also supported by UMTS.

In 2007 and Release 8 *TSG SA WG1* (SA1) working on the 3GPP system architecture published TR 22.868 *Study on Facilitating Machine to Machine Communication in 3GPP Systems* [2]. It highlights use cases such as metering and health, which are still of vital interest as 3GPP continues with the 5G specification effort. 3GPP TR 22.868 provides considerations in areas such as handling large numbers of devices, addressing of devices, and the level of security needed for machine-to-machine applications.

In 3GPP TSG SA typically initiates work for a given feature by first agreeing to a corresponding set of general service requirements and architectural considerations. In the case of MTC the SA1 work also triggered a series of Stage 1–3 activities in 3GPP Release 10 denoted *Network Improvement for Machine-Type Communications* [3]. The main focus of the work was to provide functionality to handle large numbers of devices, including the ability to protect existing networks from a flood of what could potentially be a very large number of devices. For GSM/Enhanced Data Rates for GSM Evolution (EDGE) the features *Extended Access Barring* (EAB) [4] and *Implicit Reject* (IR) [5] were fully specified as part of these Release 10 activities.

Already in the Release 99 specifications, i.e., the first 3GPP release covering GSM/EDGE, support for the *Access Class Barring* (ACB) feature is specified. It allows a network to bar devices of different *access classes* regardless of their registered *Public Land Mobile Network* (PLMN). Each device is pseudo randomly, i.e., based on the last digit of their *International Mobile Subscriber Identity* (IMSI), configured to belong to 1 of 10 normal access classes. In addition, five special access classes are defined and the device may also belong to one of these special classes. The GSM network broadcasts in its system information message, e.g., *System Information Type 1*, a bitmap as part of the *Random Access Channel (RACH) Control Parameters* to indicate if devices in any of these 15 access classes are barred. EAB is built around this functionality and reuses the 10 normal access classes. However, contrary to ACB, which applies to all devices, EAB is only applicable to the subset of devices that are configured for EAB. It also allows a network to enable PLMN-specific and domain-specific, i.e., *Packet Switched* or *Circuit Switched*, barring of devices. In GSM/EDGE, *System Information message 21* broadcasted in the network contains the EAB information. In case a network is shared among multiple operators, or more specifically among multiple PLMNs, then EAB can be configured on a per PLMN basis. Up to four additional PLMNs can be supported by a network. *System Information message 22* contains the *network sharing* information for these additional PLMNs and, optionally, corresponding EAB information for each of the PLMNs [5].

ACB and EAB provide means to protect both the radio access and core network from congestion that may occur if a multitude of devices attempt to simultaneous access a network. The 10 normal access classes allow for barring devices using these access classes with a granularity of 10% when using either ACB or EAB. Because both ACB and EAB are controlled via the system information, these mechanisms have an inherent reaction time associated with the time it takes for a device to detect that the system information has been updated and the time required to obtain the latest barring information.

The IR feature introduces an IR flag in a number of messages sent on the downlink (DL) *Common Control CHannel* (CCCH). Before accessing the network, a device configured for *Low Access Priority* [6] is required to decode a message on the DL CCCH and read the IR flag therein. Low-priority indicators are signaled by a device over the *Non Access Stratum* (NAS) interface using the *Device properties* information element [7] and over the *Access Stratum* in the *Packet Resource Request* message [6]. In case the IR flag is set to "1" the device is not permitted to access the GSM network (NW) and is required to await the expiration of a timer before attempting a new access. Because it does not require the reading of system information, IR has the potential benefit of being a faster mechanism than either ACB or EAB type-based barring. When the IR flag is set to "1" in a given DL CCCH message then all devices that read that message when performing system access are barred from NW access. By toggling the flag with a certain periodicity within each of the messages sent on the DL

CCCH, a partial barring of all devices can be achieved. Setting the flag to "1" within all DL CCCH messages sent during the first second of every 10-seconds time interval will, for example, bar 10% of all devices. A device that supports the IR feature may also be configured for EAB.

For LTE, ACB was included already in the first release of LTE, i.e., 3GPP Release 8, while the low-priority indicators were introduced in Release 10 [8]. An NAS low-priority indication was defined in the NAS signaling [15] and an *Establishment Cause* indicating delay tolerant access was introduced in the *Radio Resource Control (RRC) Connection Request* message sent from the device to the base station [9]. These two indicators support congestion control of delay tolerant MTC devices. In case the RRC connection request message signals that the access was made by a delay tolerant device, the base station has the option to reject the connection in case of congestion and via the RRC *Connection Reject* message request the device to wait for the duration of a configured *extended wait timer* before making a new attempt.

In Release 11 the MTC work continued with the work item *System Improvements for MTC* [10]. In TSG RAN EAB was introduced in the LTE specifications. A new *System Information Block 14* (SIB14) was defined to convey the EAB-related information [9]. To allow for fast notification of updates of SIB14 the paging message was equipped with a status flag indicating an update of SIB14. As for TSG GERAN, barring of 10 different access classes is supported. In case of network sharing a separate access class bitmap can, just as for GSM/EDGE, be signaled per PLMN sharing the network. A device with its low-priority indicator set also needs to support EAB.

Table 2.1 summarizes the GSM/EDGE and LTE 3GPP Release 99−11 features designed to provide overload control. It should be noted that ETSI was responsible for the GSM/EDGE specifications until 3GPP Release 99 when 3GPP took over the responsibility for the evolution and maintenance of GSM/EDGE. ACB was, for example, part of GSM/EDGE already before 3GPP Release 99.

Table 2.1 3GPP features related to MTC overload control		
Release	**GSM**	**LTE**
99	Access Class Barring	−
8	−	Access Class Barring
10	Extended Access Barring Implicit Reject Low priority and access delay tolerant indicators	Low priority and access delay tolerant indicators
11	−	Extended Access Barring

2.2.2 SMALL DATA TRANSMISSION

In Release 12 the work item for *Machine-Type Communications and other mobile data applications Communications* [11] triggered a number of activities going beyond the scope of the earlier releases that to a considerable extent were focused on managing large numbers of devices. It resulted in TR 23.887 *Study on Machine-Type Communications (MTC) and other mobile data applications communications enhancements* [12] that introduces solutions to efficiently handle small data trans-missions and solutions to optimize the energy consumptions for devices dependent on battery power.

MTC devices are to a large extent expected to transmit and receive small data packets, especially when viewed at the application layer. Consider, for example, street lighting controlled remotely where turning on and off the light bulb is the main activity. On top of the small application layer payload needed to provide the on/off indication, overhead from higher-layer protocols, for example, *User Datagram* and *Internet Protocols* and radio interface protocols need to be added thereby forming a complete protocol stack. For data packets ranging up to a few hundred bytes the protocol overhead from layers other than the application layer constitutes a significant part of the data transmitted over the radio interface. To optimize the power consumption of devices with a traffic profile characterized by small data transmissions it's of interest to reduce this overhead. In addition to the overhead accumulated over the different layers in the protocol stack, it is also vital to make sure various pro-cedures are streamlined to avoid unnecessary control plane signaling that consumes radio resources and increases the device power consumption. Figure 2.2 shows an overview of the message flow associated with an LTE *Mobile Originated* (MO) data transfer where a single uplink (UL) and single DL data packet are sent. It is clear from the depicted signaling flow that several signaling messages are transmitted before the actual data packets are sent.

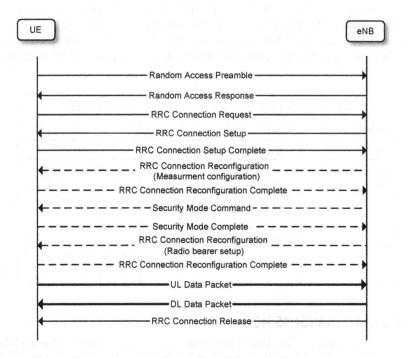

FIGURE 2.2

LTE message transfer associated with the transmission a single UL and single DL data packet. Messages indicated with dashed arrows are eliminated in the *RRC Resume procedure* solution [9].

One of the most promising solutions for support of small data transmission is the *RRC Resume* procedure [9]. It aims to optimize, or reduce the number of signaling messages, that is needed to set up a connection in LTE. Figure 2.2 indicates the part of the connection setup that becomes redundant in the RRC resume procedure, including the Security mode command and the RRC connection reconfiguration messages. The key in this solution is to resume configurations established in a previous connection. Part of the possible optimizations is to also suppress the RRC signaling associated with measurement configuration. This simplification is justified by the short data transfers expected for MTC. For these devices measurement reporting is less relevant compared to when long transmissions of data are dominating the traffic profile. In 3GPP Release 13 this solution was specified together with the *Control plane CIoT EPS Optimization* [9] as two alternative solutions adopted for streamlining the LTE setup procedure to facilitate small and infrequent data transmission [13]. These two solutions are highly important to optimize latency and power consumption for LTE-M and NB-IoT evaluated in Chapters 6 and 8, respectively.

2.2.3 DEVICE POWER SAVINGS

The 3GPP Release 12 study on MTC and other mobile data applications communications enhancements introduced two important solutions to optimize the device Power Consumption, namely PSM and eDRX. PSM was specified both for GSM/EDGE and LTE and is a solution where a device enters a power saving state in which it reduces its power consumption to a bare minimum [14]. While in the power saving state the mobile is not monitoring paging and consequently becomes unreachable for *Mobile Terminated* (MT) services. In terms of power efficiency this is a step beyond the typical *Idle Mode* behavior where a device still performs energy consuming tasks such as neighbor cell measurements and maintaining reachability by listening for paging messages. The device leaves PSM when higher layers in the device triggers an MO access, e.g., for an UL data transfer or for a periodic Tracking Area Update/Routing Area Update (TAU/RAU). After the MO access and the corresponding data transfer have been completed, a device using PSM starts an *Active Timer*. The device remains reachable for MT traffic by monitoring the paging channel until the Active Timer expires. When the Active timer expires the device reenters the power saving state and is therefore unreachable until the next MO event. To meet MT reachability requirements of a service a GSM/EDGE device using PSM can be configured to perform a periodic RAU with a periodicity in the range of seconds up to a year [7]. For a LTE device the same behavior can be achieved through configuration of periodic TAU [15]. Compared to simply turning off a device, PSM has the advantage of supporting the mentioned MT reachability via RAU or TAU. In PSM the device also stays registered in the network and maintains its connection configurations. As such, when leaving the power saving state in response to a MO event the device does not need to first attach to the network and setup the connection, as it would otherwise need to do when being turned on after previously performing a complete power off. This reduces the signaling overhead and optimizes the device power consumption.

Figure 2.3 depicts the operation of a device configured for PSM when performing periodic RAUs and reading of paging messages according to the idle mode DRX cycle applicable when the Active Timer is running. The RAU procedure is significantly more costly than reading of a paging message as indicated by Figure 2.3. During the Active timer the device is in idle mode and is required to operate accordingly. After the expiry of the Active Timer the device again reenters the energy-efficient PSM.

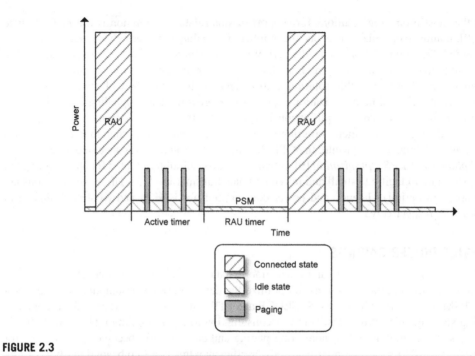

FIGURE 2.3

Illustration of operation in PSM including periodic RAUs.

In Release 13 also eDRX was specified for GSM and LTE. The general principle for eDRX is to extend the DRX cycles to allow a device to remain longer in a power saving state between *Paging Occasions* and thereby minimize its energy consumption. The advantage over PSM is that the device remains periodically available for MT services without the need to first perform e.g., a RAU or TAU to trigger a limited period of reachability. The *Study on power saving for Machine-Type Communication (MTC) devices* [16] considered, among other things, the energy consumption of devices using eDRX or PSM. The impacts of using PSM and eDRX on device battery life, assuming a 5 Watt-hour (Wh) battery, were characterized as part of the study. More specifically, the battery life for a device was predicted for a range of triggering intervals and reachability periods. A trigger may e.g., correspond to the start of a MT data transmission wherein an application server requests a device to transmit a report. After reception of the request the device is assumed to respond with the transmission of the requested report. The triggering interval is defined as the interval between two adjacent MT events. The reachability period is, on the other hand, defined as the period between opportunities for the network to reach the device using a Paging channel. Let us, for example, consider an alarm condition that might only trigger on average once per year, but when it does occur there is near real-time requirement for an application server to know about it. For this example the ongoing operability of the device capable of generating the alarm condition can be verified by the network sending it a Page request message and receiving a corresponding Page response. Once the device operability is verified, the network can send the application layer message that serves to trigger the reporting of any alarm condition that may exist.

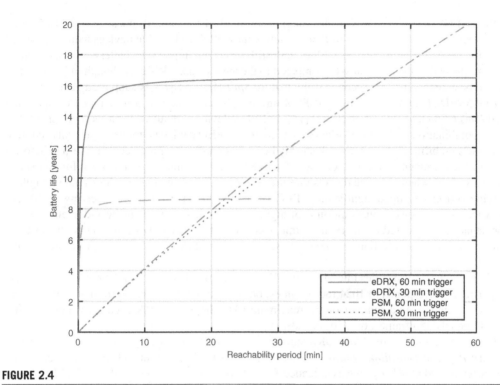

FIGURE 2.4

Estimated power consumption for a GSM/EDGE device configured to use PSM or eDRX. [16].

Figure 2.4 presents the estimated GSM/EDGE device battery life when using PSM or eDRX. Reachability for PSM was achieved by the device by performing a periodic RAU, which initiates a period of network reachability that continues until the expiration of the Active Timer. Both in case of eDRX and PSM it was here assumed that the device, before reading a Page or performing a RAU, must confirm the serving cell identity and measure the signal strength of the serving cell. This is to verify that the serving cell remains the same and continues to be *suitable* from a signal strength perspective. When deriving the results depicted in Figure 2.4 the energy costs of confirming the cell identity, estimating the serving cell signal strength, reading a Page, performing a TAU, and finally transmitting the report were all taken from available results provided within the *Study on power saving for MTC devices* [16]. A dependency both on the reachability period and on the triggering interval is seen in Figure 2.4. For eDRX a very strong dependency on the triggering interval is seen. The reason behind this is that the cost of sending the report is overshadowing the cost of being paged. For PSM the cost of performing a RAU is in this example of similar magnitude as sending the actual report so the reachability period becomes the dominating factor, while the dependency on triggering interval becomes less pronounced. For a given triggering interval Figure 2.4 shows that eDRX is outperforming PSM when a shorter reachability is required, while PSM excels when the reachability requirement is in the same range as the trigger interval.

In the end GSM/EDGE eDRX cycles ranging up to 13,312 51-multiframes, or roughly 52 min, were specified. A motivation for not further extending the eDRX cycle is that devices with an expected reachability beyond 1 hour may use PSM and still reach an impressive battery life, as seen in Figure 2.4. The eDRX cycle can also be compared to the legacy max DRX cycle length of 2.1 seconds, which can be extended to 15.3 seconds if the feature *Split Paging Cycle* is supported [17].

For GSM/EDGE 3GPP went beyond PSM and eDRX and specified a new mode of operation denoted *Power Efficient Operation* (PEO) [18]. In PEO a device is required to support either PSM or eDRX, in combination with relaxed idle mode behavior. A PEO device is, for example, only required to verify the suitability of its serving cell shortly before its nominal paging occasions or just before a MO event. Measurements on a reduced set of neighbor cells is only triggered for a limited set of conditions such as when a device detects that the serving cell has changed or the signal strength of the serving cell has dropped significantly. PEO is mainly intended for devices relying on battery power where device power consumption is of higher priority than, e.g., *mobility* and latency, which may be negatively impacted by the reduced Idle Mode activities. Instead of camping on the best cell the aim of PEO is to assure that the device is served by a cell that is good enough to provide the required services.

For LTE, Release 13 specifies Idle Mode eDRX cycles ranging between 1 and 256 hyperframes. As one hyperframe corresponds to 1024 radio frames, or 10.24 seconds, 256 hyperframes correspond to roughly 43.5 minutes. As a comparison the maximum LTE Idle Mode DRX cycle length used before Release 13 equals 256 frames, or 2.56 seconds.

LTE does in addition to Idle Mode DRX support *Connected Mode* DRX to relax the requirement on reading the *Physical Downlink Control Channel* (PDCCH) for DL assignments and UL grants. The LTE connected mode DRX cycle was extended from 2.56 to 10.24 seconds in Release 13 [9].

Table 2.2 summarizes the highest configurable MT reachability periodicities for GSM and LTE when using Idle Mode eDRX, Connected Mode DRX, or PSM. For PSM the assumption is that the MT reachability periodicity is achieved through the configuration of periodic RAUs/TAUs.

Table 2.2 Mobile terminated reachability periodicities for GSM and LTE when using idle mode eDRX, connected mode eDRX, or PSM with RAU/TAU-based reachability

	GSM	LTE
Idle mode eDRX	52 min, 13.44 s	43 min, 41.44 s
Connected mode eDRX	–	10.24 s
PSM	>1 year (RAU)	>1 year (TAU)

In general it is expected that the advantage of eDRX over PSM for frequent reachability periods is reduced for LTE compared with what can be expected in GSM/EDGE. The reason is that a typical GSM/EDGE device uses 33 dBm output power, while LTE devices typically use 23 dBm output power. This implies that the cost of transmission and a RAU/TAU in relation to receiving a Page is much higher for GSM/EDGE than what is the case for LTE.

Table 2.3 summarizes the features discussed in this section and specified in Release 12 and 13 to optimize the mobile power consumption.

Table 2.3 3GPP Release 12 and 13 features related to device power savings

Release	GSM	LTE
12	Power Save Mode	Power Save Mode
13	Extended DRX Power Efficient Operation	Extended DRX

Chapters 3, 5, and 7 will further discuss how the concepts of Paging, Idle, and Connected Mode DRX and eDRX have been designed for EC-GSM-IoT, LTE-M and NB-IoT. It will then be seen that the DRX cycles mentioned in Table 2.2 have been further extended for NB-IoT to support low power consumption and long device battery life.

2.2.4 STUDY ON PROVISION OF LOW-COST MTC DEVICES BASED ON LTE

LTE uses the concept of *device categories* referred to as *User Equipment* (UE) categories (Cat) to indicate the capability and performance of different types of equipment. As the specifications have evolved, the number of device categories has increased. Before Release 12, Category 1 (Cat-1), developed in the first LTE Release, was considered the most rudimentary device category. Since it was originally designed to support mobile broadband services with data rates of 10 Mbps for the DL and 5 Mbps for the UL, it is still considered too complex to compete with GPRS in the low-end MTC segment. To change this, 3GPP TSG RAN initiated the Release 12 study item called *Study on Provision of low-cost MTC UEs based on LTE* [19], here referred to as the *LTE-M study item*, with the ambition to study solutions providing lower device complexity in combination with improved coverage.

A number of solutions for lowering the complexity and cost of the *radio frequency* (RF) and baseband parts of an LTE *modem* were proposed in the scope of the LTE-M study item. It was concluded that a reduction in transmission and reception bandwidths and peak data rates in combination with adopting a single RF receive chain and *half-duplex* operation would make the cost of an LTE device modem comparable to the cost of an EGPRS modem. A reduction in the maximum supported transmission and reception bandwidths and adopting a single RF receive chain reduces the complexity in both the RF and the baseband because of, e.g., reduced RF filtering cost, reduced sampling rate in the *analog-to-digital and digital-to-analog conversion* (ADC/DAC), and reduced number of baseband operations needed to be performed. A peak data rate reduction helps reduce the baseband complexity in both demodulation and decoding parts. Going from *full-duplex* operation as supported by Cat-1 devices to half duplex allows the *duplex filter(s)* in the RF front end to be replaced with a less costly switch. Furthermore, a reduction in the transmission power can also be considered, which relaxes the requirements on the RF front-end *power amplifier* and may support integration of the power amplifier on the chip that is expected to reduce device complexity and manufacturing costs. Table 2.4 summarizes the findings recorded in the LTE-M study item for the individual cost reduction techniques and also indicates the expected impact on coverage from each of the solutions. Refer to

Table 2.4 Overview of measures supporting an LTE modem cost reduction [19]

Objective	Modem cost reduction [%]	Coverage impact
Limit FDD operation to half duplex	7−10	None
Peak rate reduction through limiting the maximum transport block size (TBS) to 1000 bits	10.5−21	None
Reduce the transmission and reception bandwidth for both RF and baseband to 1.4 MHz	39	1−3 dB DL coverage reduction due to loss in frequency diversity
Limit RF front end to support a single receive branch	24−29	4 dB DL coverage reduction due to loss in receive diversity
Transmit power reduction to support power amplifier integration	10−12	UL coverage loss proportional to the reduction in transmit power

Table 6.15 for cost estimates for combinations of multiple cost reduction techniques because the cost savings are not additive in all cases. The main impact on DL coverage is caused by going to a single RF chain, i.e., one receive antenna instead of two. If a lower transmit power is used in UL, this will cause a corresponding coverage loss in UL. Reducing the signal bandwidth to 1.4 MHz may cause coverage loss due to reduced frequency diversity. This can however be partly compensated for by use of frequency hopping.

Besides studying means to facilitate low device complexity the LTE-M study item [19] provided an analysis of the existing LTE coverage and presented means to improve it by up to 20 dB. Table 2.5 summarizes the frequency-division duplex (FDD) LTE Maximum Coupling Loss (MCL) presented in TR 36.888 assuming that the eNB supports two transmit and two receive antennas. The reference LTE device was assumed to be equipped with a single transmit and two receive antennas. The results were obtained through simulations assuming DL *Transmission Mode 2 (TM2)*, i.e., DL transmit diversity. It is seen that the *Physical Uplink Shared Channel* (PUSCH) is limiting the LTE coverage to a MCL of 140.7 dB.

The initial target of the LTE-M study item [19] was to provide 20 dB extra coverage for low-cost MTC devices leading to a MCL of 160.7 dB. After investigating the feasibility of extending the coverage of each of the channels listed in Table 2.5 to 160.7 dB through techniques such as *transmission time interval (TTI) bundling*, *Hybrid Automatic Repeat Request (HARQ) retransmissions* and repetitions, it was finally concluded that a coverage improvement of 15 dB leading to a MCL of 155.7 dB was appropriate to target for low-complexity MTC devices based on LTE.

The LTE-M study item triggered a 3GPP Release 12 work item [20] introducing an LTE device category (Cat-0) of low-complexity and a Release 13 work item [21] introducing an LTE-M device category (Cat-M1) of even lower-complexity for low-end MTC applications and the needed functionality to extend the coverage for LTE and LTE-M devices. Chapters 5 and 6 will present in detail the design and performance of LTE-M that is the result of these two work items.

Table 2.5 Overview of LTE MCL performance [19]

#	Physical channel name	PUCCH Format 1A	PRACH	PUSCH	PDSCH	PBCH	PSS/SSS	PDCCH Format 1A
1	Data rate [kbps]			20	20			
Transmitter								
2	Total Tx power [dBm]	23	23	23	46	46	46	46
3	Power boosting [dB]	–	–	–	0	0	0	0
4	Actual Tx power [dBm]	23	23	23	32	36.8	36.8	42.8
Receiver								
5	Thermal noise [dBm/Hz]	–174	–174	–174	–174	–174	–174	–174
6	Receiver noise figure [dB]	5	5	5	9	9	9	9
7	Interference margin [dB]	0	0	0	0	0	0	0
8	Channel bandwidth [kHz]	180	1080	360	360	1080	1080	4320
9	Effective noise power [dBm] $= (5) + (6) + (7) + 10\log_{10}(8)$	–116.4	–108.7	–113.4	–109.4	–104.7	–104.7	–98.6
10	Required SINR [dB]	–7.8	–10.0	–4.3	–4.0	–7.5	–7.8	–4.7
11	Dual antenna receiver sensitivity [dBm] $= (9) + (10)$	–124.24	–118.7	–117.7	–113.4	–112.2	–112.5	–103.34
12	MCL [dB] $= (4) - (11)$	147.2	141.7	140.7	145.4	149.0	149.3	146.1

2.2.5 STUDY ON CELLULAR SYSTEM SUPPORT FOR ULTRA-LOW COMPLEXITY AND LOW THROUGHPUT INTERNET OF THINGS

In 3GPP Release 13 the study item on *Cellular System Support for Ultra-Low Complexity and Low Throughput Internet of Things* [1], here referred to as the *Cellular IoT study item*, was started in 3GPP TSG GERAN. It shared, as discussed in Section 1.2.3, many commonalities with the *LTE-M study item* [19], but it went further both in terms of requirements and in that it was open to GSM backward compatible solutions as well as to nonbackward compatible radio access technologies. The work attracted considerable interest, and 3GPP TR 45.820 Cellular system support for ultra-low complexity and low throughput IoT capturing the outcome of the work contains several solutions based on GSM/EDGE, on LTE and nonbackwards compatible solutions, so-called *Clean Slate* solutions.

Just as in the LTE-M study item improved coverage was targeted, this time by 20 dB compared to GPRS. Table 2.6 presents the GPRS reference coverage calculated by 3GPP. It is based on the minimum GSM/EDGE *BLock Error Rate* (BLER) performance requirements specified in 3GPP TS 45.005 *Radio transmission and reception* [22]. For the DL the specified device receiver *Sensitivity* of −102 dBm was assumed to be valid for a device Noise Figure (NF) of 9 dB. When adjusted to a NF of 5 dB, which was assumed suitable for IoT devices, the GPRS *Reference Sensitivity* ended up at −106 dBm. For the UL 3GPP TS 45.005 specifies a GPRS single antenna base station sensitivity of −104 dBm that was assumed valid for a NF of 5 dB. Under the assumption that a modern base station supports a NF of 3 dB the UL sensitivity reference also ended up at −106 dBm. To make the results applicable to a base station supporting receive diversity a 5 dB processing gain was also added to the UL reference performance.

Table 2.6 Overview of GPRS MCL performance [8]			
#	Link direction	DL	UL
Transmitter			
1	Total Tx power [dBm]	43	33
Receiver			
2	Thermal noise [dBm/Hz]	−174	−174
3	Receiver noise figure [dB]	5	3
4	Interference margin [dB]	0	0
5	Channel bandwidth [kHz]	180	180
6	Effective noise power [dBm] $= (2) + (3) + (4) + 10\log_{10}(5)$	−116.4	−108.7
7	Single antenna receiver sensitivity according to 3GPP TS 45.005 [dBm]	−102 at NF 9 dB	−104 at NF 5 dB
8	Single antenna receiver sensitivity according to 3GPP TR 45.820 [dBm]	−106.0 at NF 5 dB	−106 at NF 3 dB
9	Required SINR [dB] $= (8) − (6)$	10.4	12.4
10	Receiver processing gain [dB]	0	5
11	MCL [dB] $= (1) − ((8) − (10))$	149	144

The resulting GPRS MCL ended up at 144 dB because of limiting UL performance. As the target of the CIoT study item was to provide 20 dB coverage improvements on top of GPRS, this led to a stringent MCL requirement of 164 dB.

After the CIoT study item had concluded, normative work began in 3GPP Release 13 on EC-GSM-IoT [23] and NB-IoT [24]. Chapters 3, 4, 7, and 8 go into detail and present how EC-GSM-IoT and NB-IoT were designed to meet all the objectives of the CIoT study item.

When comparing the initially targeted coverage for EC-GSM-IoT, NB-IoT, and LTE-M, it is worth to notice that Tables 2.5 and 2.6 are based on different assumptions, which complicates a direct comparison between the LTE-M target of 155.7 dBs MCL and the EC-GSM-IoT and NB-IoT target of 164 dB. Table 2.5 assumes, e.g., a base station NF of 5 dB, while Table 2.6 uses an NF value 3 dB. If those assumptions had been aligned, the LTE reference MCL had ended up at 142.7 dB and the LTE-M initial MCL target at 157.7 dB. As described in Section 6.2, if one takes into account that the LTE-M coverage target is assumed to be fulfilled for 20-dBm LTE-M devices, but that all the LTE-M coverage enhancement techniques are also available to 23-dBm LTE-M devices, the difference between the LTE-M coverage target and the 164-dB target shrinks to 3.3 dB. The actual coverage performance of EC-GSM-IoT, LTE-M, and NB-IoT is presented in Chapters 4, 6, and 8, respectively.

REFERENCES

[1] Third Generation Partnership Project, Technical Report 45.820 v13.0.0, Cellular System Support for Ultra-Low Complexity and Low Throughput Internet of Things, 2016.

[2] Third Generation Partnership Project, Technical Report 22.868 V8.0.0, Study on Facilitating Machine to Machine Communication in 3GPP Systems, 2016.

[3] Third Generation Partnership Project TSG SA WG2, SP-100863 Update to Network Improvements for Machine Type Communication, 3GPP TSG SA Meeting #50, 2010.

[4] Third Generation Partnership Project, Technical Specification 22.011 v14.0.0, Service Accessibility, 2016.

[5] Third Generation Partnership Project, Technical Specification 44.018 v14.00, Mobile Radio Interface Layer 3 Specification; Radio Resource Control (RRC) Protocol, 2016.

[6] Third Generation Partnership Project, Technical Specification 23.060 v14.0.0, General Packet Radio Service (GPRS); Service Description; Stage 2, 2016.

[7] Third Generation Partnership Project, Technical Specification 24.008 v14.0.0, Mobile Radio Interface Layer 3 Specification; Core Network Protocols; Stage 3, 2016.

[8] Third Generation Partnership Project, Technical Specification 23.401 v14.0.0, General Packet Radio Service (GPRS) Enhancements for Evolved Universal Terrestrial Radio Access Network (E-UTRAN) Access, 2016.

[9] Third Generation Partnership Project, Technical Specification 36.331 v14.0.0, Evolved Universal Terrestrial Radio Access (E-UTRA); Radio Resource Control (RRC); Protocol Specification, 2016.

[10] Third Generation Partnership Project TSG SA WG3, SP-120848 System Improvements for Machine Type Communication, 3GPP TSG SA Meeting #58, 2012.

[11] Third Generation Partnership Project TSG SA WG3, SP-130327 Work Item for Machine-Type and Other Mobile Data Applications Communications Enhancements, 3GPP TSG SA Meeting #60, 2013.

[12] Third Generation Partnership Project, Technical Report 23.887 v12.0.0, Study on Machine-Type Communications (MTC) and Other Mobile Data Applications Communications Enhancements, 2016.

[13] Third Generation Partnership Project, Technical Specification 36.300 v14.0.0, Evolved Universal Terrestrial Radio Access (E-UTRA) and Evolved Universal Terrestrial Radio Access Network (E-UTRAN); Overall Description; Stage 2, 2016.

[14] Third Generation Partnership Project, Technical Specification 23.682 v14.0.0, Architecture Enhancements to Facilitate Communications with Packet Data Networks and Applications, 2016.

[15] Third Generation Partnership Project, Technical Specification 24.301 v14.0.0, Non-Access-Stratum (NAS) Protocol for Evolved Packet System (EPS); Stage 3, 2016.

[16] Third Generation Partnership Project, Technical Report 43.869, GERAN Study on Power Saving for MTC Devices, 2016.

[17] Third Generation Partnership Project, Technical Specification 45.002 v14.0.0, Multiplexing and Multiple Access on the Radio Path, 2016.

[18] Third Generation Partnership Project, Technical Specification 43.064 v14.0.0, General Packet Radio Service (GPRS); Overall Description of the GPRS Radio Interface; Stage 2, 2016.

[19] Third Generation Partnership Project, Technical Report 36.888 v12.0.0, Study on Provision of Low-Cost Machine-Type Communications (MTC) User Equipments (UEs) Based on LTE, 2013.

[20] Vodafone Group, et al., RP-140522 Revised Work Item on Low Cost & Enhanced Coverage MTC UE for LTE, 3GPP RAN Meeting #63, 2014.

[21] Ericsson, et al., RP-141660 Further LTE Physical Layer Enhancements for MTC, 3GPP RAN Meeting #65, 2014.

[22] Third Generation Partnership Project, Technical Specification 45.005 v14.0.0, Radio Transmission and Reception, 2016.

[23] Ericsson, et al., GP-151039 New Work Item on Extended Coverage GSM (EC-GSM) for Support of Cellular Internet of Things, 3GPP TSG GERAN Meeting #67, 2015.

[24] Qualcomm Incorporated, et al., RP-151621 Narrowband IOT, 3GPP TSG RAN Meeting #60, 2015.

EC-GSM-IoT

CHAPTER OUTLINE

Cellular Internet of Things. http://dx.doi.org/10.1016/B978-0-12-812458-1.00003-4
Copyright © 2018 Elsevier Ltd. All rights reserved.

Abstract

This chapter presents the design of *Extended Coverage Global System for Mobile Communications Internet of Things* (EC-GSM-IoT). The initial section describes the background of the GSM radio access technology, highlighting the suitability of an evolved GSM design to support the Cellular IoT (CIoT) core requirements, which includes ubiquitous coverage, ultra-low-device cost, and energy efficient device operation. The following sections builds from the ground up, starting with the physical layer, going through fundamental design choices such as frame structure, modulation, and channel coding. After the physical layer is covered, the basic procedures for support of full system operation are covered, including, for example, system access, paging functionality, and for EC-GSM-IoT-improved security protocols. The reader will not only have a good knowledge of EC-GSM-IoT after reading the chapter but will also have a basic understanding of what characteristics a system developed for CIoT should possess. At the end, a look at the latest enhancements of the EC-GSM-IoT design is presented.

3.1 BACKGROUND
3.1.1 THE HISTORY OF GSM

The GSM technology was first developed in Europe in 1980s, having its first commercial launch in 1991. In current writing (2017), the technology has already turned 26 years. Despite this it still maintains global coverage and is still among the most widely used cellular technologies. This is even true in the most mature markets, such as western Europe. Compared to previous analog cellular technologies, GSM is based on digital communication that allows for an encrypted and more spectrally efficient communication. The group developing the GSM technology was called Group Spécial Mobile (GSM) but with the global success of the technology it is now referred to as Global System for Mobile Communications (GSM).

The initial releases of the GSM standard were only defined for *circuit switched* (CS) services, including both voice and data calls. A call being CS implies that a set of radio resources are occupied for the duration of the call. Even if the transmitter is silent in a CS call, the resources are occupied from a network perspective and cannot be allocated to another user.

In 1996 work was started to introduce *packet switched* services (PS). A PS service no longer occupies the resources for the full duration of the call, but instead intends to only occupy resources when there are data to send. The first PS service was called General Packet Radio Service (GPRS) and was launched commercially in 2000. Following the success of GPRS, the PS domain was further enhanced by the introduction of Enhanced GPRS (EGPRS), also known as EDGE (Enhanced Data Rates for GSM Evolution), supporting higher end user data rates, primarily by the introduction of higher order modulation and improved protocol handling. In current GSM/EDGE networks, CS services are still used for speech calls, and the PS service is predominantly used for providing data services.

Since the deployment of the first GSM network in 1991, the technology has truly become the global cellular technology. It is estimated that it today reaches over 90% of the world's population [1] as it is deployed in close to all countries in the world. In all of these networks, voice services are supported and in the vast majority of networks there is also support for GPRS/EDGE.

3.1.2 CHARACTERISTICS SUITABLE FOR IoT
3.1.2.1 Global Deployment

Considering that GSM has been deployed since 25 years, it is only natural that the characteristics of GSM/EDGE networks globally are vastly different. In countries where GSM/EDGE is deployed together with 3G and/or 4G, it is typically used as one of the main carriers of speech services, while serving as a fallback solution for data services. However, it is not always the case that devices are capable of 3G or 4G, even if the networks in many cases are. For example, in the Middle East and Africa around 75% of subscriptions 2015 were GSM/EDGE-only [1]. Furthermore, looking at the global subscription base of cellular technologies, GSM/EDGE-only subscriptions account for around 50% of the total subscription base [1]. For the remaining 50%, the vast majority of subscriptions supporting 3G and/or 4G technologies will also include GSM/EDGE capability as fallback when coverage is lost. This global presence is beneficial for all radio access technologies, including those providing Machine-Type Communication (MTC), e.g., not only to support roaming but also to reach a high level of economy of scale.

3.1.2.2 Number of Frequency Bands

An important aspect of the global success of GSM lies also in the number of frequency bands it is deployed in. Although the GSM specifications support a wider range of frequency bands, the deployment of GSM/EDGE is limited to four global frequency bands: 850, 900, 1800, and 1900 MHz. This global allocation of just four bands spectrum is much more aligned compared with other cellular technologies. The spectrum regulations used in different parts of the world typically pair the use of 900 MHz with 1800 MHz and 850 MHz with 1900 MHz. This means that in most regions only two of the frequency bands are allowed for GSM operation. For a device manufacturer, this aspect is important because to provide a single band or *dual band* device (one low band and one high band combination), which will already cover a large part of the world population, while a *quad band* device supporting all four bands truly provides global coverage. Having to support a lower number of frequency bands means less *radio frequency* (RF) components needed in the device, less optimization of components that are to be operable over multiple frequency bands, and in the end a lower *Bill of Material* and overall development cost.

Another important characteristic for cellular systems, especially for those targeting IoT, is to have extensive coverage to reach remote locations, such as deep indoor. Operation of GSM, as well as 3G and 4G technologies, is today defined from a few hundreds of MHz to a couple of GHz in the radio spectrum. It is a well known fact that the propagation loss is dependent on the carrier frequency, and that it is important to keep the carrier frequency low to minimize the path loss due to signal propagation and by that maximize the coverage provided. Because GSM is globally available either on the 900 MHz band or the 850 MHz band, it is well suited for a technology attempting to maximize coverage because the loss due to propagation is low.

Another important aspect of frequency band support in a device is roaming, which is also an important feature for CIoT. Take, for example, a tracking device where a module is integrated in a container, shipped around the world. If such a device would have to support a large variety of frequency bands just to be able to operate in the country it is shipped to, it will drive device complexity and cost.

In summary, the number of frequency bands deployed is an important factor when it comes to avoiding market fragmentation, reaching economies of scale, and allow for global roaming of devices.

3.1.2.3 Small Spectrum Deployment

Although GSM has a truly global footprint, 3G and 4G compete for the same scarce spectrum as GSM. With the advent of a fifth generation cellular system even more pressure will be put on the spectrum resources. Deployment of new technologies not only changes what technologies are available for the end consumer but will also impact how the traffic between the technologies is distributed. As the traffic shifts from GSM to 3G and/or 4G, it is possible to release parts of the spectrum used for GSM/EDGE and make room for the increased spectrum needs for improved end user experience and system capacity coming from the new technologies. This is referred to as *spectrum refarming*.

Refarming of GSM spectrum was started several years ago and will continue in the future. However, even in mature cellular markets there is one thing to reduce the spectrum operation of GSM and another to turn the GSM network completely off. Even if there are only, let's say, a few tenths of thousand devices in the network there might be contracts/subscriptions with the operator that are not possible to end prematurely. This is especially true for the *Machine-to-Machine* (M2M) market where, for example, devices can be placed in remote locations with contracts that can last for several decades. So, the spectrum allocated for GSM services is expected to shrink with time, in many markets the GSM networks will live for still a long time to come. To make future GSM deployments attractive for the operator, it is of interest to deploy the network as spectrally efficient as possible, in an as low spectrum allocation as possible.

3.1.2.4 Module Price

The low price on GSM/EDGE devices is probably one of the more important reasons why GSM is the main M2M cellular carrier in networks today. In Figure 3.1 [2], the M2M module price is estimated for a range of cellular technologies up to 2016. As can be seen, the GSM/EDGE module cost is considerably lower than other technologies reaching a global selling price of around USD 6−7. To this selling price there are of course regional variations, and, for example, the estimated average selling price in China for the same module is USD 4 [2]. There are several reasons why such low-average selling price can be achieved of which most have already been mentioned, such as a global deployment providing economies of scale, a mature and relatively low-complex technology that has been optimized in product implementations over the last 25 years, low number of frequency band-specific RF components due to the low number of frequency bands used for the technology.

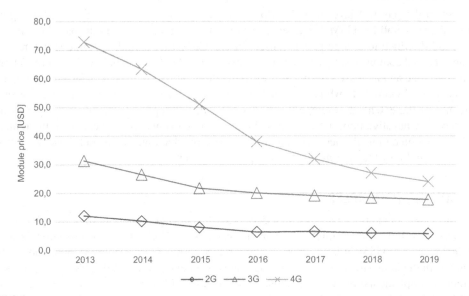

FIGURE 3.1

M2M module price by technology.

From IHS Markit, Technology Group, Cellular IoT Market Tracker − 1Q 2017.

3.1.3 ENHANCEMENTS UNDERTAKEN BY 3GPP

It was based on the knowledge above that Third Generation Partnership Project (3GPP) in its 13th release decided to further evolve GSM to cater for the requirements coming from the IoT. Beyond the requirements mentioned in Sections 1.2.3 and 2.5 to increase coverage, secure low-device cost, and high battery lifetime the work included the following:

- Provide a means to operate the technology in an as tight spectrum allocation as 600 kHz for an operator to deploy a GSM network completely removing, or at least minimizing, a conflict in spectrum usage with other technologies. In such small spectrum allocation, it can even be possible to deploy the GSM network in the guard band of wideband technologies such as 3G and 4G.
- Improve end user security to an (Long-Term Evolution) LTE/4G-grade security level to remove any security concerns that may exist in current GSM deployments.
- Ensure that all changes brought to the GSM standard by the introduction of EC-GSM-IoT ensure backward compatibility with already existing GSM deployments, to allow a seamless and gradual introduction of the technology sharing resources with existing devices.
- Ensure support of a massive number of CIoT devices in the network.

The remainder of this chapter will in detail outline how the above design guidelines have been followed by redesigning the GSM standard, and in Chapter 4 the performance evaluation of the technology will show how the performance objectives are fulfilled.

3.2 PHYSICAL LAYER

In this section the physical layer of EC-GSM-IoT is described. The section starts with the guiding principles in the overall EC-GSM-IoT design. The reader is then introduced to the basic physical layer design of GSM, which to a large part is reused for EC-GSM-IoT, including frame structure, modulation, and channel coding. The two main parts of this section are then in detail looking into techniques to extend coverage and increase system capacity. Extending the coverage is the new mode of operation introduced by EC-GSM-IoT, while the need to increase system capacity can more be seen as a consequence of the extended coverage where devices will take up more resources to operate in the system, having a negative impact on capacity. When the main concepts are described, the remaining parts of this section outline the network operation in terms of what logical channels have been defined, their purpose, and how they are mapped to the physical channels.

3.2.1 GUIDING PRINCIPLES

When redesigning a mature technology, first deployed 25 years ago, it is important to understand any limitations that current products might place on the design. For EC-GSM-IoT to build upon the already global deployment of GSM, it is of uttermost importance to be able to deploy it on already existing base stations in the field. However, it is also important for an existing GPRS/EGPRS device to be able to upgrade its implementation to either replace the GPRS/EGPRS implementation or to implement EC-GSM-IoT in addition to a GPRS/EGPRS implementation. In both cases, it is important for the design of EC-GSM-IoT to have a common design base as GPRS/EGPRS as possible.

A base station can be considered to consist of two main parts, a digital unit (or a baseband unit) and a radio unit. Supporting EC-GSM-IoT, in addition to GSM/EDGE, will mean that additional implementation effort is needed, at least on the digital unit to implement the new protocol stack and new physical layer designs. However, the support of the new feature cannot imply that hardware upgrades are needed, which would mean that products in the field need to be replaced. Assume, for example, that the base station's required processing power or sampling rate is increased to an extent where it is no longer possible to support it on existing digital units. This would imply a major investment for the operator to replace the base station hardware. In addition, for a technology that aims for extremely good coverage, it would result in a spotty network coverage until most/all of the base stations have been upgraded. Similarly, the radio unit has been designed with the current GSM signal characteristics in mind and, significantly, changing those such as increasing the signal dynamics could mean that the technology can no longer be supported by already deployed infrastructure.

A similar situation exists on the device side where a change to the physical layer that requires a new device platform to be developed would mean a significant investment in research, development, and verification. If in contrast, the basic principles of the physical layer are kept as close to GSM as possible, an already existing GPRS/EDGE platform would be able to be updated through a software upgrade to support EC-GSM-IoT, and if the additional complexity from EC-GSM-IoT is kept to a minimum, the same platform would be able to also support GPRS/EDGE operation. This, of course, does not prevent development of an EC-GSM-IoT-specific platform, more optimized in, for example, energy consumption and cost than the corresponding GPRS/EGPRS platform. However, if existing platforms can be used, a gradual introduction of the feature onto the market without huge investment costs is possible.

In addition to the aspects of product implementation mentioned above, also the network operation of EC-GSM-IoT needs to be considered. GSM networks have been operable for many years and are planned to be operable for many years to come as discussed in Section 3.1. The technology has hence been designed to be fully backward compatible with existing GSM/EDGE deployments and network configurations to seamlessly be able to multiplex GPRS/EDGE traffic with EC-GSM-IoT traffic on the same resources. By pooling the resources in this way, less overall resources will be consumed in the GSM network, and less spectrum resources will, in the end, be required for network operation.

Following these guiding principles will naturally mean that the physical layer to a large extent will be identical to already existing GSM/EDGE. At the same time, changes are required to meet the design objectives listed in Section 3.1.3.

The following section will, to some extent, be a repetition of the physical layer of GSM to serve as a basis for understanding EC-GSM-IoT. Focus will, however, be on new designs that have been added to the specifications by the introduction of EC-GSM-IoT, and details of the GSM physical layer that is not relevant for the understanding of EC-GSM-IoT have intentionally been left out.

3.2.2 PHYSICAL LAYER NUMEROLOGY

3.2.2.1 Channel Raster

GSM is based on a combination of frequency division multiple access and time division multiple access (TDMA).

The channels in frequency are each separated by 200 kHz, and their absolute placement in frequency, the so-called *channel raster*, is defined also in steps of 200 kHz. This means that the placement of the 200 kHz channels in frequency is not completely arbitrary, and that the center frequency of the channel (f_c) needs to be divisible by 200 kHz. Because the placements in frequency, in a given frequency band, are limited by the channel raster, the channels can be numbered for easier reference and are referred to as *absolute radio frequency carrier number* (*ARFCN*). That is, for a given frequency band and absolute carrier frequency the ARFCN value is fixed.

3.2.2.2 Frame Structure

In time a frame structure is defined. Each, so-called *TDMA* frame, is divided into eight timeslots (TSs). To reference a specific point in time that exceeds the duration of a TDMA frame, the TDMA frames are grouped into hierarchical frame structure including *multiframes*, *superframes*, and *hyperframes*. The time reference in the overall frame structure is within a hyperframe accuracy, being roughly 3.5 h long.

To start with, the TDMA frame is grouped into one of two multiframes, either a 51 multiframe or a 52 multiframe.

The 51 multiframe carries channels that are used for time and frequency synchronization (Frequency Correction CHannel (FCCH), SCH, and EC-SCH), (common control channels (CCCH) and Extended Coverage Common Control CHannel (EC-CCCH), see Section 3.2.6 for more details), and (broadcast channels (BCCH) and Extended Coverage Broadcast CHannel (EC-BCCH), see Section 3.2.6 for more details) and are always mapped onto the broadcast carrier (the BCCH carrier).

The 52 multiframe is used by the packet traffic channels (Packet Data Traffic CHannel (PDTCH) and Extended Coverage Packet Data Traffic CHannel (EC-PDTCH), see Sections 3.2.6 and 3.2.7 for the downlink (DL) and uplink (UL), respectively) and their associated control channels (*Packet Associated Control CHannel* (PACCH) and Extended Coverage Packet Associated Control CHannel (EC-PACCH), see Sections 3.2.6 and 3.2.7 for the DL and UL, respectively).

The use of two different multiframes has its explanation in that traditionally a GSM device assigned resources on a packet traffic channel (52 multiframe) would still need to continuously monitor its surrounding environment by synchronizing to neighboring cells and to acquire cell-specific information (System Information (SI)). By ensuring that the two multiframes are drifting relative to each other over time, the channels of interest in the 51 multiframe will drift over time relative to a certain position in the 52 multiframe, and hence will not overlap over time, which would prevent acquisition of the traffic channel in the serving cell, and information from neighbor cells. For EC-GSM-IoT, there is no requirement to synchronize to or to acquire SI from neighboring cells, while being assigned packet traffic channel resources. Hence, there is no requirement from that perspective for a 51 multiframe to be used. However, since the existing FCCH, which is already mapped to a 51 multiframe, also has been chosen to be used for EC-GSM-IoT, the use of the 51 multiframe also for some of the EC-GSM-IoT logical channels is natural.

A set of 26 51 multiframes or 25.5 52 multiframes form a superframe. The superframes are in their turn grouped in sets of 2048, each set forming a hyperframe.

The overall frame structure is shown in Figure 3.2.

Now that the frame structure is covered, let us turn our attention to the slot format. As stated above, and as shown in Figure 3.2, each TDMA frame consists of eight TSs.

When using blind physical layer transmissions (see Section 3.2.8.2), which is a new transmission scheme introduced by EC-GSM-IoT, it is of importance for the receiver to be able to coherently combine the transmissions to be able to maximize the received signal-to-interference-plus-noise power

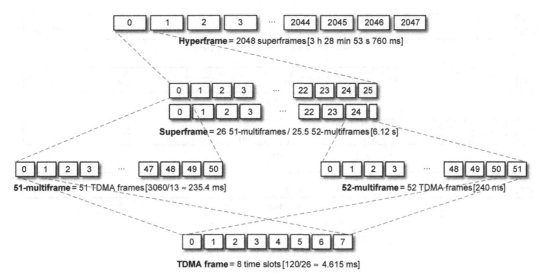

FIGURE 3.2

Frame structure in GSM.

ratio (SINR). Even a fractional symbol offset in the timing of the transmitter and receiver will cause suboptimum combination and, hence, loss in performance. Therefore, integral symbol length TS is defined for EC-GSM-IoT, as shown in Figure 3.3. Considering that in GSM, it is allowed for each slot to have a duration of 156.25 symbols, the slot length need to alternate between 156 symbols and 157 symbols, to fit within the same TDMA frame duration.

3.2.2.3 Burst Types

Each time slot is carrying a burst, which is the basic physical transmission unit in GSM. Different *burst types* are used depending on the logical channel and its use. For EC-GSM-IoT, five different burst types are used: *frequency correction bursts* (FB), *synchronization bursts* (SB), *access bursts* (AB), *dummy bursts* (DB), and *normal bursts* (NB).

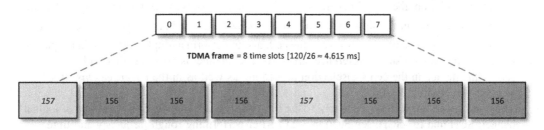

FIGURE 3.3

Integral symbol length slot structure.

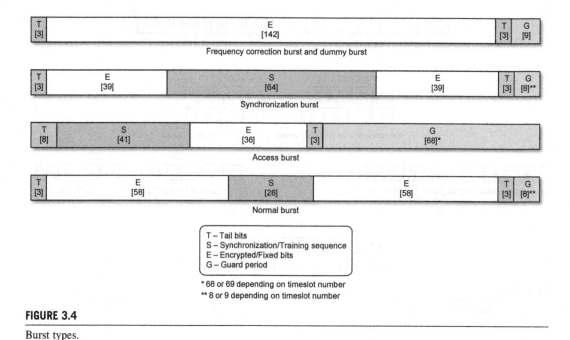

FIGURE 3.4

Burst types.

Common to all burst types is that they per definition all occupy a full slot, even if active transmission is not required in the full burst. That is, different guard periods are used for the different burst types to extend the burst to the full slot. Common to the SB, AB, and NB is that they all contain tail bits, a training sequence/synchronization sequence, and encrypted bits (payload). Both the FB and the DB consist only of a field of fixed bits, apart from the tail bits. All burst types are shown in Figure 3.4 with the number of bits per burst field in brackets.

As can be seen, all bursts occupy 156 or 157 symbols (depending on the TS they occupy, see Figure 3.3). The burst types can be shortly described as:

- **Frequency correction burst**: The burst type is only used by the FCCH, see Section 3.2.6, and consists of 148 bits of state "0." Due to the properties of *Gaussian Minimum Shift Keying* (GMSK) modulation, the signal generated will be of sinusoidal waveform over the burst length, giving rise to a spike in the signal spectrum at 67.7 kHz $\left(\frac{f_s}{4}\right)$. The burst is used by the device to identify a GSM frequency, synchronize to the cell, and to perform a rough alignment in frequency and time with the base station reference structure. The channel is also, together with EC-SCH, used by the device in the cell (re)selection procedure, as well as in the *coverage class* (CC) selection, see Section 3.3.1.4.
- **Synchronization burst (SB)**: Similarly, to the FB, the SB is also only used by one type of logical channel; SCH and EC-SCH, see Section 3.2.6. After performing rough frequency and time synchronization using the FCCH, an EC-GSM-IoT device attempts to acquire the EC-SCH to more accurately get synchronized in time and frequency. This process also includes acquiring the

frame number. Considering that the device will only be roughly synchronized to the base station reference after FCCH synchronization, the synchronization sequence of the EC-SCH has been designed longer than, for example, for the NB (which is a burst type only monitored by the device after fine synchronization) to provide a more reliable acquisition.

- **Access burst (AB)**: The AB is used when accessing the network on the Random Access CHannel/ Extended Coverage Random Access CHannel (RACH/EC-RACH) after synchronization to the DL frame structure has been obtained. To support synchronization to the UL frame structure it has been designed with a longer guard period to be able to support a wide range of cell radii. The AB also contains a longer synchronization sequence because the base station does not know where to expect the AB due to the propagation delay (the burst will arrive at different positions within the slot depending on the distance between the device and the base station). Hence, a longer synchronization window in the base station receiver is needed compared to, for example, receiving a NB. The payload of the access burst typically contains an access cause and a random reference to avoid contention with other simultaneous accesses, see Section 3.3.1.7. The AB can also be used by the EC-PACCH channel when the network wants to estimate the *timing advance* (TA) of the device.
- **Normal burst**: The NB is the burst type used for most logical channels, including the EC-BCCH, EC-CCCH (DL), EC-PDTCH, and EC-PACCH. It consists of a 26-symbol-long training sequence code (TSC) and 58 payload symbols on each side of the TSC.
- **Dummy burst**: The DB could be seen as a special type of NB where the TSC and the payload part are replaced by a fixed bit pattern. It is only used on the Broadcast carrier (BCCH) when no transmission is scheduled for the network to always transmit a signal. The signal is used by devices in the network to measure on serving and neighboring cells.

3.2.3 TRANSMISSION SCHEMES
3.2.3.1 Modulation
The symbol rate, f_s, for GSM and EC-GSM-IoT is $13 \times 10^6/48$ symbols/s or roughly 270.83 ksymbols/s, resulting in a symbol duration of roughly 3.69 μs. With a symbol rate exceeding the 200 kHz channel bandwidth there is a trade-off between the spectral response of the modulation used and by that the protection of the neighboring channels and the *intersymbol interference* (ISI), caused by the modulation. In general, a contained duration in time of the modulation means less ISI, but results in a relatively wider frequency response, compared with a modulation with longer duration.

The basic modulation scheme used in GSM, and also by EC-GSM-IoT, is GMSK. GMSK modulation is characterized by being a constant envelope modulation. This means that there are no amplitude variations in the modulated signal, and the information is instead carried in the phase of the waveform. GMSK modulation is furthermore characterized by the *BT-product*, where B is the half-power, or -3 dB, Bandwidth and T is the symbol Time duration. The BT-product in GSM is 0.3 meaning that the double-sided -3 dB bandwidth is 162.5 kHz ($2 \times 0.3 \times 13 \times 10^6/48$). A BT-product of 0.3 also gives rise to an ISI of roughly 5 symbol periods (the duration of the modulation response where a nonnegligible contribution of other symbols can be observed). That is, already in the modulation process at the transmitter, distortions of interference from other symbols are added to the transmitted waveform. This result in a relatively spectrally efficient transmission with contained spectral characteristics, at the expense of increased complexity at the receiver that need to resolve the introduced ISI. The channel propagation may result in additional ISI and the filtering of the signal at the receiver. With a matched filter implementation in the receiver, i.e., the same filter response used as in the transmitter, the GMSK modulation provides roughly an 18 dB suppression of interference coming from adjacent channels. For more details on GMSK modulation and how to generate the modulated signal, see for example [3].

It is possible to make a linear decomposition of the nonlinear GMSK modulation, as shown by P. Laurent [4]. By using the main pulse response of the decomposition, a good approximation of GMSK can be obtained by the convolution of the main pulse with a $\pi/2$ rotated Binary Phase Shift Keying (BPSK) modulation. The same pulse shape is also the one defined for 8PSK modulation used by EGPRS and EC-GSM-IoT, and is referred to as a "Linearized GMSK pulse." Before pulse shaping, the 8PSK constellation is rotated by $3\pi/8$ radians to minimize the peaks in signal level caused by the ISI and also to avoid zero-crossings in the IQ-diagram. The knowledge of the modulation-specific rotation is also used by the receiver for the purpose of modulation detection.

In Figure 3.5 the IQ-trace for both GMSK and 8PSK modulation is exemplified together with the amplitude response over the modulated bursts (mapped to a TS in the TDMA frame).

As can be seen, the GMSK modulation exhibits a constant envelope over the full burst, except for the start and stop of the burst, where the signal is ramping up and down in power, respectively. For 8PSK the situation is different with significant amplitude variations also within the burst. If one would look at the long-term characteristics of the 8PSK signal, one would see that peak-to-average-power ratio of the signal, i.e., how the absolute peak power relates to the average power of the signal, is 3.2 dB. That is, the peak signal power is roughly twice that of the average power. In addition, worth noting is that there are no zero-crossings of the signal in the IQ-diagram and this would also hold for the long-term characteristics, because of the symbol rotation angle used.

For EC-GSM-IoT, the modulation supported by the device can either be GMSK only or GMSK and 8PSK. The reason to allow only GMSK modulation to be supported is to allow for an ultra-low-cost device, for more information see Section 4.7. In terms of RF characteristics, the constant envelope modulation will allow for a power efficient implementation where the power amplifier (PA), in simplistic terms, can be optimized for a single operating point. In contrast, a PA used to support 8PSK modulation would have to be dimensioned for relatively rarely occurring peaks with twice the amplitude as the average signal. Although there are techniques that help bridge the gap in power efficiency, such as tracking the envelope of the signal, a difference is still there. In addition, considering the simple signal characteristics, only carrying information using the phase of the signal, a GMSK PA can, to a larger extent, distort the amplitude of the signal without impacting the performance of the radio link. Hence, apart from being more power efficient, a GMSK implementation can also be more cost and energy effective.

3.2.3.2 Blind Transmissions

Blind transmissions, also referred to as *blind repetitions*, are the means by which an EC-GSM-IoT transmitter, instead of transmitting a block only once, transmits a predefined number of transmissions without any feedback from the receiving end indicating to the sending end that the block is erroneous. This is one of the main mechanisms used to extend the coverage of EC-GSM-IoT. Its usefulness is in detailed elaborate upon in Section 3.2.8.2.

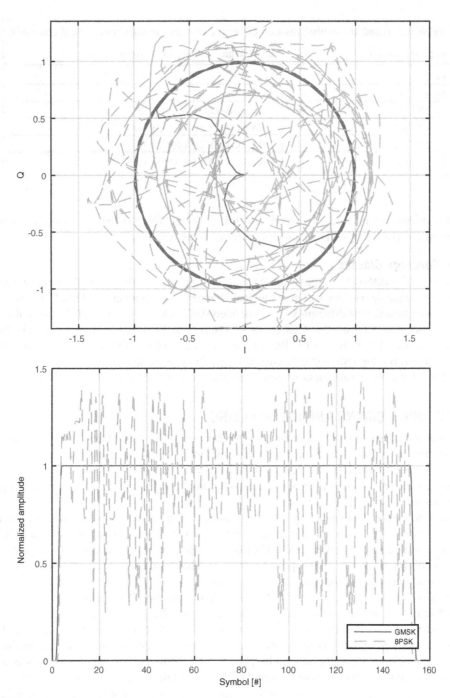

FIGURE 3.5

Examples of IQ-trace (top) and amplitude response (bottom) of a GMSK and 8PSK modulated bursts.

Table 3.1 Blind transmissions and coverage classes for different logical channels

Logical Channel	Blind Transmissions and Coverage Classes
EC-SCH	28
EC-BCCH	16
EC-CCCH/D	1 (CC1), 8 (CC2), 16 (CC3), 32 (CC4)
EC-CCCH/U	1 (CC1), 4 (CC2), 16 (CC3), 48 (CC4)
EC-PACCH	1 (CC1), 4 (CC2), 8 (CC3), 16 (CC4)
EC-PDTCH	1 (CC1), 4 (CC2), 8 (CC3), 16 (CC4)

A similar concept is used to extend the link coverage for NB-IoT and *LTE Machine-Type Communications* (LTE-M) but is then simply referred to as *repetitions*.

3.2.3.3 Coverage Classes

To keep the EC-GSM-IoT feature implementation simple, at most four different numbers of blind repetitions are defined for any given logical channel. Each number is referred to as a coverage class (CC) and, hence, four different CCs are defined (CC1, CC2, CC3, and CC4). Logical channels using the concept of CCs are EC-CCCH, EC-PDTCH, and EC-PACCH. For synchronization and broadcast channels, EC-SCH and EC-BCCH, only a single set of blind transmissions are defined, dimensioned to reach the most extreme coverage conditions expected in the cell.

In Table 3.1 the blind transmissions defined and the associated CCs (where applicable) is shown.

3.2.4 CHANNEL CODING AND INTERLEAVING

This chapter intends to give an introduction to the channel coding procedures for EC-GSM-IoT. The channel coding procedures here mirror the steps contained within the channel coding specification, 3GPP TS 45.003 [5], which also includes interleaving and mapping of the encoded bits onto bursts. The interested reader is referred to the specification for more details, considering that the description of these procedures is kept relatively short.

The first step in the channel coding is typically to add a number of bits to the payload for detecting errors induced in the transmission. The error detecting code consists of a number of parity bits added before the forward error correction (FEC). The number of bits differs depending on the logical channel and/or the coding scheme. Generally speaking, the length has been chosen considering the implication of a block being erroneously interpreted by the receiving end.

After the error detecting capability has been added to the payload, a FEC code is applied. The FEC used in EC-GSM-IoT is based on convolutional codes that are fully reused from the EGPRS channel coding design. Two different mother codes are used, defined by either a 1/2 or a 1/3 code rate that are followed by optional puncturing/rate matching of encoded bits to reach the final code rate. The puncturing is typically defined by puncturing schemes (PSs), which basically is a list of bit positions identifying the ones to be removed from the encoded bit stream.

Both tail biting and zero padded convolutional codes are used. For coding schemes where the block length is limited, tail biting is typically used, which reduces the overhead from channel coding at the

Table 3.2 Channel coding details for EC-SCH, EC-CCCH, EC-BCCH, and EC-PACCH

Logical Channel	Uncoded Bits	Parity Bits	Mother Code	Code Rate	Tail Biting	Interleaver	Burst Type	Bursts per Block
EC-SCH	30	10	1/2	0.56	No	No	SB	1
EC-CCCH/D	88	18	1/3	0.91[a]	Yes	No	NB	2
EC-CCCH/U	11	6	1/2	0.58	No	No	AB	1
EC-BCCH	184	40	1/2	0.50	No	Yes	NB	4
EC-PACCH/D	80	18	1/3	0.86[b]	Yes	No	NB	4
EC-PACCH/U	64	18	1/3	0.71[b]	Yes	No	NB	4

[a]Code rate of a single block, which is then at least repeated once (for CC1).
[b]Code rate of a single block, which is then at least repeated three times (for CC1).

expense of an increased decoding complexity. Instead of a known starting and ending state of the decoder (as for the case of zero padding), the start and end state can only be assumed to be the same, but the state itself is not known.

The encoding process is typically followed by interleaving, which is simply a remapping of the bit order using a 1-to-1 mapping table.

After interleaving the bits are mapped onto the burst type associated with the logical channel, see Figure 3.4. Different burst types are associated with different number of encrypted bits, and there can also be different number of bursts comprising the full block, depending on the logical channel, see Tables 3.2 and 3.3.

The general channel coding procedure is illustrated in Figure 3.6.

The EC-PDTCH is different from the other logical channels in that the full block constitutes a number of separately encoded fields, see Table 3.3.

Compared with EC-PDTCH, which follows the EGPRS design of PDTCH, the EC-PACCH is a new block format for EC-GSM-IoT. Because Uplink State Flag (USF)-based scheduling is not used in EC-GSM-IoT, as explained in see Section 3.3.2.1, the USF in the EC-PACCH is included in a rather unconventional way by simply removing some of the bits for the payload and replacing them by USF bits, effectively increasing the code rate of EC-PACCH control message content. This is referred to as bit stealing for USF and is only applicable to the EC-PACCH block on the DL. The EC-GSM-IoT receiver is, however, not aware of if a USF is included in the EC-PACCH block or not and will treat the block in the same way in each reception. It can be noted that the inclusion USF is solely included for the purpose of scheduling other PS devices on the UL.

The channel coding procedure for each logical channel is summarized in Tables 3.2 and 3.3.

3.2.5 MAPPING OF LOGICAL CHANNELS ONTO PHYSICAL CHANNELS

In Sections 3.2.6 and 3.2.7 the DL and UL logical channels will be described. Each logical channel is mapped onto one or more basic physical channels in a predefined manner. A *basic physical channel* is defined by a set of resources using the one and the same TS in each TDMA frame over the full hyperframe. For packet logical channels the basic physical channel is also referred to as a *Packet Data*

Table 3.3 Channel coding details for EC-PDTCH

Modulation and Coding Scheme	Modulation	SF Code Rate	USF Code Rate	RLC/MAC Header			RLC Data Block		Burst Type	Bursts per Block
				Header Type	Parity Bits	Code Rate[a]	Parity Bits	Code Rate		
MCS-1	GMSK	1/4	1/4	3	8	0.53 (DL) 0.49 (UL)	12	0.53	NB	4
MCS-2	GMSK			3				0.69	NB	4
MCS-3	GMSK			3				0.89	NB	4
MCS-4	GMSK			3				1.00	NB	4
MCS-5	8PSK	1/2	1/12	2		0.33 (DL) 0.33 (UL)		0.38	NB	4
MCS-6	8PSK			2				0.50	NB	4
MCS-7	8PSK	1/2		1				0.78	NB	4
MCS-8	8PSK			1				0.92	NB	4
MCS-9	8PSK			1				1.00	NB	4

[a]Tail biting.

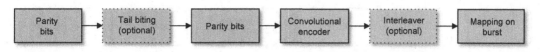

FIGURE 3.6

Channel coding procedure for EC-GSM-IoT.

CHannel (PDCH) and is, in this case, always mapped to a 52 multiframe [6]. The logical channels that are mapped to PDCHs are EC-PDTCH and EC-PACCH.

To understand the mapping of the logical channels onto physical channels, it is good to understand the two dimensions in time used by the GSM frame structure (for more details on the frame structure, see Section 3.2.2), that is the TSs that a logical channel is mapped to, and the TDMA frames over which the logical channel spans.

For illustration purposes, this book uses the principle of stacking consecutive TDMA frames horizontally, as shown in the bottom part of Figure 3.7 (where the arrows show the time direction in each TDMA frame). This illustration will be used in the following chapters when illustrating the mapping of the logical channels onto the physical frame structure.

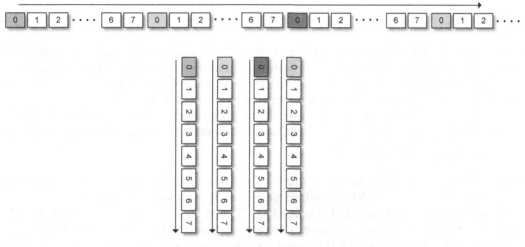

FIGURE 3.7

TDMA frame illustration.

3.2.6 DOWNLINK LOGICAL CHANNELS

The set of logical channels used in the DL by EC-GSM-IoT are shown in Figure 3.8.

The purpose of each channel, how it is used by the device, and the mapping of the channel onto the physical resources will be shown in the following chapters. The notation of "/D" for some of the channels, indicate that the channel is defined in both downlink and uplink, and that the downlink is here referred to.

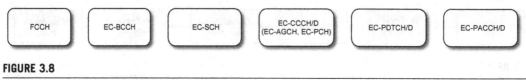

FIGURE 3.8

Downlink logical channels used in EC-GSM-IoT.

3.2.6.1 FCCH

TS	0
TDMA frames	0, 10, 20, 30, 40 (see Figure 3.9)
Mapping repetition period	51 TDMA frames
Multiframe	51
Burst type	Frequency correction burst
Block size	–
Carrier	BCCH

The Frequency Correction Channel (FCCH) is the only channel that is the same as used by GSM devices not supporting EC-GSM-IoT. When the device is scanning the frequency band, the acquisition of FCCH will identify if a certain frequency belongs to a GSM network. The device also uses it to perform a rough frequency (to the base station reference) and time alignment.

As already mentioned in Section 3.2.2.3, the burst carrying the FCCH consists of 148 bits of state "0," which will generate a signal of sinusoidal character over the burst length, giving rise to a spike in the signal spectrum at 67.7 kHz $\left(\frac{f_c}{4}\right)$.

The FCCH is mapped onto the 51 multiframe of TS0 of the BCCH carrier. This is the only basic physical channel, and the only RF channel, where the FCCH is allowed to be mapped.

As shown in Figure 3.9 the FCCH is mapped onto TDMA frames 0, 10, 20, 30, and 40 in the 51 multiframe. This implies that the separation between the frequency correction bursts is 10, 10, 10, 10, 11 TDMA frame. Although the FCCH does not carry any information, the irregular structure of the FCCH mapping can be used by a device to determine the overall 51 multiframe structures.

How well the device is synchronized after the FCCH acquisition is very much dependent on the algorithm used and also the time spent to acquire the FCCH. The device oscillator accuracy is for typical oscillator components used in low-complexity devices equals 20 ppm (parts per million), i.e., for a device synchronizing to the 900 MHz frequency band, the frequency offset prior to FCCH acquisition can be up to 18 kHz (900e6 × 20 × 10^{-6}). After FCCH acquisition a reasonable residual frequency and time offset is up to a few hundred Hz in frequency and a few symbol durations in time.

The next step for the device is to decode the EC-SCH channel.

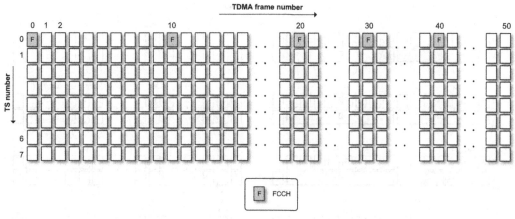

FIGURE 3.9

FCCH mapping.

3.2.6.2 EC-SCH

TS	1
TDMA frames	See Figure 3.10
Mapping repetition period	204 TDMA frames
Blind physical layer transmission	28
Multiframe	51
Burst type	Synchronization burst
Block size	1 burst
Carrier	BCCH

After the rough frequency and time correction by the use of the FCCH, the device turns to the Extended Coverage Synchronization Channel (EC-SCH). As mentioned in Section 3.2.6.1 after acquiring the FCCH the device will have knowledge of the frame structure to a precision of the 51 multiframe but will not know the relation to the overall frame structure on a superframe and hyperframe level (see Section 3.2.2). The EC-SCH will assist the device to acquire the knowledge of the frame structure to a precision of a quarter hyperframe. There is no reason for the device at this point to know the frame structure more precisely. Instead, the missing piece of the puzzle is provided in the assignment message, where 2 bits will convey which quarter hyperframe the assignment message is received in.

The frame number on a quarter hyperframe level is communicated partly through information contained in the payload part of the EC-SCH. However, considering that the interleaving period of the EC-SCH is four 51 multiframes (i.e., the payload content will change after this period), see Figure 3.10, and that the device will start its acquisition of the EC-SCH in any of these multiframes, a signaling, indicating which multiframe out of four is required. To solve this, a cyclic shift of the

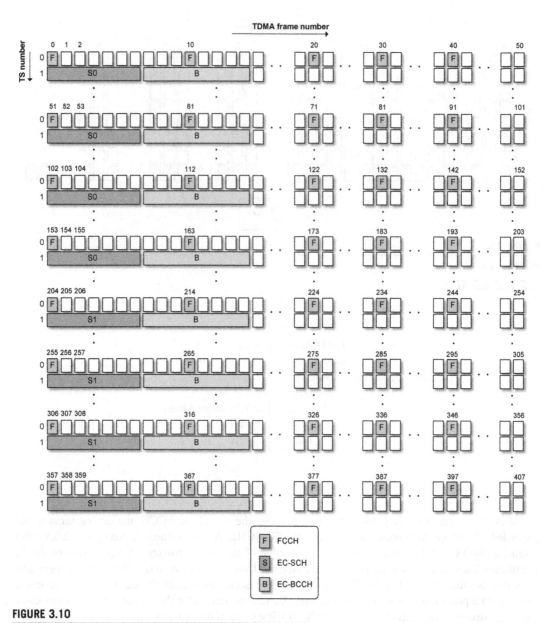

FIGURE 3.10

Mapping of FCCH, EC-SCH, and EC-BCCH.

encoded bits is applied specific to each multiframe. The same cyclic shift is applied to all seven repetitions within the multiframe. This allows the receiver to accumulate, on IQ level, all bursts within each multiframe to maximize processing gain (following the same principle as illustrated in Figure 3.21). The device would then need to decode the EC-SCH using up to four different hypotheses (one for each cyclic shift) before the block is decoded.

After EC-SCH acquisition the device should be synchronized in time and frequency, fulfilling the requirements of the specification on ½ symbol synchronization accuracy and 0.1 ppm frequency accuracy (for the 900 MHz band this corresponds to residual time and frequency offset of at most 1.8 µs and 90 Hz, respectively).

The device is mandated to decode the EC-SCH each time, it is attempting to initiate a connection to receive on the DL or transmit on the UL. This makes the EC-SCH a powerful channel for indicating information to the device. It can be compared with the NB-IoT and LTE-M Master Information Block. Hence, apart from the remaining frame number that is communicated by the channel, the EC-SCH also communicates:

- The cell ID, i.e., the *Base Station Identity Code* (BSIC).
- System overload control (see Section 3.3.1.7).
- Load balancing on the random access channel (see Section 3.3.1.7).
- System Information (SI) change (see Section 3.3.1.2)

The BSIC is a 9-bit field, which can address 512 different cell identities. In traditional GSM the BSIC is a 6-bit field but to accommodate future IoT deployments of GSM, 3GPP took the initiative to expand the space of possible cell identities from 64 to 512.

This was mainly motivated by three factors

- **Tighter frequency reuse**: The frequency domain that provides orthogonality between cell IDs will be reduced.
- **Increased use of network sharing**: With IoT, operators are expected to more frequently share the networks that reduce the possibility to use the *Network Color Code*, which is part of the BSIC, as a means to separate cell IDs.
- **Reduced idle mode measurements**: With the reduced measurements by the device, it can wake up in a cell that uses the same BSIC as when previously was awake and with the reduced measurement activity, it can take up to 24 h before it will realize it.

The mapping of the EC-SCH channel is shown in Figure 3.10 together with the FCCH and EC-BCCH, where TS 2−7 of each TDMA frame has been omitted to simplify the figure. It should be noted that the repetition period of the EC-SCH channel only spans four 51 multiframes, using 28 blind transmissions, and hence two EC-SCH blocks are (0 and 1) are depicted in the figure.

3.2.6.3 EC-BCCH

TS	1
TDMA frames	See Figure 3.10
Mapping repetition period	408 TDMA frames
Blind physical layer transmissions	16
Multiframe	51
Burst type	Normal burst
Block size	4 bursts
Carrier	BCCH

After EC-SCH acquisition the device may potentially continue with acquiring the EC SI, which is transmitted on the Extended Coverage Broadcast Control Channel (EC-BCCH). As the name implies the EC SI is system-specific information to EC-GSM-IoT and will convey not only information related to the specific cell, it is transmitted in but can also provide information on other cells in the system.

The coding scheme used by the EC-BCCH channel is CS-1, which is also the modulation and coding scheme used by GPRS/EGPRS for control signaling purposes. After channel coding and interleaving, the block consists of four unique bursts that are mapped onto one TS over four consecutive TDMA frames. Each *EC SI message instance* (where each *EC SI message* can consist of multiple instances) is mapped onto in total 16 block periods, using two consecutive block periods in each 51 multiframe, mapped over in total eight 51 multiframes. The EC SI messages, and the multiple EC SI message instances (if any), are mapped onto the physical frame in ascending order. The sequence of EC SI messages and their associated EC SI message instances can be seen as an EC SI cycle that repeats itself over time.

By defining the EC SI placement in this static manner, the device will always know where to look for EC SI message instances, once the frame number has been acquired from the EC-SCH. In addition, the network will not know which devices are in a specific cell, or which conditions the devices are deployed in, it will have to always repeat each message the same and maximum number of times to be certain to reach all devices.

Figure 3.10 shows TS0 and TS1 of the TDMA frame with TS2−TS7 omitted (showed by a dashed line), illustrating the mapping of FCCH, EC-SCH, and EC-BCCH. It can be noted that the FCCH bursts are all identical (there is no distinction between them in the figure), while EC-SCH is transmitted in two unique blocks (S0 and S1) over the eight 51 multiframes because the repetition length is four 51 multiframes as explained in Section 3.2.6.2. Only one EC-BCCH block is transmitted (B) because the repetition period is eight 51 multiframes (see above).

3.2.6.4 EC-CCCH/D (EC-AGCH, EC-PCH)

TS	1, and optionally 3, 5, 7
TDMA frames	See Figure 3.11
Mapping repetition period	51 (CC1), 102 (CC2), 102 (CC3), 204 (CC4)
Blind physical layer transmissions	1 (CC1), 8 (CC2), 16 (CC3), 32 (CC4)
Multiframe	51
Burst type	Normal burst
Block size	2 bursts
Carrier	BCCH

The Extended Coverage Common Control Channel (EC-CCCH) is used by the network to reach one or more devices on the DL using either Extended Coverage Access Grant CHannel (EC-AGCH) or Extended Coverage Paging CHannel (EC-PCH). From a physical layer point of view, the two logical channel types are identical and they differ only in which TDMA frames the channels can be mapped onto. In some TDMA frames only EC-AGCH can be mapped, whereas in other TDMA frames both EC-AGCH and EC-PCH are allowed. The reason for having a restriction on how to map the EC-PCH is related to how paging groups are derived, and how cells in the network are synchronized (see Section 3.3.1.5 for more information).

Whether the message sent is mapped onto EC-AGCH or EC-PCH is conveyed through a message type field in the message itself. That is, it is only after decoding the block that the device will know whether the message sent was carried by EC-AGCH or EC-PCH. In case of EC-AGCH, only one device can be addressed by the message sent, whereas for EC-PCH up to two devices can be addressed by the same message.

Compared with FCCH, EC-SCH, and EC-BCCH that have been described in Section 3.2.6, the EC-CCCH/D channel makes use of CCs introduced in Section 3.2.8, to be able to reach users in different coverage conditions effectively.

Each block for each CC is mapped onto predefined frames in the overall frame structure. In Figure 3.11 the mapping of the EC-CCCH/D blocks are shown.

As can be seen, the CC1 blocks are mapped to two TDMA frames, whereas in case of CC4 32 blind transmissions are used, spread over four 51 multiframes, to reach devices in extreme coverage conditions. To spread the transmissions over several multiframes instead of transmitting them consecutively in time, will provide time diversity, improving the reception of the block. It can be noted that this is also the case for completely stationary devices, as long as the surrounding environment provides time variations in the radio propagation (for example, cars driving by, leaves in trees caught by wind etc.).

One of the main objectives with serving the IoT segment is not only to support challenging coverage conditions but also to provide an energy efficient operation reaching up to 10 years of battery lifetime as evaluated in Section 4.5. Here, the EC-CCCH plays an important role. Each time a device accesses the system it will have to monitor the EC-CCCH to provide the device with dedicated resources. Hence, an energy efficient operation on the EC-CCCH can contribute considerably to the overall battery lifetime. This is specifically true for devices that may only monitor the paging channel during its whole lifetime, where the device is only triggered in exceptional cases (for example, a fire break-out). To provide an energy efficient operation, three design considerations have been taken related to physical layer processing:

- **Single block decoding**: The EC-CCCH/D is at least transmitted using two bursts (for CC1), but each burst transmitted is identical, and hence self-decodable. This means that in case the device is in good enough radio condition, it only needs to decode a single burst before the block can be decoded.
- **Downlink coverage class indication 1**: In each EC-CCCH/D block, the DL CC by which the block is transmitted is indicated in the message. Because the EC-CCCH/D block is mapped to a predefined set of TDMA frames, a device that has selected, for example, CC1 knows that the network is mandated to transmit a CC1 block to that device, and hence the device is not interested to decode CC2, CC3, or CC4 blocks. Because the CC is indicated in the message, a device in CC1 could, for example, decode a CC4 message after only a single burst (see single block decoding) and sleep for the remaining 63 bursts, see Figure 3.12.
- **Downlink coverage class indication 2**: To include the DL CC in the message will only help devices in better coverage than the recipient of the block is intended for. It is however more important to save energy for devices that are in more challenging coverage conditions (i.e., CC2, CC3, and CC4). To enable a more efficient energy saving for these devices, different training sequences are used in the EC-CCCH/D block, depending on if all devices addressed by the block are in CC1 or not. That is, as long as one or more devices addressed by the block has selected

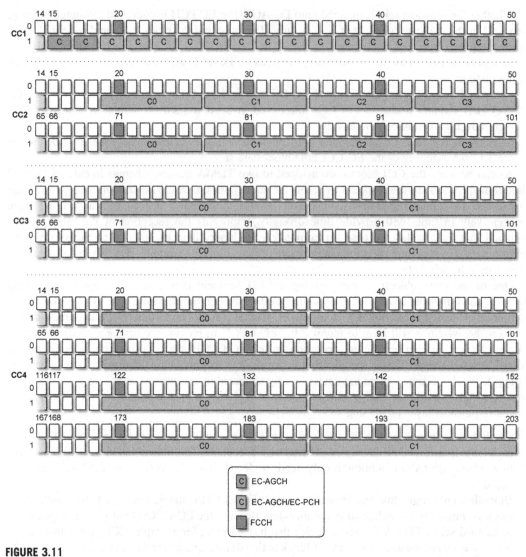

FIGURE 3.11

Mapping of EC-CCCH/D for different Coverage Classes.

CC2, CC3, or CC4, an alternative TSC shall be used by the network. Because the training sequence detector can operate correctly at lower SNR than where the EC-CCCH/D block is decodable, the device can make an early detection of a block sent with CC1, and by that determine, for example, that a CC4 block will not be transmitted on the remaining bursts of the CC4 block. Simulations have shown [7] that roughly 80%–85% of the energy can be saved in the monitoring of the DL EC-CCCH/D by early detection of the TSC transmitted. Figure 3.13

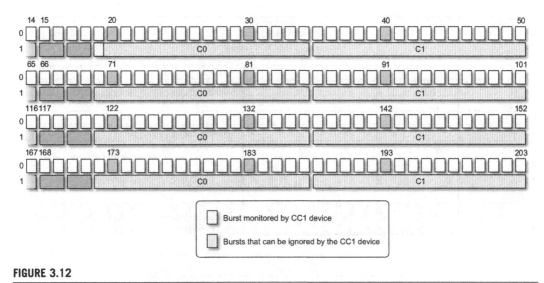

☐ Burst monitored by CC1 device

▨ Bursts that can be ignored by the CC1 device

FIGURE 3.12

Downlink coverage class indicator—in message content.

illustrates the savings in DL monitoring for a CC4 device detecting the TSC indicating CC1 block, six bursts into the CC4 block. In this case 91% is saved of the DL monitoring.

3.2.6.5 EC-PDTCH/D

TS	Any
TDMA frames	See Figure 3.15
Mapping repetition period	52
Blind physical layer transmissions	1 (CC1), 4 (CC2), 8 (CC3), 16 (CC4)
Multiframe	52
Burst type	Normal
Block size	4 bursts
Carrier	Any

The Extended Coverage Packet Data Traffic Channel (EC-PDTCH) is one of the logical channels for EC-GSM-IoT, which is almost identical to the corresponding logical channel used in EGPRS. It is used for carrying payload from the network to the device or, in case of UL transmission, from the device to the network. The EC-PDTCH block consists of four different bursts mapped to the same TS over four consecutive TDMA frames. The block is referred to as a *radio block* and is transmitted over a *Basic Transmission Time Interval (BTTI)* of 20 ms. The EC-PDTCH is mapped onto one or more physical channels referred to as *PDCH (Physical Data CHannel)*.

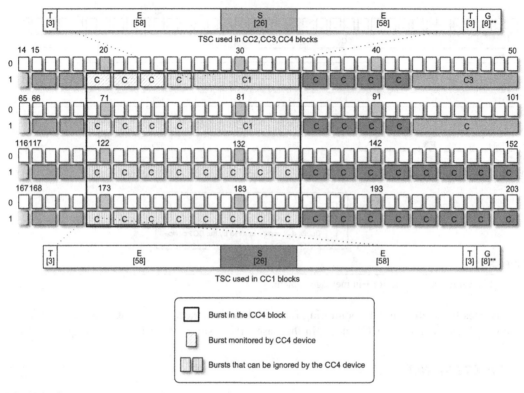

FIGURE 3.13

Downlink coverage class indicator—by training sequence.

To adapt to more extreme coverage conditions, CCs are also used for EC-PDTCH, see Table 3.1. When monitoring the DL EC-PDTCH (EC-PDTCH/D), it is up to the device to blindly detect the modulation scheme used in the block. This is done by detecting the rotation angle of the predefined training sequence. After synchronization, channel estimation and modulation detection, the device equalizes the symbols of the burst to provide a set of estimated probabilities that a certain bit has been transmitted. This probability is denoted a "soft bit." The higher the magnitude of the soft bit, the more certain the receiver is that a specific bit value was transmitted. Because different *Radio Link Control* (RLC)/*Medium Access Control* (MAC) headers can be used for the same modulation scheme the next step is to read the Stealing Flags (see Section 3.2.4), to determine the header type used in the block. After the RLC/MAC header has been decoded, the device will have knowledge of the modulation and coding scheme used in the block and also the redundancy version or PS, used in the *Hybrid Automatic Repeat Request* (HARQ) process, see Section 3.3.2. The procedure is illustrated in Figure 3.14.

As with the EC-CCCH channel, the EC-PDTCH also includes an indication of the DL CC in each block to allow energy efficient operation. Figure 3.12 illustrates this mechanism for EC-CCCH.

In case of CC1, the PDTCH block is mapped onto one PDCH, over four consecutive TDMA frames. For CC2, CC3, and CC4 the blind repetitions of the block are mapped onto four consecutive TSs and over 4, 8, or 16 TDMA frames, respectively.

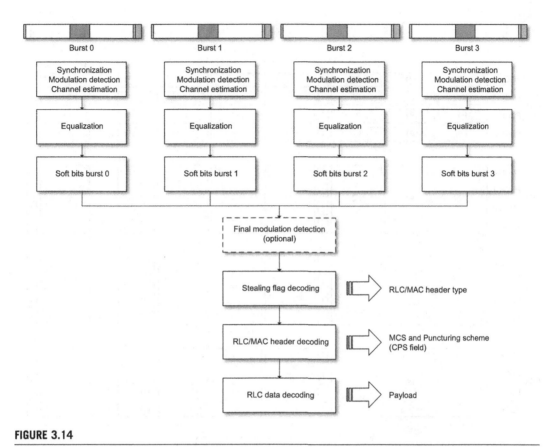

FIGURE 3.14

Illustration of procedure to equalize an EC-GSM-IoT radio block.

Increasing the mapping for CC1 from one to four PDCHs was decided based on multiple reasons. The more TSs that are used in the TDMA frame, the better it is for the receiver that can rely on coherent transmissions within the TDMA frame, see Section 3.2.8, improving the processing gain compared to mapping the blind repetitions over a smaller number of PDCHs (and instead more TDMA frames). Hence, from a performance point of view, it would be best to map the repetitions overall eight TSs in the TDMA frame. This poses potential problems of PA heat dissipation and resource management restrictions. If the EC-PDTCH would take up all 8 TSs, the only channel possible to transmit in addition on that carrier would be the associated control channel, the EC-PACCH, as well as the traffic channel and associated control channel for non-EC-GSM-IoT traffic (the PDTCH and PACCH). Hence, from a network deployment point of view, a shorter allocation over TSs in a TDMA frame is required. As a trade-off between network deployment flexibility, radio link level performance, and device implementation, four consecutive TSs have been chosen for all higher CCs, although a second option is introduced in Release 14 introduced in Section 3.4.

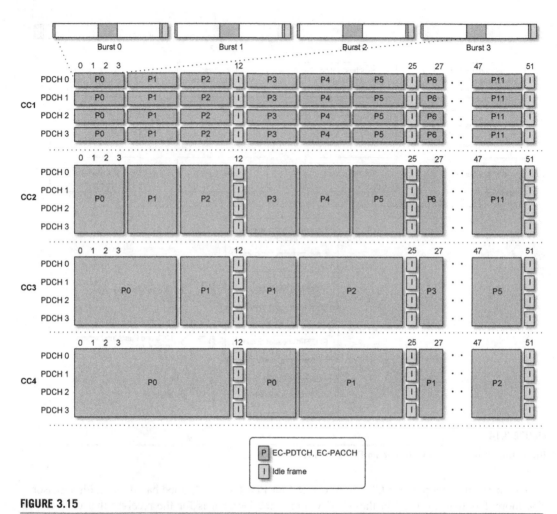

FIGURE 3.15

Mapping of EC-PDTCH and EC-PACCH for different Coverage Classes.

The mapping of the EC-PDTCH is shown in Figure 3.15. Although CC1 is only mapped to one PDCH, four consecutive PDCHs are shown to illustrate the same amount of resources, irrespective of the CC. It can be noted that there are no restrictions on the mapping over the four consecutive PDCHs as long as the block is kept within the TDMA frame, i.e., "PDCH 0" in Figure 3.15 can be mapped to TS index 0, 1, 2, 3, or 4 (in which, for the latter case, the block is mapped to TS4, TS5, TS6, TS7).

3.2.6.6 EC-PACCH/D

The Extended Coverage Packet Associated Control Channel (EC-PACCH) is, as the name implies, an associated control channel to the EC-PDTCH. This means that it carries control information related to the EC-PDTCH operation, e.g., output power level to be used (power control), resource (re)assignment, and Ack/Nack bitmap (for the UL HARQ process).

The EC-PACCH/D information is encoded onto a single normal burst, carrying 80 information bits, which is always repeated four times to construct an EC-PACCH/D block. The reason for this is twofold to ensure an energy efficient operation for users in good radio conditions, and to allow the device to correct its frequency to the base station reference.

Ensuring an energy efficient operation implies here that a device in good radio conditions can monitor the DL channel for a single burst, attempts to decode the block, and if successfully decoded, go back to sleep in similarity to the EC-CCCH operation presented in Section 3.2.6.4.

The second benefit is related to the stability of the frequency reference in the device over time. As stated in Sections 3.2.6.1 and 3.2.6.2, the device will align its time and frequency base with the base station reference during FCCH and EC-SCH acquisition. However, once a device has been assigned resources for its data transfer, it would be more suitable if it can fine tune its reference on the monitored dedicated channels. By designing the EC-PACCH as a single burst block, the device is able to estimate the frequency offset by correlating and accumulating the multiple bursts transmitted. Since at least one EC-PACCH block is transmitted after each set of EC-PDTCH blocks assigned in the UL, the frequency reference can continuously be updated by the device.

A USF need to be included in every EC-PACCH block because non-EC-GSM-IoT devices monitor the DL for potential UL scheduling (assuming a block could contain the USF).

3.2.7 UPLINK LOGICAL CHANNELS

The set of logical channels used in the UL by EC-GSM-IoT are shown in Figure 3.16.

FIGURE 3.16

Uplink logical channels used in EC-GSM-IoT. The notation of "/U" for some of the channels, indicate that the channel is defined in both downlink and uplink, and that the uplink is here referred to.

3.2.7.1 EC-CCCH/U (EC-RACH)

TS	0, 1, 3, 5, 7 and [0,1], [2,3], [4,5], [6,7]
TDMA frames	See Figures 3.17 and 3.18
Mapping repetition period	51 (CC1), 51 (CC2), 51 (CC3), 102 (CC4)
Blind physical layer transmissions	1 (CC1), 4 (CC2), 16 (CC3), 48 (CC4)
Multiframe	51
Burst type	Access burst
Block size	1 burst
Carrier	BCCH

Before a connection between the device and the network can be setup, the device needs to initiate an access on the random access channel. The initiation of the system access request can either be triggered by the device referred to as *Mobile Originated* (MO) traffic or by the network referred to as *Mobile Terminated* (MT) traffic, initiated, for example, when paging the device. Before using the random access channel, the device needs to be synchronized to the network in time and frequency and not being barred for access (see Sections 3.2.6 and 3.3.1).

The device sends an AB to initiate the communication, which contains 11 bits, containing, for example, an indication of DL CC. In addition, the UL CC is implicitly indicated to the network by the choice of the TSC used.

The device will transmit the Packet Channel Request using either one or two TSs per TDMA frame. This is referred to as a 1 or 2 TS mapping of the Extended Coverage Random Access Channel (EC-RACH). At first glance, using two different mappings for transmitting the same information seems unnecessary complex. There are, however, reasons for this design. The 2 TS mapping extends coverage further than the 1 TS mapping. The reason is that the device needs to keep the coherency of the transmission over the two TSs in the TDMA frame and hence, the processing gain at the base station is improved. However, it comes at a cost of having to multiplex system accesses from EC-GSM-IoT devices with the access of non-EC-GSM-IoT devices accessing on TS 0. Hence, an operator would, depending on the load in the network from non-EC-GSM-IoT devices, and the need for extending the coverage, allow the use of 2 TS mapping by the EC-GSM-IoT devices. Only one of the mapping options will be used at a specific point in time in the cell.

The performance gain of the 2 TS EC-RACH mapping has been evaluated [8] where it was seen that up to 1.5 dB performance gain is achieved for the users in the worst coverage.

Since the device accessing the network only knows the timing of the DL frame structure (from the synchronization procedure, see Sections 3.2.6.1 and 3.2.6.2), it will not know the distance from the device position to the base station receiver. To accommodate for different base stations to device distances (propagation delays) in the cell, the AB has been designed to be shorter than a regular NB

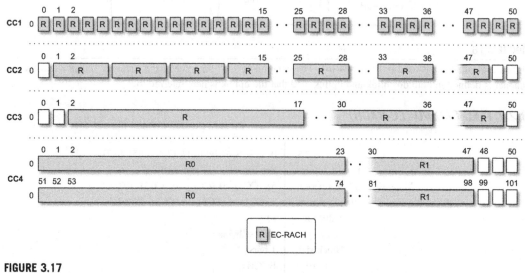

FIGURE 3.17

EC-CCCH/U EC-RACH, 1 TS mapping.

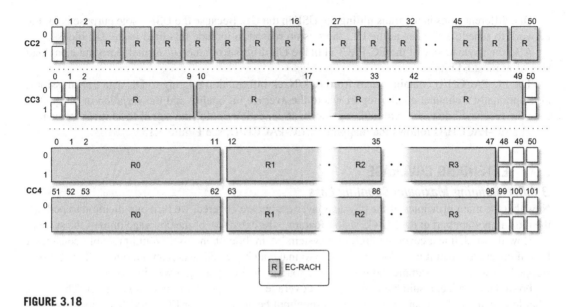

FIGURE 3.18

EC-CCCH/U EC-RACH, 2 TS mapping.

(see Section 3.2.2.2 for the burst structures). The shorter burst duration provides a 68-symbol-long guard period, which allows for roughly a 35 km cell radius (3e8 × (T_{symb} × 68)/2 ≈ 35 km). Because the DL synchronization will also be affected by the propagation delay, the UL transmission will have to be shifted in time to take into account a propagation delay twice the distance between the device and the base station. The propagation delay is estimated by the base station (comparing its reference timing with the timing of the received AB from the device) and the amount of TA to be used by the device is communicated to it in the message sent on the access grant channel (see Section 3.2.6.4).

The mapping of the logical channel onto the physical resources is shown in Figures 3.17 and 3.18.

3.2.7.2 EC-PDTCH/U

As for the DL traffic channel, the EC-PDTCH/U is almost identical to the corresponding channel used in EGPRS. Hence, the description of EC-PDTCH/D applies equally well to the EC-PDTCH/U with some exceptions. The differences lie in, for example, the exclusion of the USF in the overall block structure (as for EC-PACCH/U compared with EC-PACCH/D, see Section 3.2.7.3). In addition, the RLC/MAC header format will not be the same comparing the UL and DL EC-PDTCH blocks.

The EC-PDTCH/U follows the mapping of EC-PDTCH/D, see Figure 3.15.

3.2.7.3 EC-PACCH/U

From a physical layer point of view, the design of the EC-PACCH in the UL is very similar to the corresponding DL channel presented in Section 3.2.6.6.

The EC-PACCH/U block is constructed of a single burst repeated at least four times, using the same number of blind block repetitions as for the DL for the different CCs.

One difference lies in not transmitting the USF in the UL. Because the USF's sole purpose is for the network to schedule devices in the UL, there is no reason to include this field in the UL blocks.

Also the payload size of the EC-PACCH in the UL differs from the DL format, carrying a payload size of at most 64 bits.

The EC-PACCH/U contains, apart from Ack/Nack information relating to DL data transmissions, also optionally a channel quality report where the average bit quality and the variation in bit quality over the received bursts are reported, together with an estimation of the signal level measurements.

The EC-PACCH/U follows the mapping of EC-PACCH/D, see Figure 3.15.

3.2.8 EXTENDING COVERAGE

3.2.8.1 Defining Maximum Coupling Loss

Now that the main principles of the physical layer design are covered, we turn our attention to perhaps the most important part of the EC-GSM-IoT feature—the extension of the coverage limit of the system.

How do we define a coverage limit of a system? A mobile phone user would probably consider a loss of coverage when it is no longer possible to make a phone call, when an ongoing call is abruptly stopped, or when, for example, no web page is loading when opening a web browser.

For a data service, a suitable definition of coverage is at a specified throughput target. That is, at what lowest signal level can still X bps of throughput be achieved. For EC-GSM-IoT, this target was set to 160 bps. In addition to the traffic channel reaching a certain minimum throughput, also the signaling/control channels need to be operable at that point for the device to synchronize to the network, acquire the necessary associated control channels etc.

When designing a new system, it is important that the coverage is balanced. When comparing the coverage of different logical channels, it is easily done in the same direction (UL or DL) using the same assumptions of transmit power and thermal noise levels (see below) at the receiver. When we are to balance the UL and DL, assumption need to be made on device and base station transmit levels, and also what level of thermal noise that can typically be expected to limit the performance.

Multiple factors impact the thermal noise level at the receiver, but they can be separated into a dependency of:

- **Temperature**: The higher the temperature, the higher the noise level will be.
- **Bandwidth**: Because the thermal noise is spectrally white (i.e., the same spectral density irrespective of frequency) the larger the bandwidth of the signal, the higher the absolute noise power level will be.
- **Noise Figure (NF)**: The overall NF of a receiver stem from multiple sources, which we will not cover in this book, but one can see the NF as an increased noise level after the receiver chain, compared to the ideal thermal noise without any imperfections. The NF is expressed in decibel and is typically defined at room temperature.

Lowering the NF would improve the coverage in direct proportion to the reduction in NF, but a lower NF also implies a more complex and costly implementation, which is directly in contrast to one of the main enablers of a CIoT.

Another simple means to improve coverage is to increase the transmit power in the system, but this would imply an increase in implementation complexity and cost. Also the bandwidth needs to be kept as in GSM to minimize implementation and impact to existing deployments.

An alternative way to express coverage is in terms of its *Maximum Coupling Loss* (MCL). The MCL defines the maximum loss in the radio link that the system can cope with between a transmitter and a receiver, before losing coverage.

The MCL can be defined as,

$$MCL(dB) = P_{Tx} - (NF + SNR - 174 + 10\log_{10}(B)),\qquad(3.1)$$

where P_{Tx} is the output power [dBm], B is the bandwidth [Hz], 174 is the ideal thermal noise level expressed as [dBm/Hz] at temperature 300 K, NF is the Noise Figure [dB], SNR is the signal-to-noise ratio [dB].

We have already concluded that P_{Tx} (the output power), B (the signal bandwidth), and NF (the noise figure) are not quantities suitable to use in improving coverage for EC-GSM-IoT. Hence, only the SNR is the quantity left to improve.

A simple means for improving the experienced SNR at the receiver and at the same time reusing existing system design and ensuring backward compatibility are to make use of blind repetitions (see Section 3.2.8.2).

The feature also relies on an improved channel coding for control channels (see Section 3.2.8.3), a more efficient HARQ retransmissions for the data channels (see Sections 3.2.8.4 and 3.3.2.2), and an increased allowed acquisition time (see Section 3.2.8.5).

3.2.8.2 Maximizing the Receiver Processing Gain

In case of extended coverage, the receiver will, in worst case, not even be able to detect that a signal has been received. This problem is solved by the introduction of blind repetitions where the receiver first can accumulate the IQ representation of blindly transmitted bursts to a single burst with increased SNR before synchronizing to the burst and performing channel estimation. After the SNR is increased the rest of the receiver chain can be kept identical to that of GSM. This can be seen in Figure 3.19.

To maximize SNR, the signals should be coherently, i.e., aligned in amplitude and phase, combined. Because the white noise is uncorrelated it will always be noncoherently combined, resulting in a lower rise in the noise power than in the wanted signal. Taking this to a more generic reasoning, when combining N number of transmissions, the SNR compared to a single transmission can, in the ideal case, be simply expressed as in Eq. (3.2).

$$SNR_N(dB) = 10\log_{10}(N) + SNR_1\qquad(3.2)$$

However, in reality a combination of two signals is rarely perfect, and instead signal impairments will result in an overall processing gain lower than expressed in Eq. (3.2).

To assist the receiver in the coherent combination, the blind repetitions should also be coherently transmitted. This means that multiple signals from the transmitting antenna will follow the same phase

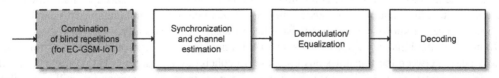

FIGURE 3.19

EC-GSM-IoT receiver chain.

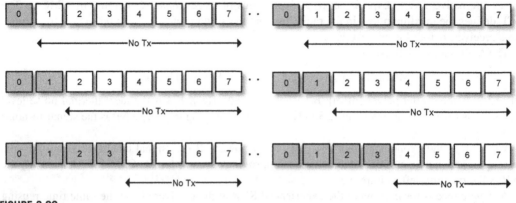

FIGURE 3.20

Different ways to blindly perform repetitions over multiple TDMA frames.

trajectory. As long as no other impairments are added to the signal, and as long as the propagation channel is stationary, the receiver could simply blindly accumulate the signal and achieve the expected processing gain. As we will see later, reality is not as forgiving as the simplest form of theory. Furthermore, to ensure coherent transmissions the phase reference in the transmitter need to be maintained between all blind repetitions. For the TDMA structure in GSM, where blocks are typically transmitted in one or a few TSs, the components in the transmitter would have to be active to keep the phase reference over the full TDMA frame. This not only consumes more energy, specifically of interest in the device, but might not be easily supported by existing GSM devices because before the introduction of EC-GSM-IoT there is no requirement on coherent transmissions and the phase is allowed to be random at the start of every burst.

Figure 3.20 illustrates three different ways in the EC-GSM-IoT specification to transmit blocks, using one, two, or four bursts in each TDMA frame.

In all cases there will be a substantial time where no transmission occurs, and when the device can turn off several components and basically only keep the reference clock running. What can also be seen from Figure 3.20 is that when transmitting on consecutive TSs there is no interruption in the transmission, and hence coherency can more easily be kept. This is also the approach that the 3GPP specifications have taken, i.e., those transmissions need only be coherent in case bursts are transmitted over consecutive TSs. Between TDMA frames any phase difference could hence be expected.

To maximize the processing gain, and hence the experienced SNR by the receiver, preferably all blind repetitions are to be combined before calling the demodulator including the channel estimator. This can be achieved by first blindly combining the transmissions within a TDMA frame. Blind in this regard means that the bursts are not compensated in anyway before the combination. This can be done because the receiver knows that the bursts have been transmitted coherently.

To combine the blind repetitions across TDMA frames, the receiver would have to estimate the random phase between the transmissions (because no coherency is ensured). This combination will result in a potential error in that case the processing gain from the combination will not follow Eq. (3.2).

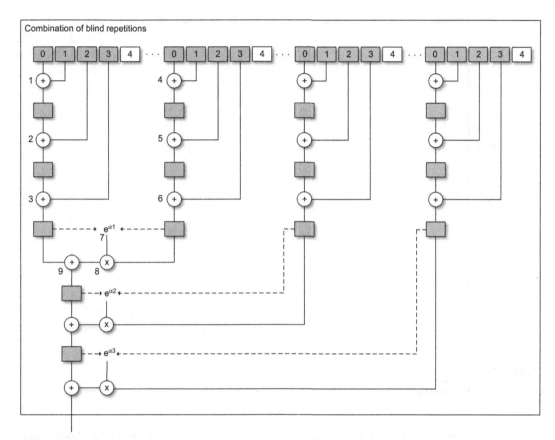

FIGURE 3.21

Combination of blind repetitions.

An example of a receiver implementation (an expansion of the dashed block in Figure 3.19) over four consecutive TSs and four consecutive TDMA frames, following the procedure described above, is shown in Figure 3.21.

The combinations are sequentially numbered in time starting with the blind accumulation over the four blind transmissions in the first TDMA frame (**1−3**), followed by the blind accumulation in the second TDMA frame (**4−6**). The phase shift between the two accumulated bursts is then estimated (**7**) and compensated for (**8**) before accumulation (**9**). The two last TDMA frames of the block follow the same sequence and procedure.

To illustrate the nonideal processing gain from a nonideal combination of transmissions, simulations have been carried out with blind transmissions of 2^N with N ranging from 1 to 10. For each increment of N (doubling of the number of transmissions), a 3 dB gain is expected from the coherent combination. However, as can be seen in Figure 3.22, the ideal gain (solid line) according to Eq. (3.2) can only be reached without any visible degradation between 2 and 16 blind repetitions. At higher number of repetitions, the experienced gain is lower than the ideal gain. Four times more transmissions

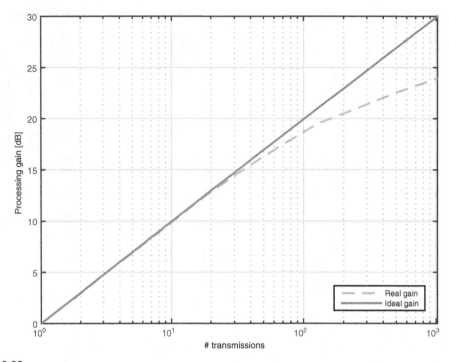

FIGURE 3.22

Real versus ideal processing gain at different number of blind repetitions.

are, for example, required to reach 24 dB gain (1024 in real gain vs. 256 with ideal gain), leading to a waste of radio resources.

It can be noted that in the 3GPP specifications, blind repetitions are referred to as *blind physical layer transmissions*. In this book when discussing EC-GSM-IoT, blind transmissions or blind repetitions are simply referred to.

3.2.8.3 Improved Channel Coding

The compact protocol implementation of EC-GSM-IoT compared with GPRS/EGPRS opens up for a reduced message size of the control channels. The overall message size of the control channels have, for example, been reduced from 23 octets as defined for GPRS/EGPRS to 8, 10, and 11 octets for EC-PACCH/U, EC-PACCH/D, and EC-CCCH/D, respectively. In addition to, the already mentioned blind transmissions, the improved channel coding from the reduced payload space also contributes to the coverage extensions on these logical channels.

3.2.8.4 More Efficient HARQ

HARQ type II was introduced with EGPRS and is also used for EC-GSM-IoT. More details are provided in Section 3.3.2.2 on the HARQ operation for EC-GSM-IoT, but one can note here that for UL operation, the use of *Fixed Uplink Allocation* (FUA) for allocation/assignment of resources will (which is more elaborated upon in Section 3.3.2.1) allow the receiver to operate at a higher BLock Error Rate (BLER) on the UL, because no detection of the sequence number is required. A higher BLER of the traffic channel will effectively increase coverage (as long as the targeted minimum throughput is reached).

3.2.8.5 Increased Acquisition Time

To see an increased acquisition time as an extension of coverage might seem strange at an initial thought but this could actually be a means that can improve coverage for all type of channels. For EC-GSM-IoT, the only channel that is extended in coverage by an increased acquisition time is the channel used for initial synchronization to the network/cell. At initial detection of the network the device will have to scan the RFs to find a suitable cell to camp on (for more details on the cell selection and cell reselection procedures, see Section 3.3.1.2). In EC-GSM-IoT the channel used for this purpose is the FCCH, see Section 3.2.6.1 for more details. Having a device synchronize to a cell during a longer period will not have a negative impact on the network performance but will have an effect on the latency in the system. For the IoT applications targeted by EC-GSM-IoT, however, the service/application is usually referred to as delay tolerant.

3.2.9 INCREASING SYSTEM CAPACITY

One of the more profound impacts to the system by serving an extensive set of users in extreme coverage situations is the impact on the system capacity. That is, when users are positioned in extreme coverage locations, they need to blindly repeat information (see Section 3.2.8.2) and by this eat up capacity that otherwise could serve other users. To combat this negative effect, EC-GSM-IoT has been designed using what is referred to as overlaid Code Division Multiple Access (CDMA). The code is applied across the blind physical layer transmissions (i.e., on a burst-per-burst basis), and, consequently, has no impact to the actual spectral properties of the signal.

The basic principle is to assign multiple users different codes that allow them to transmit on the UL, simultaneously. The codes applied are orthogonal to allow the base station to receive the super-positioned signal from the (up to four) users and still separate the channel from each user. The code consists of shifting each burst with a predefined phase shift, according to the assigned code. The codes are picked from the rows of a 4×4 *Hadamard* matrix, see Eq. (3.3), where "0" implies that no phase shift is performed, whereas a "1" means a 180 degrees (π radians) phase shift.

$$H = \begin{bmatrix} 0 & 0 & 0 & 0 \\ 0 & 1 & 0 & 1 \\ 0 & 0 & 1 & 1 \\ 0 & 1 & 1 & 0 \end{bmatrix} \tag{3.3}$$

A 180 degrees phase difference is the same as having the signal completely out of phase, and hence the total signal will be canceled out if adding a 180 degrees phase shifted signal to the original signal.

To receive the signal from the multiple users the base station simply applies the code of the user it will try to decode. In this way all paired users are cancelled out since the codes are orthogonal to the

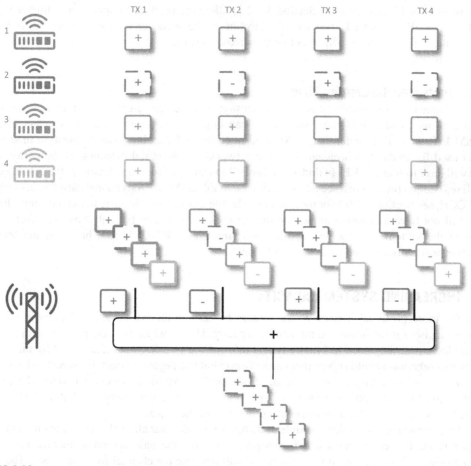

FIGURE 3.23

Illustration of overlaid CDMA.

applied code, whereas the signal for the user of interest still will benefit from the processing gain of adding the repeated bursts together.

The procedure is illustrated in Figure 3.23 where four devices are, simultaneously, transmitting on the UL, each device using a different code. The codes are illustrated as "+" (no change of phase) and "−" (180 degrees phase shift) on a per burst basis.

Compared to having a single user transmitting on the channel, the same processing gain is achieved, i.e., the SNR is increased by 6 dB ($10 \log_{10}(4)$). Hence, the channel capacity has increased by a factor of four!

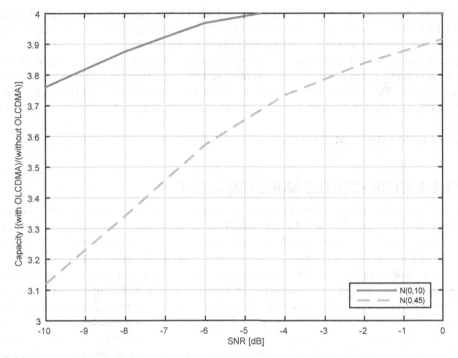

FIGURE 3.24

Capacity increase from overlaid CDMA (OLCDMA).

In reality the ideal capacity increase is not always reached, and this mainly depends on the following:

1. Not being able to pair users that, simultaneously, want to use the channel using blind repetitions.
2. Impairments in the signal, e.g., frequency offsets, that partly destroys the orthogonality.
3. Large power imbalances between the received signal that makes it sensitive to signal imperfections.
4. An increased power used on the UL channel that can cause interference to other devices.

Considering (1) this is not really a problem for the technique itself because the main target is to increase capacity, and if there are not enough users to pair to increase the capacity, the need for the extra capacity is not that evident, (4) is also not considered a problem since it will only be users in bad coverage, i.e., with a low signal level received at the serving base station, that will make use of the technique. Hence, any contribution of increased interference to other cells is expected to be small. It is thus (2) and (3) that are of main concern regarding reaching a suboptimum channel capacity.

In Figure 3.24 [9] the increase in channel capacity from multiplexing four users on the same channel, compared with allocating a single user to the same resources is shown. For a specific realization of power imbalance, the device can be up to 20 dB lower in received power compared to each paired device. The residual frequency offset (i.e., the remaining error in frequency after the device has synchronized to the base station reference) is picked from a normal distribution of mean 0 Hz and a standard deviation of either 10 or 45 Hz. It can be noted that N(0,10) can be considered a more typical frequency offset, whereas N(0,45) can be considered a more pessimistic scenario.

As can be seen, at the targeted MCL of 164 dB (20 dB improved coverage compared to GPRS/EGPRS), the capacity ranges from roughly 3.5 to 4.0 in capacity increase, reaching higher values when the coupling loss (CL) is reduced (SNR is increased).

The codes used for the technique are always of length four and are only applied over blind repetitions sent with a coherent phase, i.e., over the four blind repetitions in a TDMA frame introduced in Section 3.2.8.2. This implies that it can be used for EC-PACCH and EC-PDTCH transmissions for CC2, CC3, and CC4. For more details of the resource mapping and different UL channels, see Section 3.2.7.

Overlaid CDMA (OLCDMA) is only applied on the UL.

3.3 IDLE AND CONNECTED MODE PROCEDURES

In Section 3.2 the physical layer of EC-GSM-IoT was described. In this chapter, the procedures related to the physical layer operation will be outlined. That is, apart from knowing how information is transferred over the physical layer, this chapter will describe the related behavior of the device and the network in different modes of operation. This chapter reflects the sequential order, of events that typically occur when a device attempts a system access to initiate a data transfer. First the idle mode operation is described, with all its related activities from the device, including descriptions of how the network reaches the device when in idle mode, by, for example, paging the device. The idle mode description is followed by the system access procedure where the device initiates a connection with the network. After this follows the assignment of resources and how resources can be managed during the data transfer. The chapter will also look into how to protect the network from overload and how the security of the network has been improved to protect devices from different type of attacks.

The description of the higher layers is limited to the layers directly above the physical layer. To understand the full GSM protocol stack, and how different protocols maps to different nodes in the architecture, the interested reader is referred to Reference [10].

3.3.1 IDLE MODE PROCEDURES

Idle mode operation, or simply *idle mode*, is the state of the device when a device is not connected to the network. For example, an electricity meter that is not providing a report to the server will, to save power, not constantly stay connected to the network and instead operate in idle mode. For a device in idle mode there is still some level of activity, e.g., to support mobility and paging-based reachability, but substantially less than when connected to the network, in that case the device is in *packet transfer mode*.

There are six main idle mode procedures for which the behavior of an EC-GSM-IoT device differs from that of a GSM/EDGE device:

- Cell selection
- Cell reselection
- SI acquisition
- CC selection
- Paging
- Power Saving Mode

3.3.1.1 Cell Selection

For a legacy GSM device to always be connected to the most suitable cell, near continuous measurements are typically performed of the surrounding radio environment. This means measuring its currently selected cell as well as the surrounding neighboring cells, or if no cell has been selected yet, a scan of the supported frequency band(s) is required. Two important modes of cell selection are defined: *cell selection* and *cell reselection*.

When performing cell selection, the device will, in worst case, not have knowledge of the surrounding radio environment and hence need to acquire knowledge of what cells are around it and which cell is most suitable to connect to. This activity is hereafter referred to as *normal cell selection*. If a device would not perform this task and, for example, simply connect to the first cell detected, it would typically mean a less than optimum cell would be used, thereby resulting in high interference levels in the network because the device would have to use a higher power level to reach the base station (and the base station in its turn a higher power level to reach the device). Apart from causing interference to other devices in the network, this would also mean an increased level of packet resource usage (e.g., due to data retransmission), as well as energy consumption for the device.

Cell selection is, for example, performed at power on of the device, or when the device is roaming and the radio environment is not previously known. The task in normal cell selection is to perform a full scan of the frequency band supported by the device and connect to what is referred to as a *suitable cell*. A suitable cell shall fulfill certain criteria [11] among which the most important ones are that the cell is allowed to be accessed (for example, that it is not barred), and that the path loss experienced by the device is sufficiently low (this criterion is referred to as *C1* in the GSM/EDGE specifications). After finding a suitable cell, the device is allowed to *camp* on that cell.

An alternative to the normal cell selection described above is the use of a *stored list cell selection*; in that case the device uses stored information to limit the scan of the frequency band to a stored set of frequencies. These set of frequencies can, for example, be based on EC SI, see Section 3.3.1.3, acquired from previous connections to the network and stored at power off of the devices. Using a stored list cell selection will speed up the process of finding a suitable cell as well as help reduce energy consumption.

The scan of the frequency band is, to a larger extent, left up to implementation, but there are certain rules a device needs to follow. One important difference between a legacy GSM device and an EC-GSM-IoT device is that the latter can be in extended coverage when performing the search, which means that scanning a certain ARFCN will take longer time (see Section 3.2.8.4) and by that consume more energy. If a device is not in extended coverage, a GSM frequency can be detected by simply scanning the frequency for energy and acquiring a *received signal strength indication (RSSI)*, referred to in the specification as *RLA_C*. Cells are then ranked according to their RSSI. For legacy GSM, the cell scan stops here and a cell selection is made. For EC-GSM-IoT however, the RSSI is not enough. Because only the total energy level of the signal is considered, the measurement will also include contributions from interference and thermal noise at the receiver. Hence, to get a reasonably correct RSSI estimate, the SINR should be sufficiently high. In GSM the cell edge has traditionally been assumed to have an SINR of roughly 9 dB. This means that noise and interference at most contribute to an error in the signal estimate of 0.5 dB.

Considering the refarming of GSM frequencies (see Section 3.1) and that EC-GSM-IoT is designed to be able to operate in rather extreme coverage together with a system frequency allocation down to 600 kHz, both interference and high levels of thermal noise (compared to the wanted signal level) become more of a problem for EC-GSM-IoT than in traditional GSM networks. The EC-GSM-IoT system is, instead of designed to operate at +9 dB SINR, roughly aiming at an operation at −6 dB SINR (see Chapter 4). If only measuring RSSI the signal level estimation will be off by around 7 dB (because the total power will be dominated by the thermal noise level, 6 dB higher than the wanted signal).

Hence, after the RSSI scan, an EC-GSM-IoT device is required to perform a second scan wherein more accurate measurements are made on the strongest set of cells identified by the RSSI scan. The number of cells to include in the second scan is determined by the difference in signal strength to the strongest cell. The maximum difference is set by the network and indicated within the information devices acquire from the EC SI. This refined measurement intends to only measure the signal of a single cell, and exclude the contribution from interference and noise, and is referred to as *RLA_EC*. The device is only allowed to estimate RLA_EC using FCCH and/or EC-SCH. This is in contrast to the RSSI scan, which could measure any of the physical channels on the BCCH carrier. Requiring only FCCH and EC-SCH to be measured will, for example, allow the network to down-regulate the other channels on the BCCH carrier (except the EC-BCCH) and by this lower the overall network interference. Especially for a tight frequency reuse deployment, see Chapter 4, this will have a large impact on the overall network operation since only 5% of all bursts on the BCCH carrier are required be transmitted with full power. Figure 3.25 shows that when EC-GSM-IoT-only is supported in a given cell, resources in a 51 multiframe except FCCH, EC-SCH, and EC-BCCH can be down-regulated. It can be noted that in case also non-EC-GSM-IoT devices are supported in the cell, also the SCH, CCCH, and BCCH are transmitted with full power, in that case 16% of the resources cannot be down-regulated.

In case the device itself is in an extreme coverage situation, it could be that the RSSI scan does not provide a list of frequencies for a second RLA_EC scan (because the cell signals are all below the noise floor in the receiver). In this case, a second scan over the band using the RLA_EC measurement is been performed. Each ARFCN will at most be scanned for 2 s before a device is required to select another ARFCN to scan.

After the second scan is completed, the device attempts to find a suitable cell (from the set of cells for which RLA_EC measurements were performed) by acquiring the EC-SCH (identifying the BSIC, frame number, and additional access information regarding the cell) and the EC SI. In both EC-SCH and EC SI information regarding barring of devices can be found. In EC SI further details related to camping of the cell can be found, such as, minimum allowed signal level measured to access the cell, see Section 3.3.1.3. If a suitable cell is found, the device can camp on it.

3.3.1.2 Cell Reselection

After a cell has been selected, a legacy GSM the device will continuously monitor the cell and its surrounding neighbor cells to always camp on the most suitable cell. The procedure when a device changes the cell it is camping on is referred to as cell reselection. Compared to legacy GSM devices

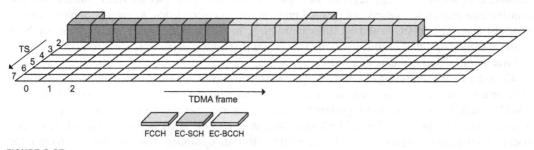

FIGURE 3.25

Power down-regulation on the BCCH carrier.

that are almost continuously monitoring the surrounding environment, an EC-GSM-IoT device need only perform measurements of the neighboring cells under certain conditions. The reason for performing continuous measurement is to ensure that the best cell is camped on and thereby guarantee e.g., good call quality should call establishment be triggered at any point in time, as well as quick mobility (handover or cell reselection) between cells. In case of CIoT the mobility of the devices is expected to be limited, the frequency of data transmission is expected to be low, and many of the applications are expected to be delay tolerant. This allows the requirements to be relaxed. In addition, considering that continuous measurements have a large impact on the battery lifetime, a minimization of these activities should be targeted.

To allow a reasonable trade-off between camping always on the most suitable cell, and saving battery, the measurements for cell reselection are only triggered by certain events, of which the most important ones are as follows:

- **Signaling failure**: If the device fails to decode the EC-SCH within 2.5 s.
- **A change in BSIC is detected**: For example, if the device wakes up to reconfirm the previously camped on cell and detects when acquiring EC-SCH that the BSIC is different from previously recorded. This implies that it has moved during its period of sleep and that it needs to reacquire the surrounding radio environment.
- **Failed path loss criterion**: The camped on cell is no longer allowed to camp on (no longer considered suitable) because the abovementioned path loss criterion (C1) fails (i.e., the measured signal level of the cell is below the minimum allowed level)
- **24 h**: More than 24 h have passed since the last measurement for cell reselection was performed
- **Change in signal level**: If the measured signal level of the camped on cell drops more than a certain threshold value compared to the best signal level measured on that cell. The threshold value, C1_DELTA, is set to the difference between the C1 value of the selected cell and the strongest neighbor determined at the most recent instance of performing the cell reselection measurement procedure.

An illustration of the fifth criterion is provided in Figure 3.26. At time N the device has selected cell A as its serving cell after performing measurements for cell reselection. The strongest neighbor cell recorded, at this point in time, is cell B, and the difference in signal strength between the two cells, *C1_DELTA*, is recorded. At a later point in time (M) the device wakes up to reconfirm cell A as a suitable cell to remain camping on. It measures a signal strength (A'') therein and determines it has degraded by more than C1_DELTA compared with the previously measured signal level (A'). This triggers measurements for cell reselection and based on the measurements, the device detects cell B to be the one with the strongest signal level and hence reselects to it. It should be noted that the above criteria only trigger *measurements* for cell reselection. It might well be so that after the measurements are performed, the device remains camping on the same serving cell.

The metric used when performing measurements for cell reselection is the same as used for cell selection, i.e., RLA_EC. To get an RLA_EC estimate for a specific cell, at least 10 samples of the signal strength of the cell, taken over at least 5 s, need to be collected. This is to ensure that momentary variations of the signal level are averaged out. It should be noted that the RLA_EC measurements performed for a set of cells need not be performed in serial but could done in parallel to minimize the overall measurement period.

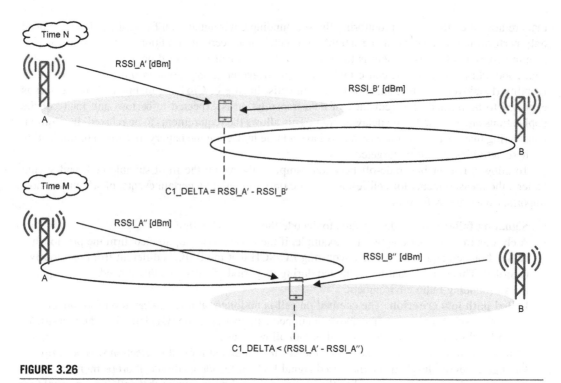

FIGURE 3.26

Triggering of measurements for cell reselection.

Other means to save the battery lifetime consist of only reacquiring EC SI if the EC-SCH indicates that the EC SI content has changed. Because the EC SI is not expected to frequently change, this is an easy way to keep SI reading to a minimum. The network will toggle the states of a change indicator information in the EC-SCH at the time of EC SI change, which informs the device that EC SI needs to be reacquired. However, the change indicator information indicates which EC SI message that has been changed, and hence it need not acquire the full EC SI. A complete reacquisition of the EC SI is required if more than 24 h has passed since the last reading.

3.3.1.3 Extended Coverage System Information (EC SI)

The EC SI contains, as the name indicates, information needed to operate in the system. The SI messages are sent on the BCCH with a certain period, see the EC-BCCH description in Section 3.2.6.3.

There are currently four messages defined, EC SI 1, EC SI 2, EC SI 3, and EC SI 4. Each message can vary in size depending on the parameters transmitted in the cell, and hence the number of EC SI message instances per EC SI message can also vary. Each EC SI message instance is mapped to an EC-BCCH block mapped to the physical resources as described in Section 3.2.6.3.

For the device to save energy and to avoid acquiring EC SI too frequently, the synchronization channel (EC-SCH) will assist the device by changing the EC-BCCH CHANGE MARK field, thereby indicating that one or more EC SI messages has changed. The device then acquires any

EC SI message and reads the EC SI_CHANGE_MARK field included therein. This field includes 1 bit for each of four EC SI messages where a change of state for any of these 4 bits (relative to the last time EC SI_CHANGE_MARK field was acquired) indicates to the device that the corresponding EC SI message must be reacquired. The EC SI_CHANGE_MARK field also includes one overflow bit, which can be toggled to indicate to the device that all EC SI messages need to be reacquired. There are rules associated with how often bit states are allowed to be changed by the network, see Reference [12], and the specification recommends that changes to the EC SI information do not occur more frequently than 7 times over a 24-h period. In addition, if a device has not detected any changes to EC-BCCH CHANGE MARK over a period of 24 h, the full EC SI message set is reacquired.

The following principles apply for the EC SI messages regardless of how many message instances are required to transmit any given EC SI message:

- The EC SI messages comprising a cycle of EC SI information are sent in sequence in ascending order.
- In case any given EC SI message contains multiple message instances (up to four) they are also sent in ascending order.
- The sequence of EC SI messages and associated message instances are repeated for each cycle of EC SI information.

The transmission of the first three EC SI messages is mandatory. The fourth message is related to network sharing and is only required if network sharing operation is activated in the network.

The content of each EC SI message is shortly described in Figure 3.27.

3.3.1.4 Coverage Class Selection

Cell selection and cell reselection are activities performed both for EC-GSM-IoT and in legacy GSM operation. CC selection, however, is something specific to EC-GSM-IoT.

The most important improvement to the system operation introduced with EC-GSM-IoT is the ability to operate devices in more challenging coverage conditions (i.e., extended coverage) than what is supported by GPRS/EGPRS. To accommodate the extended coverage and still provide a relatively efficient network operation, CCs have been introduced, as described in Section 3.2.3.3. In case a logical channel makes use of CCs, a CC is defined by a certain number of blind repetitions used, each repetition being mapped onto the physical resources in a predetermined manner, see Sections 3.2.7 and 3.2.8 for more details for DL and UL channels, respectively.

When a device is in idle mode the network will have no knowledge of the whereabouts of the device or in what coverage condition it is in. Hence, in EC-GSM-IoT it is under device control to select the CC, or more correctly put, it is under device control to perform measurements that form the basis for the CC selection. The network provides, through EC SI (broadcasting to all devices in the cell), information used by a device to determine which CC to select, given a certain measured signal.

The procedure is illustrated in Figure 3.28.

In Figure 3.28 the device measures a signal level of -110 dBm. This in itself will not allow the device to select the appropriate CC but it needs information from the network to determine the CC to be selected. In this case, -110 dBm implies that CC3 should be selected.

Two different CC selection procedures are supported by the specification in the DL, either a signal level-based selection (RLA_EC) or an SINR-based selection (*SLA*). Which measurement method to be

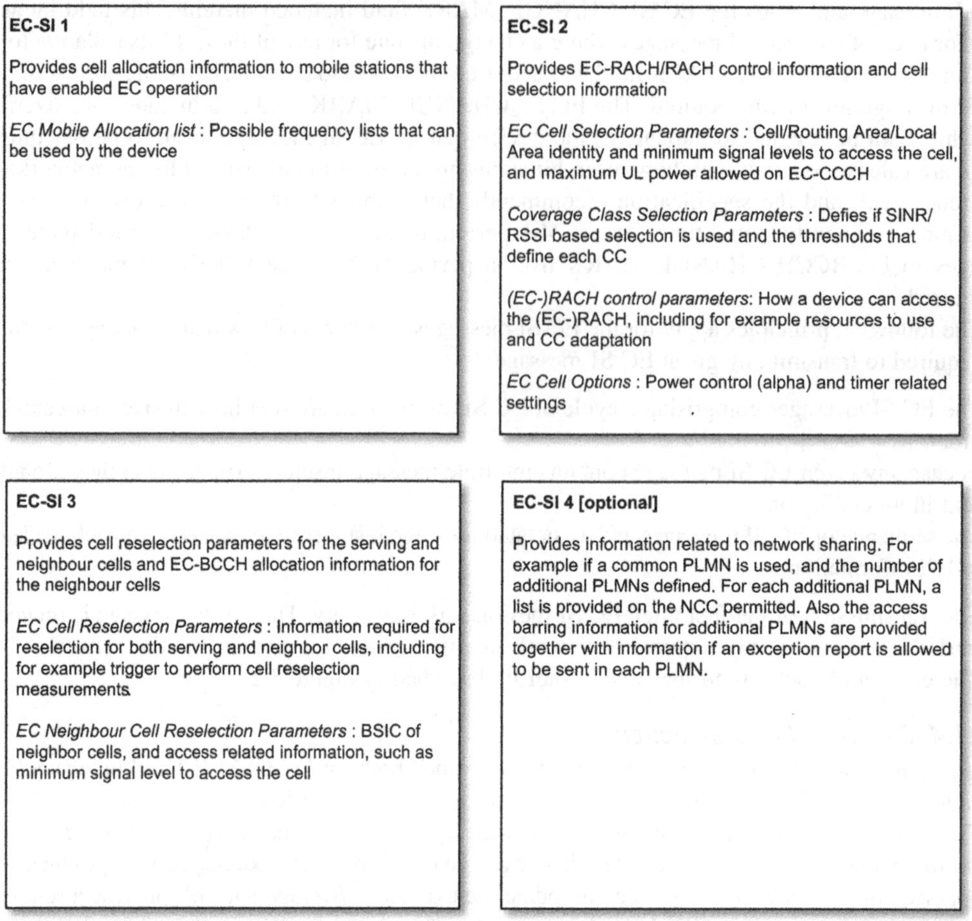

EC-SI 1
Provides cell allocation information to mobile stations that have enabled EC operation
EC Mobile Allocation list : Possible frequency lists that can be used by the device

EC-SI 2
Provides EC-RACH/RACH control information and cell selection information
EC Cell Selection Parameters : Cell/Routing Area/Local Area identity and minimum signal levels to access the cell, and maximum UL power allowed on EC-CCCH
Coverage Class Selection Parameters : Defies if SINR/ RSSI based selection is used and the thresholds that define each CC
(EC-)RACH control parameters: How a device can access the (EC-)RACH, including for example resources to use and CC adaptation
EC Cell Options : Power control (alpha) and timer related settings

EC-SI 3
Provides cell reselection parameters for the serving and neighbour cells and EC-BCCH allocation information for the neighbour cells
EC Cell Reselection Parameters : Information required for reselection for both serving and neighbor cells, including for example trigger to perform cell reselection measurements
EC Neighbour Cell Reselection Parameters : BSIC of neighbor cells, and access related information, such as minimum signal level to access the cell

EC-SI 4 [optional]
Provides information related to network sharing. For example if a common PLMN is used, and the number of additional PLMNs defined. For each additional PLMN, a list is provided on the NCC permitted. Also the access barring information for additional PLMNs are provided together with information if an exception report is allowed to be sent in each PLMN.

FIGURE 3.27

EC SI message content (nonexhaustive).

used by the device is broadcasted by the base station in the EC SI. Using a SINR-based selection can greatly improve the accuracy by which a device determines the applicable CC. In contrast, for the case where the CC selection is based on signal level, any interference would be ignored by the device and essentially the network would have to compensate for a high interference level by setting the CC thresholds in the network more conservatively, meaning that a device would start using blind repetitions earlier (at higher signal levels). This in turn will increase the interference in the network, and consume more energy in the device (longer transmission time).

The two procedures have been evaluated by system level simulations where interference from other devices and the measurement performed by the device is in detail modeled [13].

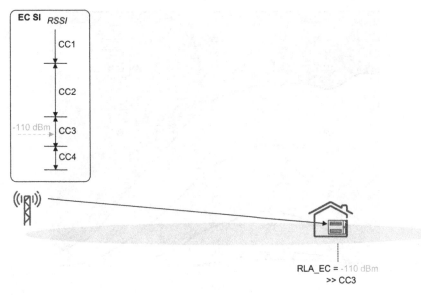

FIGURE 3.28

Coverage class selection.

As can be seen in Figure 3.29, the distributions of CC over SINR are much more concentrated in the bottom plot than in the upper one. Ideally only a single CC should be selected at a given SINR, but due to different imperfections, this will not happen. The device will, for example, not perfectly measure the SINR during the measurement period and the SINR in the system will not be the same during the measurements as when the actual transmission takes place (although the thermal noise can be considered stable over time, the interference is difficult to predict). One can note that more than four curves are seen in the figure, which is related to the possibility for a device to report the signal level above the signal level where the device switches from CC1 to CC2, which is explained in more detailed in Section 3.3.2 (here this is only referred to as CC1, a, b, and c respectively). It can further be noted that a specific reading of the percentage value for a specific CC should not be the focus when interpreting the figures. For the purpose of understanding the benefits with SINR-based CC selection, the overall shape of the curves is what matters.

Although the device is under control of the CC selection, it is important that the network knows the CC selected when the device is in idle mode, in case the network wants to reach the device using paging (see Section 3.3.1.2). There are a set of specific conditions specified where the device need to inform the network about a change in CC. One of them is, for example, if the selected CC is increased, the device experiences worse coverage than before. For the opposite case (lowering the selected CC) the device needs to only inform the network when experiencing an extreme coverage improvement, switching from CC4 to CC1. Based on this, there could, for example, be situations where the selected CC last time communicated to the network is CC3, while the device currently selected CC is CC2. In this case the device is not required to, but could, report the updated selection of CC2. Section 3.3.1.2 provides more details on why this type of operation is allowed.

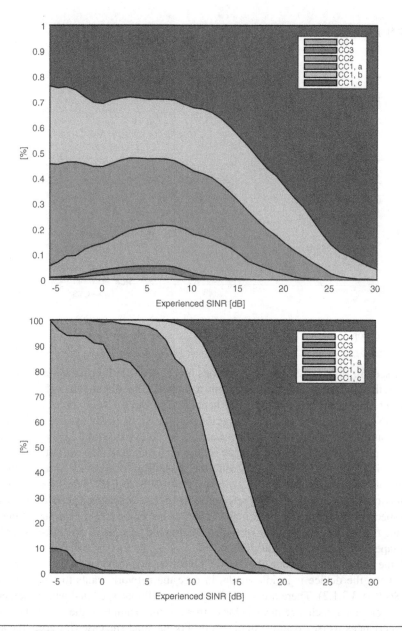

FIGURE 3.29

CC distribution over SNR (top: signal level-based measurements; bottom: SINR-based measurements).

Although all measurements for CC selection are performed on the DL, the device also needs to estimate its UL CC. This is done based on the estimated CL in the DL, which can be assumed to be the same in both directions. To help devices estimate their CL the base station sends in EC SI the output power level (P_{out}) used by the base station on FCCH and EC-SCH, and hence by knowing the received signal power (P_{rx}), the CL can be derived (see Eq. 3.4)

$$CL[dB] = P_{out}[dBm] - P_{rx}[dBm] \qquad (3.4)$$

Because the device is aware of its own output power capability, the received power at the base station can be derived using Eq. (3.4) with CL assumed the same in UL and DL, and P_{out} being the device output power.

It can be noted that only signal level-based CC selection is supported on the UL since the reciprocity of the CL does not apply to the interference (i.e., knowledge of the DL interference level does not provide accurate information about the UL interference level).

3.3.1.5 Paging

For an EC-GSM-IoT device, reachability from the network (i.e., means by which the network initiates communication with the device) is supported either by paging when Discontinuous Reception (DRX) operation is used or in conjunction with Routing Area Updates (RAUs) in the case of Power Saving Mode (PSM) (see Section 3.3.1.3).

In case of paging-based reachability, the functionality is significantly different than what is traditionally used in legacy GSM, where all devices in a cell make use of the same DRX cycle configured using the SI. In this case, the DRX setting from the network is a trade-off between having a fast reachability and enabling energy savings in the device.

The same principle applies for EC-GSM-IoT, but in this case the used DRX cycle is negotiated on a per device basis. In case a DRX cycle is successfully negotiated (during Non Access Stratum (NAS) signaling), the device can consider that DRX cycle to be supported in all cells of the Routing Area (a large set of cells defined by the network operator), and that the same negotiated DRX cycle therefore applies upon reselection to other cells in the same Routing Area.

A DRX cycle is the time period between two possible paging opportunities when the network might want to (but is not required to) reach the device. In-between these two opportunities the device will not be actively receiving and can be considered to be in a level of sleep, hence the reception is discontinuous over time. The functionality is illustrated in Figure 3.30 by the use of two possible DRX cycle configurations. It can be noted that if two devices share the same DRX cycle, they will still most probably not wake up at the same time given that they will typically use different paging opportunities

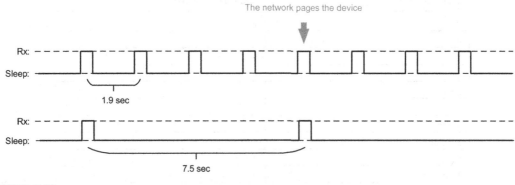

FIGURE 3.30

Discontinuous reception.

within that DRX cycle. In other words, it is only the periodicity with which the devices wake up that is the same between them.

The full set of DRX cycles supported for EC-GSM-IoT is shown in Table 3.4.

As can be seen, there is a large range of reachability periods supported, from around 2 s up to 52 min. This can be compared to the regular DRX functionality in legacy GSM that only allows DRX cycles to span a few seconds at most.

The negotiation of the DRX cycle is performed between the core network and device using NAS signaling.

The paging group to monitor for a specific DRX cycle is determined using the mobile subscription ID, International Mobile Subscriber Identity (*IMSI*). This provides a simple means to spread out the load on the paging channel over time because the IMSI can be considered more or less a random number between different devices. The shorter the DRX cycle, the fewer are the number of different nominal paging groups available within that DRX cycle. For example, for CC1 using the lowest DRX cycle of 2 s, 128 different nominal paging groups are supported therein, while for the DRX cycle of 52 min, around 213,000 different nominal paging groups are supported. The higher the CC the smaller number of nominal paging groups supported for each DRX cycle because each nominal paging group consists of a set of physical resources used to send a single paging message for a given CC. For example, using a DRX cycle of 2 s, the number of paging groups for CC4 is 4 (compared to 128 for CC1). To increase paging capacity, the network can configure up to four physical channels in the cell to be used for access grant and paging. In addition, up to two pages can be accommodated in the same paging group, in that case the paging message needs to be sent according to the highest CC of the two devices multiplexed in the same paging message, see Section 3.2.6.4.

The determination of the nominal paging group selected by a device for a given DRX cycle has been designed so that the physical resources used by that device for a lower CC are always a subset of the physical resources used by the same device for a higher CC (i.e., the resources overlap). This is to ensure that if the network assumes a device to be in a higher CC than the device has selected, the device

Table 3.4 DRX cycles	
Approximate DRX cycle length	**Number of 51-MF per DRX cycle**
2 s	8
4 s	16
8 s	32
12 s	52
25 s	104
49 s	208
1.6 min	416
3.3 min	832
6.5 min	1,664
13 min	3,328
26 min	6,656
52 min	13,312

will still be able to receive the paging message. The network will in this case send additional blind transmission that will not be monitored by the device, which can be seen as a waste of network resources. The gain with this approach is that instead the device need not continuously inform the network when going from a higher CC to a lower CC, which saves energy (device transmission is much more costly compared to reception, which is more elaborated upon in Chapter 4). The principle of the overlapping paging groups for different CCs is illustrated in Figure 3.31. In this case, the paging groups for CC1 occur over TDMA frame 27 and 28 in 51 multiframe N. If the negotiated DRX cycle

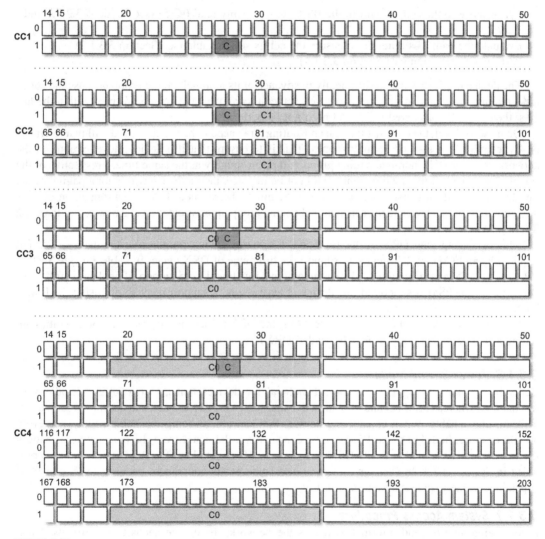

FIGURE 3.31

Overlapping paging blocks between CCs.

would be eight 51 multiframes (see Table 3.4), the same group would occur in 51 multiframe N + 8, N + 16, etc. As can be seen, TDMA frame 27 and 28 are also contained in the paging group for CC2 that spans TDMA frames 27 to 34 over 51-multiframe N and N + 1. The same principle follows for higher CCs as well, e.g., for a given device and DRX cycle, the physical resources of the nominal paging group used for CC3 are fully contained within the physical resources used for the nominal paging group for CC4 for that device.

Before each time the device wakes up to monitor its nominal paging group, it needs to perform a number of tasks:

- Synchronize to the cell, reconfirm the BSIC, and possibly read EC SI (see Sections 3.2.7.1 and 3.2.7.2)
- Evaluate the path loss criterion to ensure the cell is still suitable (see Section 3.3.1.2)
- Perform CC selection (see Section 3.3.1.4)

The device needs to wake up sufficiently in advance of its nominal paging group because the above tasks require some time to perform. However, if the tasks above have been performed within the last 30 s, they are still considered valid, and need not be performed.

On network level, the cells within a given Routing Area need to be synchronized within a tolerance of 4 s regarding the start of 51 multiframes. This is to ensure that if a device moves around in the network, the time of its nominal paging group will occur roughly at the same time, irrespective of the actual serving cell. If this requirement would not be in place, then a device with a long paging cycle, say 26 min, would, in worst case, have to wait 52 min to be reached if the synchronization of cells differ enough for the device to miss its nominal paging group upon reselection to a new cell in the same Routing Area.

Although using long DRX cycles is critical for energy saving purposes in the device, it is not required by an EC-GSM-IoT device to support the full range of DRX cycles in Table 3.4 (see Section 3.3.1.5). Also PSM (see Section 3.3.1.3), is an optional functionality to support. If neither the full set of DRX cycles nor PSM is supported, the device at least needs to support reachability using the lowest DRX cycle in the set, i.e., 2 s.

It can be noted that in the 3GPP specifications, the DRX cycles and the related functionalities are referred to as extended DRX, eDRX (extended DRX).

3.3.1.6 Power Saving Mode
Generally speaking, using DRX is more beneficial for devices that prefer a lower periodicity of reachability, whereas PSM becomes an attractive alternative for devices that can accept longer periods of being unreachable. When considering devices that can accept being reachable about once every 30 min or more, both DRX and PSM can be considered to be equal from the perspective of battery lifetime targets associated with EC-GSM-IoT devices. More details of the PSM functionality is provided in Section 2.2.3. In Chapter 2 there is also some evaluation shown between DRX and PSM depending on the requirement on reachability.

3.3.1.7 System Access Procedure
Before being able to send information across the network, the device needs to identify itself and establish a connection. This is done using the random access (system access) procedure that allows a device to make the transition from idle mode to packet transfer mode (connected mode).

The random access by EC-GSM-IoT devices can either be triggered using RACH or the EC-RACH channel. Which of the two channels to use is indicated to the device through EC-SCH (using the parameter *RACH Access Control*). Because the device is always required to acquire EC-SCH before performing a random access, it will always read the RACH Access Control parameter and thereby acquire the most up-to-date control information applicable to its serving cell. It is only applicable if the device is in CC1 that the RACH is allowed to be used, and hence devices that have selected CC2, CC3, and CC4 will always initiate system access using EC-RACH (irrespective of the RACH Access Control bit). The use of the RACH by CC1 devices depends on the traffic situation in the cell and because the control thereof is put in EC-SCH the RACH usage can dynamically vary over time (and can also be toggled by the network to for example offload traffic in 40% of the CC1 access attempts from EC-RACH to RACH). If allowing CC1 users to access on RACH mapped on TS0 it will share the random access resources with non-EC-GSM-IoT devices. Hence, in case of an already loaded RACH channel (high traffic load) it is advisable not to offload CC1 users to the RACH. On the other hand, if they stay on the EC-RACH they are more likely to interfere with CC2, CC3, and CC4 users, which, because of operating in extended coverage, are received with a low signal level at the base station. This trade-off needs to be considered in the network implementation of the feature.

3.3.1.7.1 EC Packet Channel Request
The message sent on the (EC-)RACH contains the following:

- A random reference (3 bits)
- The number of blocks in the transmit buffer (3 bits)
- Selected DL CC/Signal level indication (3 bits)
- Priority bit (1 bit)

The random reference helps the network to distinguish a given device from other devices that happen to select the same physical resource to initiate access on. As the name implies the field is set randomly, and hence there is a risk of 1/8 that two users accessing the same (EC-)RACH resource will select the same random reference (see Section 3.3.1.7 for details on how this is resolved).

The number of blocks to transmit provides information to the network on how many blocks it should allocate in the FUA, see Section 3.3.2.1. In this context, a block is based on the smallest modulation and coding scheme (MCS) size, MCS-1, carrying 22 bytes of payload.

Since the device has performed idle mode measurements, see Section 3.3.1, before initiating system access, it will not only indicate the selected DL CC, but, in case CC1 is selected, it will also indicate the region in which its measured DL signal level falls (using a granularity set by the network) thereby allowing the network to know the extent to which the signal level is above the minimum CC1 level. This then leads to the possibility of the network down regulating the power level used for sending subsequent messages to the device in the interest of interference reduction.

The final bit is related to priority, and can be set by the device to indicate that the access is to be prioritized, for example, in an alarm situation (also referred to as sending an Exception report). However, allowing the priority bit to be arbitrarily set would open up for a misuse of the functionality. That is, why would a device manufacturer not set priority bit if he/she knows that the associated data transfer would be prioritized in the network. To avoid this abusive behavior, the priority bit is forwarded to the core network that would, for example, support a different charging mechanism for exception reports.

3.3.1.7.2 Coverage Class Adaptation

Before initiating system access, the device need to perform measurements to select an appropriate CC, see Section 3.3.1.4. However, measurements are always associated with a level of uncertainty, and there is always a risk that the CC thresholds set by the network (indicated within EC SI) are not conservative enough to provide a successful system access. In contrast, setting too conservative threshold to ensure successful system access would imply a waste of resources (devices using a higher CC than required). To combat these drawbacks, a CC adaptation can be used in the system access procedure wherein a device can send multiple access requests during a single access attempt.

This means that the device can increase its initially selected CC in case a failure is experienced during a given access attempt (i.e., an access request is transmitted but the device fails to receive an expected response from the network within a certain time frame). The number of failures experienced by a device before increasing CC during a given access attempt is controlled by the network (in EC SI). At most two increments of the CC are allowed during an access attempt. That is, if CC1 is selected initially, the device will at most send an access request using CC3 as a result of repeated failures being experienced during that access attempt (i.e., if access requests sent using CC1 and CC2 have both failed).

It can be noted that a device will not know if it is the DL or the UL CC that is the reason for a failed access request sent during a given access attempt. In addition, the failure could also be the result of an unfortunate collision with an access request sent in the same or partially overlapping TDMA frames by another device. As such, to remove as much ambiguity as possible following an access request failure both the DL and UL CC are incremented at the same time.

Simulations have shown [14] that, at the same access attempt success rate, the resource usage can be reduced between 25% and 40% by allowing CC adaptation. This does not only benefit the energy consumption of the device (less transmissions) but will also help reduce the overall interference level in the network.

3.3.1.7.3 Contention Resolution

The network will not know to which device the channel request belongs because the device only includes a "Random Reference" of 3 bits in the EC Packet Channel Request (see Section 3.3.1.7.1). This problem is resolved during the contention resolution phase.

For EC-GSM-IoT, there are two procedures that can be used for contention resolution (which one to be used is controlled by the EC SI).

The first procedure is very similar to the contention resolution used in an EGPRS access. In this case, the device will include its unique ID, the *Temporary Logical Link Identity* (*TLLI*), 4 bytes long, in each UL RLC data block until the network acknowledges the reception of the device ID. Considering the use of FUA (see Section 3.3.2.1), this becomes a relatively inefficient way of communicating the device ID to the network because essentially 4 bytes need to be reserved for TLLI in each UL RLC data block transmitted by the device during the FUA, as illustrated in Figure 3.32.

As can be seen from Figure 3.32, if a low MCS is used, the overhead from 4 bytes of TLLI in each RLC data block can be significant (18% for MCS-1, with each block carrying 22 bytes). Assume, for example, that the payload to be transferred is 62 bytes. This would fit into 3 MCS-1 blocks. However, if a TLLI of 4 bytes need to be included in each block, the device need instead to transmit 4 RLC data blocks, which is an increase of 33% in transmitted resources and energy.

To minimize overhead an alternative contention resolution phase can be used, referred to as enhanced AB procedure. In this case the TLLI is only included in the first UL RLC data block, and, in

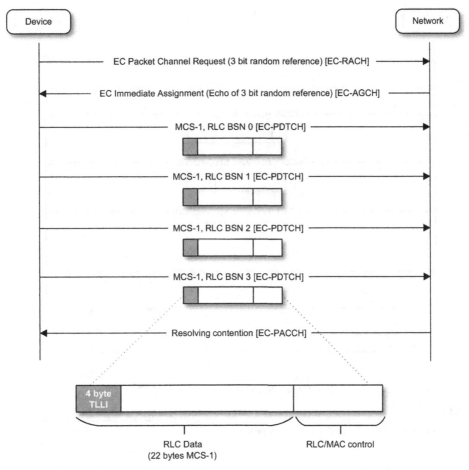

FIGURE 3.32

Contention resolution, not using enhanced access burst procedure.

addition, a reduced TLLI (rTLLI) is included in the RLC/MAC header of the subsequent RLC data blocks. The rTLLI comprises the 4 least significant bits of the full TLLI. The procedure is illustrated in Figure 3.33.

The inclusion of the rTLLI in the RLC/MAC header will not consume RLC data block payload space and hence, given the example above only 3 MCS-1 blocks would be required to transmit the 62 bytes of payload (18 + 22 + 22).

Comparing the two options for contention resolution procedure, the use of the enhanced AB procedure will result in a somewhat higher risk that two devices accessing at the same time will stay on the channel for a longer time simultaneously transmitting. This can occur only if the first RLC block has not been received by the network at the point when it sends the device an acknowledgment to the FUA. In this case only the rTLLI received by the network within the RLC/MAC header can be used to

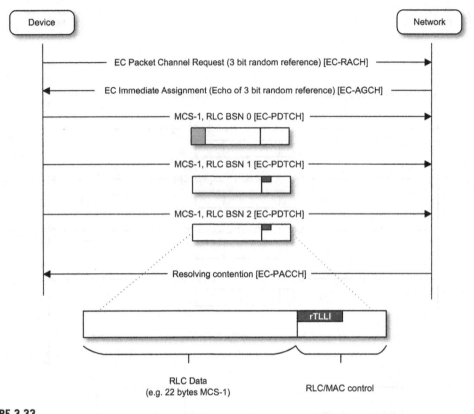

FIGURE 3.33

Contention resolution, using enhanced access burst procedure.

distinguish the users (i.e., there is a risk that the same rTLLI is used by the devices accessing), and this inability to complete contention resolution continues until the RLC data block with the full TLLI has been received. Hence, both implementation options are kept in the specification.

3.3.1.7.4 Access Control
To protect cellular networks from overload, there are typically multiple mechanisms in place. Most such mechanisms are expected to be used in exceptional situations, when, for example, turning on a network/parts of a network after a power outage. It also serves as a protection between operators, if, for example, operator A's overlapping network is shut down; in that case the devices can start camping on, and trying to access, operator B's network.

As described in Chapter 2, when 3GPP started to adapt the GSM network toward IoT by the work on MTC, the networks were enhanced to protect them from a flood of new devices. Two levels of overload control were specified: Extended Access Barring (EAB) and Implicit Reject (IR), see Chapter 2 for more details. With EC-GSM-IoT the same functionality is used but it is implemented in a somewhat different fashion.

The most important difference to legacy functionality is to make the devices aware of any potential barring of system accesses in an energy efficient manner. To achieve this, the first level of barring is placed in the EC-SCH, which is anyway required to be read by the device before any communication with the network, see Section 3.3.1.7. A two-bit *implicit reject status* flag is defined that allows for indicating either no barring, barring of all devices or two possible barring levels of roaming devices for the serving cell.

As mentioned in Section 3.1, roaming is an important functionality for CIoT devices, and for many of them, they will be deployed in networks where they are roaming for their entire life-cycle. A device using EC-GSM-IoT is not subject to the legacy EAB-based barring and instead supports a second level of barring wherein information sent in EC SI is used to determine if PLMN-specific barring is in effect for its serving cell. The PLMN-specific barring information allows for indicating whether or not barring applies to exception reports, a subset of Access Classes 0–9 or a subset of Access Classes 10–15. In addition, EC SI may indicate when a cell is barred for all access attempts (i.e., regardless of the registered PLMN used by a device).

3.3.2 CONNECTED MODE PROCEDURES

3.3.2.1 *Assignment and Allocation of Resources*
3.3.2.1.1 Downlink
The DL resource allocation for EC-GSM-IoT is very similar to the procedure used for legacy GPRS/EGPRS.

The data connection, or the *extended coverage temporary block flow (EC TBF)*, is established by the network (see Section 3.3.1.7) wherein the DL resources assigned to the device are communicated in the *EC Immediate Assignment* sent on EC-AGCH. The device is also assigned a *Temporary Flow Identity (TFI)* associated with the EC TBF.

The device monitors the assigned resources looking for an (RLC)/MAC header containing the TFI assigned to it. If the assigned TFI is found, the device considers the block to be allocated to it, and decodes the remainder of the block. The physical layer process of the DL EC-PDTCH block reception is described more in detail in Section 3.2.8.2.

3.3.2.1.2 Uplink
One of the main differences in the resource management comparing EC-GSM-IoT with previous PS allocations for GPRS/EGPRS is the use of FUA. As the name implies, it is a means to allocate resources in a fixed manner, compared with previously used allocation schemes, which were dynamic in nature.

In legacy GPRS/EGPRS a device gets resources assigned in an assignment message, and after that, resources can be dynamically allocated (scheduled) on those resources. The decision to dynamically allocate resources to a specific user is taken on a transmission time interval (TTI) basis (20 ms for BTTI and 10 ms for Reduced TTI) by the scheduler in the base station. The scheduling decision is communicated by the USF that can take up to 8 values (3-bit field). The USF is sent in each dedicated block on the DL. In order not to put restrictions on the scheduler, the USF can address any device on the UL, i.e., irrespective of which device the rest of the DL radio block is intended to. Due to this, the USF has been designed to perform with similar reliability irrespective of modulation scheme used by the base station when sending any given radio block on the DL. For example, even if the radio block is transmitted using 32QAM, the USF carried in that block is designed to be received by devices that are scheduled to use GMSK modulation.

Hence, from a perspective of increasing coverage compared to legacy GPRS/EGPRS the robust baseline performance of the USF can be viewed as a good starting point. However, looking at it more closely, improving the performance further will have rather large implications to the rest of the system if the dynamic scheduling aspects of USF are to be maintained. Considering the means to improve coverage in Section 3.2.8, any attempt to evolve the legacy USF to support extended coverage using either blind repetitions or allowing an increased acquisition time would have severe implications to the resource efficiency on the UL. Improving the channel coding performance by increasing redundancy would lead to less users possible to multiplex in the UL (using a 1-bit or 2-bit field instead of 3-bits). If anything, considering the massive number of devices expected for IoT, the addressing space currently offered by USF should be increased, not decreased.

The above realizations lead 3GPP to adopt instead the FUA approach to UL traffic channel resource management, which eliminates the need for the legacy USF-based UL scheduling. With FUA, the network allocates multiple blocks in the UL (contiguous or noncontiguous in time) based on device request. The basic procedure is as follows:

1. The device requests an UL allocation (on EC-RACH) and informs the network on the size of the UL transfer (by an integer number of MCS-1 blocks required to transmit the data in its buffer).
2. The network allocates, by an assignment message (on EC-AGCH), one or more blocks in the UL to the device.
3. After the transmission of the allocated blocks (using EC-PDTCH), the network sends an Ack/Nack report and a new assignment message (using EC-PACCH) for any remaining blocks and/or for block that are to be retransmitted. This step is repeated until all blocks have been correctly received.
4. If the device has more data in its buffer than indicated to the network in step (1), it can indicate the need to continue its UL connection by a *follow-on indicator* informing the network on the number of blocks it still has available in its buffer.

The procedure is illustrated in Figure 3.34, which illustrates the full packet flow of a UL transfer.

3.3.2.2 Hybrid ARQ

HARQ Type II using *incremental redundancy (IR)* is deployed on the EC-PDTCH. The functionality is similar as what is used for EGPRS although there are some important differences that will be outlined in this chapter.

3.3.2.2.1 EGPRS
The basic HARQ functionality in EGPRS is implemented using the following principles:

Each MCS is designed with a separately coded RLC/MAC header in addition to the RLC data block(s) transmitted in the radio block (see Section 3.2.6.5 for more details). The header is designed to be more robust than the corresponding associated data block, and hence during reception the header can be correctly received even when the data part is erroneous. In case this happens, the receiver will know the *Block Sequence Number (BSN)* of the transmitted block(s) and can store the estimated bits (soft bits) of the data block in its soft bit buffer. The receiver also needs to know the MCS and the redundancy version, or the *PS*, used to store the soft bits in the correct position. This is communicated with the *Coding and Puncturing Scheme Indicator Field (CPS)* in the RLC/MAC header. Different

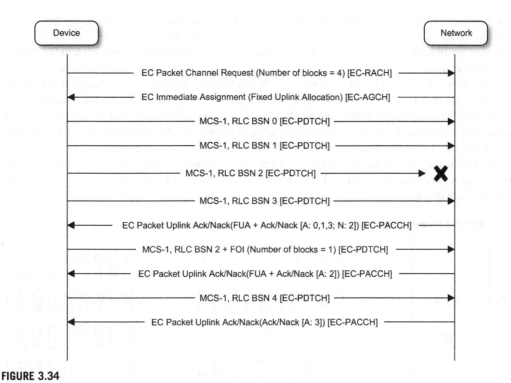

FIGURE 3.34

Fixed uplink allocation.

numbers of PSs (2 or 3) are used depending on the code rate of the MCS. The number of PSs used is determined by how many redundancy versions that are required to cover all bits from the encoder. This is illustrated by a simple example in Figure 3.35.

In this case, after receiving all three PSs the code rate of the MCS is R = 1, and hence three PSs are needed to cover all bits from the encoder, and every third bit of the encoded bit stream is added to the respective PS.

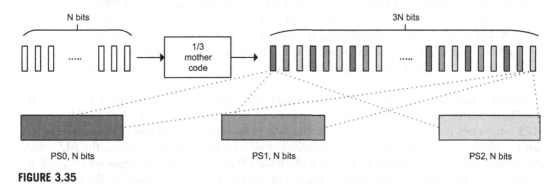

FIGURE 3.35

Puncturing schemes for HARQ.

At the receiver, the bits transmitted can be estimated and put in the soft bit buffer as long as the RLC/MAC header is received. In case no soft bit information is available for some of the bits, the decoding of the block is still possible, but those bit positions will not add any information to the overall decoding procedure.

The receiver behavior is illustrated in Figure 3.36 where the header is correctly received in both the first and second transmission (while the data was not possible to decode). In the third transmission (using PS2 at the transmitter) neither the header nor the data was decodable, and hence no additional information can be provided for further decoding attempts (since no information on BSN, MCS, and/or PS was obtained). In the fourth transmission PS0 is used again (the transmitter will cyclically loop over the PSs), and hence the soft bits associated with PS0 will increase in magnitude (the receiver will be more certain on their correct value). This additional information enables the receiver to correctly decode the data block.

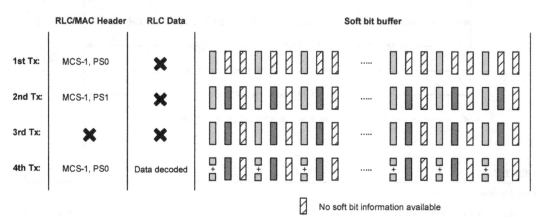

FIGURE 3.36

HARQ soft buffer handling in the receiver.

3.3.2.2.2 EC-GSM-IoT
3.3.2.2.2.1 Downlink
The principles of the DL HARQ operation for EC-GSM-IoT are the same as for EGPRS except for one important difference. To limit the device complexity when it comes to soft buffer memory, the EC-GSM-IoT RLC protocol stack only supports an RLC window size of 16 blocks (compared to 1024 for EGPRS). Although a device is not required to be able to store the soft bits for every RLC block in the RLC window, it still implies a significantly reduced requirement on soft buffer capacity in the device.

A further reduction in complexity for DL HARQ is the reduction of the number of PSs used for the GMSK-modulated MCSs. Simulations have shown that reducing the number of PS for MCS-1 and MCS-2 from two to one has no visible impact on performance [15]. This can be explained by the fact that the MCSs are already operating at a relatively low code rate, and furthermore, the RLC/MAC header is relatively close to the RLC data performance (i.e., the achievable gains with IR will be limited by the RLC/MAC header performance). For MCS-3 and MCS-4, however, the code rate is relatively high, see Section 3.2.4, and only using chase combining (every retransmission contains the same bits) will result in a link performance loss of up to 4–5 dB compared to IR using three redundancy versions [15]. If limiting the number of PSs to two instead of three, a 33% lower requirement on soft buffer memory is achieved and the performance degradation is limited.

Figure 3.37 shows the expected performance difference between the approach of using two or three redundancy versions when considering a nonfrequency hopping channel based on simulations [15]. As can be seen, for the first two transmissions (where PS0 and PS1 are used respectively in both cases) there is no performance difference (as expected). The largest performance difference (0.6 dB) is seen in the third transmission when using PS2 (if supporting three PSs) will decrease the code rate, while a second transmission of PS0 (if using two PSs) will only increase the received bit energy but not provide coding gain. All in all, the performance difference is limited and supports a reduction of the PSs. The number of PSs for the first four MCSs comparing EGPRS and EC-GSM-IoT is shown in Table 3.5.

The reduction in the number of PSs was only considered for the purpose of targeting ultra-low-cost devices supporting only GMSK modulation, and hence a similar complexity reduction for higher MCSs (MCS-5 to MCS-9) was not considered.

The same number of PSs for MCS-1 to MCS-4 applies both on the DL and UL.

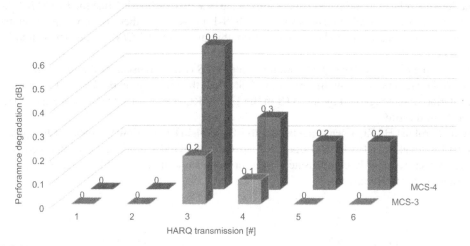

FIGURE 3.37

Performance difference comparing three or two PSs.

Table 3.5 Number of PSs		
	EGPRS	**EC-GSM-IoT**
MCS-1	2	1
MCS-2	2	1
MCS-3	3	2
MCS-4	3	2

3.3.2.2.2.2 Uplink

In the UL there is a more profound difference in the HARQ design comparing EGPRS and EC-GSM-IoT.

In typical HARQ operation the receiver would be required to decode the RLC/MAC header for the HARQ to work efficiently (see Section 3.3.2.2). For EC-GSM-IoT, due to the use of FUA (see Section 3.3.2.1), this requirement does not apply. Instead of the base station acquiring the information on BSN, MCS, and PS from the RLC/MAC header, the information can instead be provided by the Base Station Controller (BSC). Since the BSC sends the Ack/Nack information to the device, and also the allocation of the resources to be used for any given FUA, it will know how many blocks will be transmitted and over which physical resources. Furthermore, the device is required to follow a certain order in the transmission of the allocated blocks. For example, retransmissions (Nacked blocks) need to be prioritized before new transmissions, and new transmissions need to be transmission in order. Hence, for the purpose of soft buffering at the base station there is no longer a requirement to receive and correctly decode the RLC/MAC header over the air interface. The HARQ shown in Figure 3.36 instead becomes more optimized for EC-GSM-IoT as shown in Figure 3.38.

However, before the decoded data block is delivered to upper layers, also the RLC/MAC header needs to be decoded, to verify the information received by the BSC and to acquire the remaining information (not related to the HARQ operation) in the header. Considering that the RLC/MAC header will not change over multiple transmissions for MCS-1 and MCS-2 (there is only one PS supported, see Table 3.5) the receiver can also apply soft combining on the RLC/MAC header to improve its performance.

The principle of the UL HARQ operation between BTS (Base Transceiver Station, or base station) and BSC is shown in Figure 3.39. In case of EC-GSM-IoT, the information on BSN, MCS, and PS over the air interface can be ignored by the BTS in the UL HARQ operation since the same information is acquired from the BSC.

The more robust HARQ operation specific to EC-GSM-IoT will allow the receiver to operate at a higher BLER level than for HARQ operation specific to EGPRS, implying an improved resource efficiency, which will partly compensate for the increased use of resources for CC2, CC3, and CC4 when using blind repetitions.

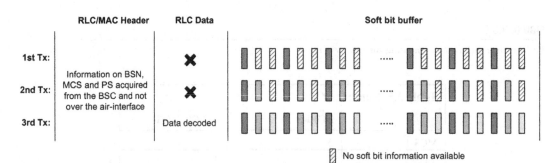

FIGURE 3.38

EC-GSM-IoT UL HARQ.

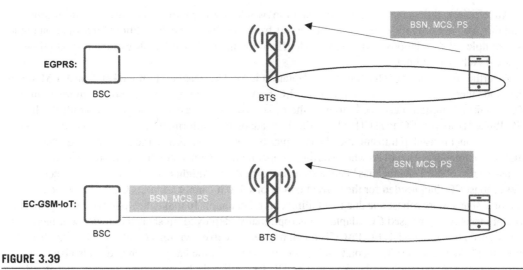

FIGURE 3.39

UL HARQ operation.

3.3.2.3 Link Adaptation

Link adaptation is the means where network adapts the MCS transmitted (DL), or instructs the device to adapt its MCSs (UL). For EC-GSM-IoT link adaptation also involves the adaptation of the CC used.

For the network to take a decision on the appropriate DL MCS, the device is required to continuously measure (and potentially report) what is referred to as a *C-value* on EC-PDTCH and EC-PACCH blocks addressed to it (where it finds a TFI value matching the assigned value). The C-value is filtered over time and can either be signal level-based (RLA_EC, see Section 3.3.1.1) or SINR-based (SLA, see Section 3.3.1.4), as indicated by the network in EC SI. From the C-value reported the network will hence get an understanding of the signal level or SINR level experienced at the receiving device.

In addition to the signal level estimation or SINR estimation the device also reports an estimation of the mean (*MEAN_BEP*) and standard deviation (*CV_BEP*) of the raw bit error rate probability of the demodulated bursts over the radio blocks transmitted on all channels (TSs) assigned to it in the DL. In case blind repetitions are used the MEAN_BEP and CV_BEP are calculated after the combination of any blind repetitions. The reason to, in addition to the mean, also estimate the standard deviation is to get an estimate of the variations of the channel. A channel with a large diversity will benefit more from using MCSs with a lower code rate, while if a low standard deviation is experienced, MCSs with higher code rate will be more suitable.

A *channel quality report* for the purpose of updating the DL link adaptation is requested from the network on a perneed basis. That is, the network can decide when, and how often, a report should be requested. On the device side, the channel quality report can be multiplexed with the Ack/Nack reporting sent on the UL EC-PACCH.

The corresponding UL link adaptation in the base station is left to implementation, and there are no requirements on measurements by the specifications. It is however reasonable to believe that a similar estimation as provided by the device, also exist in the base station.

An efficient link adaptation implementation would furthermore make use of all information available, i.e., not only the channel quality report from the device. Such additional information could, for example, be estimations of interference levels in the network and BLER estimations based on the reported Ack/Nack bitmap.

A further functionality related to link adaptation is the CC adaptation in Packet Transfer Mode (in idle mode there is also a CC adaptation on the EC-CCCH, see Section 3.3.2.2). For the more traditional MCS link adaptation described above, there is always an associated packet control channel (EC-PACCH) to the EC-PDTCH where the link adaptation information is carried. The associated control channel is used to transmit the channel quality report, and it is also the channel used when a new MCS is commanded. However, when starting to operate in coverage conditions more challenging than the associated control channel has been designed for, the device might end up in a situation where it is in a less robust CC than needed for the current conditions, but it cannot adopt its CC because the network cannot assign a more robust one due to a failing control channel. To make the system robust against this type of event, a timer-based CC adaptation is (optionally) deployed. In short the device will increment its assigned DL CC if no DL EC-PACCH block has been received within a (configurable) time from the previous FUA. The same CC adaptation is of course needed at the base station and is also based on the most recent FUA assigned to the device. The DL CC will only be incremented once, and if still no EC-PACCH is received from the network, the device will leave the EC TBF and return to idle mode.

3.3.2.4 Power Control

The UL power control for EC-GSM-IoT devices is the same as used for GPRS/EGPRS operation.

The network can choose to completely control the power level used by the device by assigning a power level reduction (Γ_{CH}). In this case the network needs to base its decision on previous feedback from the device to avoid guessing an appropriate power level to be used. It is hence a closed-loop power control. Alternatively, the network can allow the device to decide its output power based on the measured C-value (see Section 3.3.5) representing more of an open-loop power control. Still, the network is in control regarding to what extent the C-value should be weighted in the power loop equation, as determined by a parameter *alpha* (α). Alpha is broadcasted in EC SI, and hence the same value applies to all devices in the cell. However, whether to apply alpha in the power control equation is controlled by the UL assignment message sent to each device.

The resulting output power level based on different settings of α and Γ_{CH} is illustrated in Figure 3.40 (for the 900 MHz band and a device with a maximum output power of 33 dBm). It can be noted that in all possible configurations of alpha, the maximum output power of 33 dBm will always be used if Γ_{CH} (indicated by Γ in the figure legend) is set to 0 (not shown in the figure). However, by setting Γ_{CH} to a higher value (at most 62) a power reduction can be achieved. As stated above, if setting $\alpha = 0$ the DL signal strength (C-value) will not have an impact to the output power level, which can be seen by the horizontal line in the figure, while if $\alpha > 0$ there is a C-value dependency on the used output power.

The output power equation is shown in Eq. (3.5).

$$PCH = \min(\Gamma_0 - \Gamma_{CH} - \alpha \times (C + 48), PMAX)[dBm] \qquad (3.5)$$

where Γ_0 is a constant of either 39 or 36 (depending on frequency band), and PMAX is the maximum output power supported by the device. The lowest power level defined in the 900 MHz frequency band is -7 dBm, which can also be seen in Figure 3.40.

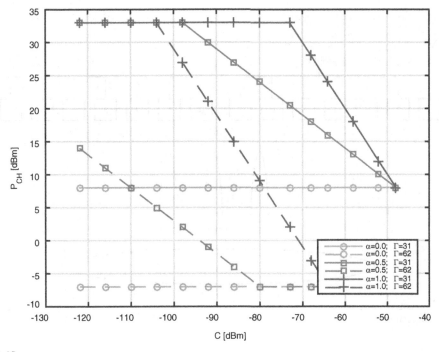

FIGURE 3.40

EC-GSM-IoT uplink Power Control.

3.3.3 BACKWARD COMPATIBILITY

As stated already in Section 3.1, backward compatibility is an important part of the EC-GSM-IoT feature. The backward compatibility applies not only to the deployment of the system but also to the implementation of it in existing device and networks. That is, for a device manufacturer and a network vendor, the feature has been designed with as much commonality as possible with legacy devices and networks, thereby allowing it to be implemented on existing device and network platforms along with existing GSM features.

For a network operator the deployment of the EC-GSM-IoT feature should consist of a software upgrade of the existing installed base. No further replanning of the network or installment of additional hardware is expected. Hence a smooth transition from a network without EC-GSM-IoT devices to a more IoT-centric network can easily be supported in a seamless fashion. The only dedicated resources required for EC-GSM-IoT operation is the new synchronization, BCCH and CCCH on TS1 of the BCCH carrier.

Figure 3.41 illustrates a cell where EC-GSM-IoT has been deployed where apart from the broadcast carrier, also two traffic (TCH) carriers are used to increase cell capacity. As can be seen, it is only TS1 of the BCCH carrier that need to be exclusively allocated to EC-GSM-IoT-related channels. In all other TSs (apart from TS0 on the BCCH carrier that need to be allocated for non-EC-GSM-IoT

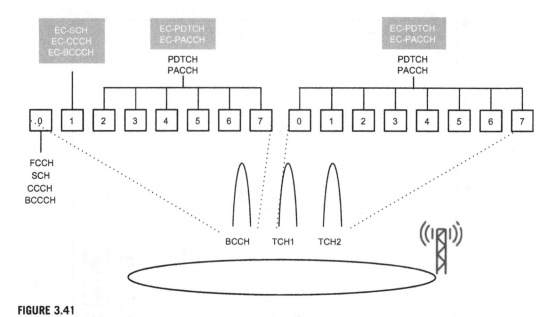

FIGURE 3.41

Channel configuration.

traffic) the PS traffic of GPRS/EGPRS can be fully multiplexed, without any restrictions, with the corresponding EC-GSM-IoT traffic.

It can be noted that in the figure it is assumed that no resources in the cell is allocated for speech services, which would not be able to share dedicated traffic resources with neither GPRS/EGPRS nor EC-GSM-IoT packet services, and hence are not of relevance for backward compatibility in this context. It should, however, be noted that there support of CS and PS services are not impacted by the introduction of EC-GSM-IoT. That is, any limitations, or possibilities, of resource multiplexing are not changed with the introduction of EC-GSM-IoT.

3.3.4 IMPROVED SECURITY

Cellular IoT devices are designed to be of ultra-low cost, and by that, to have limited system complexity. This does, however, not imply that the security of the system is not of major concern - rather the opposite. Consider, for example, a household wherein 40 devices are connected to different applications. Having a person with malicious intent being able to control of all these devices could have severe implications to personal integrity, security, and perhaps even health. Hence, a decision was taken in 3GPP to further enhance the security of GSM to state-of-the-art grade security, targeting IoT. Although the driving force behind the improvements was IoT, the enhanced security features can also be implemented in GSM/EDGE devices not supporting EC-GSM-IoT.

GSM was designed more than 20 years ago, considering, at that time, known security threats and computational power available in those days. Although the security in GSM has evolved over the years, there are still basic design principles of the system that pose threats to security, such as bidding-down attacks (an attacker acting as fake base station, requesting the device to downgrade its level of security).

Apart from providing state-of-the-art grade security, the characteristics of Cellular IoT compared to regular cellular operation need also to be considered. These include, for example, longer inactivity periods between connections to the network and attention to minimizing the impact to battery lifetime from security-related procedures. Minimizing the impact to battery lifetime includes simply minimizing the messages being transmitted and received over the radio interface related to security. In particular, it is of interest to minimize transmissions, which consumes roughly 40 times the energy of reception (see Section 4.2). The longer inactivity periods stem, for example, from the use of DRX cycles up to 52 min (see Section 3.3.1.5) or from the use of Power Saving Mode (see Section 2.2). In general, the less frequent the security context is updated the less impact on battery lifetime but the more vulnerable the security context becomes. It could, however, be argued that a less frequency reauthentication for devices supporting CIoT applications will also provide less opportunities for a potential attacker to decipher the device communication. It can also be noted that the frequency with which the authentication procedure is performed is up to the operator deployment and is not specified in 3GPP.

The improvements to the GSM security-related procedures for EC-GSM-IoT include the following:

- **Mutual authentication**: In GSM, only the device is authenticated from the network perspective, whereas the network is never authenticated at the device. This increases the risk of having what is usually referred to as a fake-BTS (an attacker acting as a network node) that the device can connect to. With EC-GSM-IoT the protocol for mutual authentication used in 3G networks has been adopted. Mutual authentication here refers to both the device and the network being authenticated in the connection establishment.
- **Improved encryption/ciphering**: The use of various 64-bit long encryption algorithms in GSM has since a few years back been shown to be vulnerable to attacks. Although 128-bit ciphering has been supported in GPRS (GEA4) for some time, a second 128-bit algorithm (GEA5, also making the use of 128-bit encryption keys mandatory) has been defined, and its use is mandatory for EC-GSM-IoT devices and networks. The same length of encryption keys is also used in 3G and 4G networks.
- **Rejection of incompatible networks**: One of the issues already mentioned with GSM is the bidding-down attack, where the attacker acts as a network node instructing the device to lower its encryption level (to something that can be broken by the attacker) or to turn the encryption completely off. With EC-GSM-IoT there is a minimum security context that the network needs to support, and in case the network is found not to support these, the device shall reject the network connection.
- **Integrity protection**: Integrity protection of control plane has been added which was earlier not supported for GPRS/EGPRS. The protection profile has been updated to match protection profiles used in 4G. Integrity protection can also optionally be added to the user plane data, which especially is of interest in countries where encryption cannot be used.

The changes introduced to EC-GSM-IoT puts the security on the same level as what is used by 3G and 4G networks today.

For more details on the security-related work, the interested reader is referred to References [16,17].

3.3.5 DEVICE AND NETWORK CAPABILITIES

To support EC-GSM-IoT for a device implies the support of all functionality described in this chapter, with the following exceptions:

- The device has two options of support when it comes to DL and UL modulation schemes, either to only support GMSK modulation scheme or to support both GMSK and 8PSK modulation (see Section 3.2.2 for more details).
- The device has also two options of supporting different output power levels. Almost all GSM devices are capable of 33 dBm maximum output power. With EC-GSM-IoT also an additional power class of 23 dBm has been added to the specifications. Since both 33 and 23 dBm devices follow the same specification (i.e., no additional blind physical layer transmissions are, for example, used by 23 dBm devices), the result of lowering the output power is a reduction in maximum UL coverage by 10 dB. At a first glance, this might seem counterproductive for a feature mainly aiming at coverage improvement. The reason to support a second output power class is the ability to support a wider range of battery technologies and to be able to integrate the PA onto the chip, which decreases implementation cost, and hence will allow an even lower price point for 23 dBm devices compared to 33 dBm devices. It should also be noted that the percentage of devices outside of a given coverage level rapidly decreases with increasing coverage. Hence, the majority of the devices outside of GPRS/EGPRS coverage will be covered by a 23 dBm implementation for which a 10 dB UL coverage improvement can be realized.
- For DRX-related functionality the device needs to only support the shortest DRX cycle and not the full range as shown in Table 3.4. Although energy efficiency is of utmost importance for many of the CIoT modules and related applications, it does not apply to all of them, in that case a DRX cycle of 2 s can be acceptable. In addition, Power Saving Mode can be implemented as an alternative to longer DRX cycles, if reduced reachability is acceptable, reaching years of battery life (see Chapter 2 for more details).
- The number of TSs that can be simultaneously allocated in UL and DL respectively is related to the *multislot class* of the device (the class to support being up to device implementation). The multislot class that EC-GSM-IoT is implemented on must, however, align with the multislot class for EGPRS (in case EGPRS is also supported by the device). The minimum supported capability is 1 TS UL and 1 TS DL.

In contrast to the device, a network implementation of EC-GSM-IoT can leave many aspects of the feature for later implementation. For example, the capacity enhancing feature described in Section 3.2.5 is not of use in a network where only a small amount of EC-GSM-IoT device is accessing the system simultaneously. Other aspects of network implementation could, for example, be to only support one of the two contention resolution options in Section 3.3.2.3 or one of the two EC-RACH mapping options described in Section 3.2.8.1. Since the main intention of the feature is to support improved coverage it is, however, also mandatory for the network to support at least CC1 and CC4, leaving CC2 and CC3 optional. This ensures that any network implementation of EC-GSM-IoT can support the intended 164 dB in MCL, exceeding the coverage limit of GPRS/EGPRS by roughly 20 dB.

3.4 RELEASE 14 IMPROVEMENTS

So far, the functionality of EC-GSM-IoT has been described for the already available Release 13 version the 3GPP specifications (with its final parts being approved in December 2016). As described in Chapter 2, the 3GPP specifications are continuously evolving, and hence Release 14 is already being discussed.

Before starting a new Work Item to introduce new functionality in the specifications, there are discussions to identify, which changes would most benefit the industry. In case of EC-GSM-IoT, three areas of improvement have been identified for the Release 14 scope:

- Improved positioning of devices
- Improved coverage for 23 dBm devices
- New TS mapping in extended coverage for improved resource handling

3.4.1 IMPROVED POSITIONING OF DEVICES

In smartphones, today the use of GPS to estimate the position of the device is commonly used in, for example, map applications or other location-based services. In low-cost cellular devices targeted for IoT, adding a GPS to the IoT module will increase the price point for it to no longer be competitive in the CIoT market. At the same time, positioning is of great importance for certain IoT applications, such as the tracking of goods.

In GSM there are multiple positioning methods specified. Several of them require, however, dedicated hardware and/or synchronized networks to perform the positioning estimation and have therefore not be deployed to a wide extent in current networks. The task of the Release 14 work on positioning enhancements [18] is, hence, to provide an improvement to device positioning without requiring network synchronization or dedicated hardware used specifically for positioning.

The reference positioning method, which does not require dedicated hardware or network synchronization, is a method that makes use of the cell ID and TA estimation (Figure 3.42). When a device accesses a cell, the position of that base station is known to the network operator. Furthermore, the access will include an estimation of the TA. The TA is basically an estimate of the radial distance between the device and the base station. That is, the device will be placed in a "donut"-like figure defined by two radii R_1 and R_2 from the base station position. The width of the donut, i.e., R_2-R_1, is defined by how accurate the base station and device can synchronize to the overall frame structure, and also how

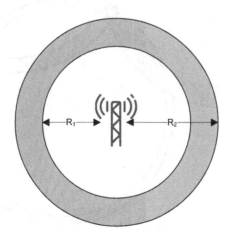

FIGURE 3.42

Cell ID together with estimated TA for positioning.

accurately the synchronization is reported (for example, if the base station uses symbol resolution on the measured TA to the network node that determines the location, the Serving Mobile Location Center).

The improvements looked at by 3GPP in Release 14 is to use the known method of multilateration to more precisely determine the device position. This means that the device will synchronize to multiple base stations to acquire the associated TA (or estimated distance), which after reported to a location server can be used to more accurately determine the device position. The angle of arrival from the base stations used for multilateration should from a device perspective be as evenly spread as possible to provide a good positioning estimate. For example, if using three base stations for the positioning attempt, they should preferably be separated with an angle of 120 degrees (360/3).

Figure 3.43 illustrates this procedure where three base stations are used for positioning. After the TA values to the three base stations are known, the network can calculate the device position as the intersection of the three "donut" (the black circle in the figure). The more base stations that are used for positioning, the smaller will the area of intersection be, and hence the better the estimate.

The accuracy of the distance to each of the base stations need not be the same (i.e., the width of the donut-like shape) because the estimation error is dependent on the synchronization performance,

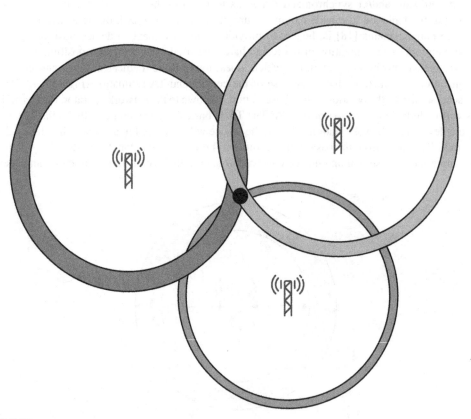

FIGURE 3.43

Multilateration.

which in turn will be SINR dependent, and hence could be quite different between different base stations.

The selection of cells can be either under network or device control. In case the network selects the base stations it would first need to know the possible base stations to select from (i.e., which base stations are reachable by the device), and in a second step, select the subset the device is to acquire TA from. If the network would have information about SINR and/or RSSI to each base station, it could consider both the geometry of the cells and the expected device TA synchronization accuracy in its selection of cells. In case the device selects the base stations, it would instead select the base stations with the highest signal level (indicating also a higher SINR), because it will have no perception of the angular location of the base stations. It could however be informed by the network which base stations are colocated, in that case it can avoid to include more than one base station from the same site in the selection process.

The positioning accuracy has been evaluated [19] by system simulations, modeling, for example, the SINR dependent synchronization accuracy. The results are presented in Figure 3.44. For the multilateration case, three base stations are used in the positioning attempt. It can be noted that the more base stations that are used, the better the accuracy, but also the more energy will be consumed by the device. From the figure it is seen that the most suitable method is where the device selects the base stations (based on descending SINR), excluding cells that are cosited (legend: "NW guided sel."). It

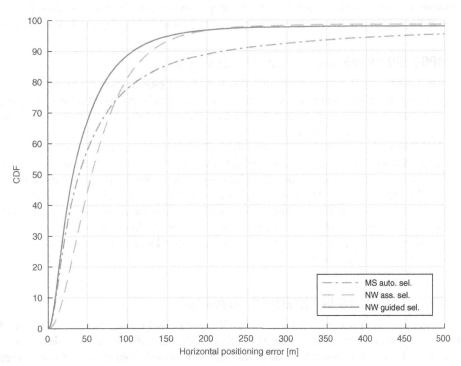

FIGURE 3.44

Positioning accuracy of multilateration.

should be noted that another method could be the optimal one for another number of base stations used in the positioning attempt [19].

A second positioning method is also considered for the Release 14 specification work where, in addition to the multilateration method, a method not relying on acquiring TA values from multiple base stations can be used under certain conditions. To understand this approach, we introduce the notion of Real Time Difference (RTD) and Observed Time Difference (OTD). RTD is the actual time difference of the frame structure between cells. OTD is the time difference observed at the device between the frame structures of different cells. The difference between them is, hence, the time it has taken for the signal to propagate from each base station to the receiving device. The propagation delay is also what the TA estimates. In simple terms, if two out of OTD, RTD, and the TAs are known, the third can be calculated.

A two-step approach is taken, where the first step involves one or more TA-based multilateration positioning attempts by one or more devices. In these attempts, not only the TA is reported from the device to the network but also the OTD. Hence, based on the previous reasoning, the RTD can be derived. The second step occurs at a later point in time where another positioning attempt is to be performed (not necessarily by the same device(s) as in the first step). If the new positioning attempts involves the same set of cells as in the first step, and if the RTD is still considered valid for these cells (the time base at the network will slowly drift over time, and hence depending on the accuracy of the network timing, the previously acquired RTD can only be considered valid for a limited time), then only the OTD need to be acquired, and the TA can instead be derived. Because the acquisition of the OTD only involves DL monitoring (i.e., reception), the method is more energy efficient than the multilateration based on TA acquisition at the cost of a more complex network implementation and lower positioning accuracy.

3.4.2 IMPROVED COVERAGE FOR 23 dBm DEVICES

A new power class for the device was introduced in Release 13 (see Section 3.3.5), which lowers the output power level by 10 dB compared with previously used levels. Allowing a lower output power class allows a more cost efficient implementation at the expense of a 10 dB loss in UL coverage. In Release 14, work has been initiated to minimize this loss. Although the full 10 dB gap is not expected to be bridged, an improvement of roughly 5 dB could still be achieved without significant impact to network operation. What is considered is to introduce a new CC, CC5, and making use of a more extensive number of blind repetitions. It is also considered to further minimize protocol overhead to allow a higher level of redundancy in the transmission. One logical channel that needs special consideration is the EC-RACH where a simple increase in the number of repetitions also would mean an increase of interference and collision-risk. Hence, means to improve the EC-RACH is also considered where, for example, a longer burst, using a longer known synchronization sequence is considered to improve the link budget.

3.4.3 NEW TS MAPPING IN EXTENDED COVERAGE

One of the implications of using CC2, CC3, and CC4 on dedicated traffic channels (EC-PDTCH and EC-PACCH) is that they all need four consecutive TSs on the same carrier to be operable. For a network operator, this results in restrictions in how the feature can be deployed. Consider, for example, a cell with only a single BCCH carrier. As seen from earlier chapters (see Section 3.2.6) TS0 and TS1 will be occupied by synchronization channels, control channels, and BCCH for GSM and EC-GSM-IoT, respectively. In addition to this, the cell might serve voice calls on CS speech channels that are not possible to multiplex with PS channels to which GPRS, EGPRS, and EC-GSM-IoT belongs. Serving voice calls are often prioritized over serving PS calls; in that case EC-GSM-IoT can only be deployed in the cell if the speech channels are not taking up more than 2 TS, see Figure 3.45.

To improve deployment flexibility, it is possible to allow CC2, CC3, and CC4 to be mapped onto, for example, 2 TS instead of 4 TS. Considering that the 4 TS mapping was chosen to maximize coverage and increase coherent reception (see Section 3.2.8) there will be a negative impact on coverage by the new mapping, but this is certainly better than not being able to operate EC-GSM-IoT using CC2, CC3, or CC4 in the cell at all. The difference of the two mapping options is illustrated in Figure 3.46 for CC2.

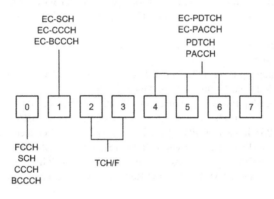

FIGURE 3.45

BCCH channel configuration including speech channels.

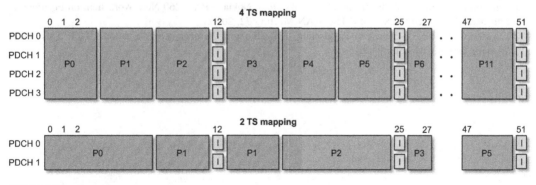

FIGURE 3.46

4 TS versus 2 TS mapping for CC2.

REFERENCES

[1] Ericsson, Ericsson Mobility Report, May 2016.

[2] IHS Markit, Technology Group, Cellular IoT Market Tracker, Q1, 2017. Results Based on IHS Markit, Technology Group, Cellular IoT Market Tracker − 1Q 2017. Information Is Not an Endorsement. Any Reliance on These Results Is at the Third Party's Own Risk, 2017. technology.ihs.com.

[3] Third Generation Partnership Project, Technical Specification 45.004 v13.0.0 Modulation, 2016.

[4] P. Laurent, Exact and approximate construction of digital phase modulations by superposition of amplitude modulated pulses (AMP), IEEE Transactions on Communications 34 (2) (February 1986) 150−160.

[5] Third Generation Partnership Project, Technical Specification 45.003 v13.0.0 Channel Coding, 2016.

[6] Third Generation Partnership Project, Technical Specification 45.002 v13.0.0 Multiplexing and Multiple Access on the Radio Path, 2016.

[7] Nokia, GP-160448 EC-PCH Enhancements for EC-GSM-IoT, 3GPP TSG GERAN Meeting 70, 2016.

[8] Ericsson LM, GP-160032 Additional EC-RACH Mapping, 3GPP TSG GERAN Meeting 69, 2016.

[9] Ericsson LM, GP-160096 EC-EGPRS, Overlaid CDMA Design and Performance Evaluation, 3GPP TSG GERAN Meeting 69, 2016.

[10] Third Generation Partnership Project, Technical Specification 43.051 v13.0.0, GSM/EDGE Overall Description; Stage 2, 2016.

[11] Third Generation Partnership Project, Technical Specification 43.022 v13.0.0 Functions Related to Mobile Station (MS) in Idle Mode and Group Receive Mode, 2016.

[12] Third Generation Partnership Project, Technical Specification 44.018 v13.0.0, Mobile Radio Interface Layer 3 Specification; GSM/EDGE Radio Resource Control (RRC) Protocol, 2016.

[13] Ericsson LM, GP-160266 Impact on PDCH for EC-GSM-IoT in a Reduced BCCH Spectrum Allocation (Including Carrier SS vs SINR Measurements), 3GPP TSG GERAN Meeting 70, 2016.

[14] Ericsson LM, GP-160035 Coverage Class Adaptation on EC-RACH, 3GPP TSG GERAN Meeting 69, 2016.

[15] Ericsson LM, GP-150139 EC-GSM, Incremental Redundancy and Chase Combining, 3GPP GERAN TSG Meeting 65, 2016.

[16] Third Generation Partnership Project, Technical Specification 33.860 v13.0.0 Study on Enhanced General Packet Radio Service (EGPRS) Access Security Enhancements with Relation to Cellular Internet of Things (IoT), 2016.

[17] Vodafone, Ericsson, Orange, TeliaSonera, SP-160203 New GPRS Algorithms for EASE, 3GPP TSG SA Meeting 71, 2016.

[18] Ericsson LM, Orange, Mediatek Inc., Sierra Wireless, Nokia, RP-161260 New Work Item on Positioning Enhancements for GERAN, 3GPP TSG RAN Meeting 72, 2016.

[19] Ericsson, R6−160235 Enhanced Positioning − Positioning Performance Evaluation, 3GPP TSG RAN Working Group 6 Meeting 2, 2016.

EC-GSM-IoT PERFORMANCE

4

CHAPTER OUTLINE

Cellular Internet of Things. http://dx.doi.org/10.1016/B978-0-12-812458-1.00004-6

Abstract

This chapter presents the performance of Extended Coverage Global System for Mobile Communications Internet of Things (EC-GSM-IoT) in terms of coverage, throughput, latency, power consumption, system capacity, and device complexity. It is shown that EC-GSM-IoT meets commonly accepted targets for these metrics in the Third Generation Partnership Project (3GPP), even in a system bandwidth as low as 600 kHz. In addition to the presented performance, an insight into the methods and assumptions used when deriving the presented results is given. This allows for a deeper understanding of both the scenarios under which EC-GSM-IoT is expected to operate and the applications EC-GSM-IoT is intended to serve.

4.1 PERFORMANCE OBJECTIVES

The design of the EC-GSM-IoT physical layer and the idle and connected mode procedures presented in Chapter 3 were shaped by the objectives agreed during the study on *Cellular System Support for Ultra-low Complexity and Low Throughput Internet of Things* [1], i.e., the support of:

- Maximum Coupling Loss (MCL) of 164 dB
- Data rate of at least 160 bits per second (bps)
- Service latency of 10 seconds
- Device battery life of up to 10 years
- System capacity of 60,000 devices/km^2
- Ultra-low device complexity

During the normative specification work item of EC-GSM-IoT [2] it was, in addition, required to introduce necessary improvements to enable EC-GSM-IoT operation in a spectrum deployment as narrow as 600 kHz.

While presenting the EC-GSM-IoT performance for each of the the just introduced objectives, this chapter will give an introduction to the methodologies used in the performance evaluations. Similar methodologies are also used when evaluating the performance of LTE-M and NB-IoT in Chapters 6 and 8, respectively.

The results and methodologies presented in this chapter are mainly collected from 3GPP TR 45.820 Cellular System Support for Ultra-low Complexity and Low Throughput Internet of Things [1] and TS 45.050 *Background for Radio Frequency (RF) requirements* [3]. In some cases, the presented results are deviating from the performance agreed by 3GPP, for example, because of EC-GSM-IoT design changes implemented during the normative specification phase subsequent to the Cellular IoT study item and the publishing of 3GPP TR 45.820. Such deviations are, however, minor and the results presented in this chapter have to a large extent been discussed and agreed by the 3GPP community.

4.2 COVERAGE

One of the key reasons for starting the work on EC-GSM-IoT was to improve the cellular coverage beyond that supported by General Packet Radio Service (GPRS). The target was to exceed the GPRS maximum coupling loss, defined in Section 2.2.5, by 20 dB to reach a maximum coupling loss of 164 dB. Coverage is meaningful first when coupled with a quality of service target and for

EC-GSM-IoT it was required that the maximum coupling loss was fulfilled at a minimum guaranteed data rate of 160 bps.

To determine the EC-GSM-IoT maximum coupling loss, the coverage performance of each of the EC-GSM-IoT logical channels is presented in this section. The section also discusses the criterias used to define adequate performance for each of the channels and the radio-related assumptions used in the evaluation of their performance. In the end of the section the achieved maximum coupling loss and data rates are presented.

4.2.1 EVALUATION ASSUMPTIONS

4.2.1.1 Requirements on Logical Channels

4.2.1.1.1 Synchronization Channels

The EC-GSM-IoT synchronization channels, i.e., the Frequency Correction CHannel (FCCH) and Extended Coverage Synchronization CHannel (EC-SCH), performance is characterized by the time taken by a device to synchronize to a cell. During the design of EC-GSM-IoT, no explicit requirement was defined on the time to synchronize to a cell, but a short synchronization time is important in most idle and connected mode procedures to provide good latency and power-efficient operation.

While the FCCH is defined by a sinusoidal waveform, the EC-SCH is a modulated waveform that contains cell-specific broadcast information that needs to be acquired by a device before initiating a connection to a base station. Beyond the time to synchronize to a cell also the achieved Block Error Rate (BLER) is a good indicator of the EC-SCH performance. A 10% BLER is considered to be a relevant target for the EC-SCH to provide adequate system access performance in terms of latency and reliability.

4.2.1.1.2 Control and Broadcast Channels

The performance of the control and broadcast channels are typically characterized in terms of their BLER. A BLER level of 10% has traditionally been targeted in the design of GSM systems [4] and is well proven. At this BLER level, the Extended Coverage Packet Associated Control Channel (EC-PACCH), Extended Coverage Access Grant CHannel (EC-AGCH), Extended Coverage Paging CHannel (EC-PCH), and the Extended Coverage BroadCast CHannel (EC-BCCH) are considered to achieve sufficiently robust performance to support efficient network operation.

For the Extended Coverage Random Access CHannel (EC-RACH), a 20% BLER is a reasonable target. Aiming for a higher BLER level on the EC-RACH than for the downlink Extended Coverage Common Control CHannels (EC-CCCH/D), which delivers the associated assignment of resources, or pages a device, reflects that this is a best effort channel where it is not critical if a device needs to perform multiple attempts before successfully accessing the system. Furthermore, because the EC-RACH is a collision-based channel, using a slightly higher BLER level operating point will allow an efficient use of the spectrum and will increase the utilization of the channel capacity.

4.2.1.1.3 Traffic Channels

For the Extended Coverage Packet Data Traffic CHannel (EC-PDTCH) data rate is a suitable design criterion. EC-GSM-IoT uses, besides blind transmissions, Hybrid Automatic Repeat reQuest (HARQ) to achieve high coupling loss, as explained in Section 3.3.2.2. To use HARQ efficiently, an average BLER significantly higher than 10% is typically targeted. To derive the achievable EC-PDTCH coupling loss and data rate, the HARQ procedure needs to be evaluated. The high-level HARQ process flow, in terms of EC-PDTCH data packets transmitted in one direction and EC-PACCH Ack/Nack control messages transmitted in the opposite direction, is depicted in Figure 4.1.

As a response to each set of transmitted EC-PDTCH blocks, an EC-PACCH control block is transmitted to positively acknowledge (Ack) or negatively acknowledge (Nack) the reception of the EC-PDTCH block. Failed blocks are retransmitted. At each stage, both total processing delays and transmission times are incremented to derive the total latency associated with the HARQ transmission of an uplink or downlink report. To generate reliable performance results, and to construct a cumulative distribution function of the EC-PDTCH HARQ latency, a large number of instances of the HARQ packet flow was simulated. The latency achieved at the 90th percentile (i.e., the time below which 90% of the simulated reports are delivered) is used in the EC-PDTCH performance evaluations presented later in this chapter.

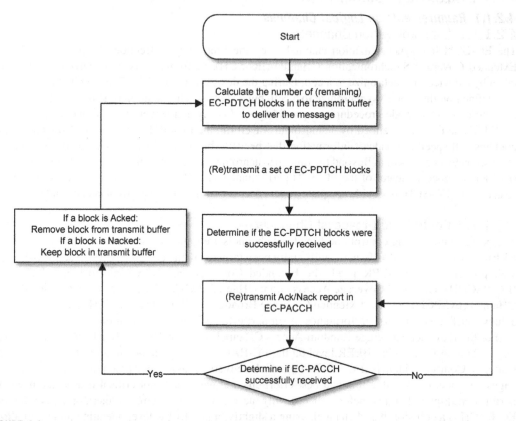

FIGURE 4.1

HARQ packet flow used in the modeling of the EC-PDTCH performance.

4.2.1.2 Radio-Related Parameters

For the EC-GSM-IoT performance evaluations presented herein, Typical Urban (TU) propagation conditions with a Rayleigh fading of 1 Hz maximum Doppler spread is assumed this is intended to model a stationary device with the fading generated by mobility of items in the spatial proximity of the device. The *root mean square delay spread* of the TU channel is around 1 µs, which is fairly challenging and not as typical as the name indicates; see Reference [6] for more details.

To derive the noise level in the receiver, the ideal thermal noise density at 300K, or 27°C (-174 dBm/Hz; see Section 3.2.8.1), in combination with a device noise figure of 5 dB and base station noise figure of 3 dB are used. These noise figures are assumed to correspond to realistic device and base station implementations although it needs to be stressed that the noise figures between different implementations can vary substantially, even between units using the same platform design. An initial frequency offset in the device of 20 ppm when synchronizing to the FCCH and EC-SCH is also assumed. The source of this initial frequency error is described in Section 4.7.1. In addition to the initial frequency error, a continuous frequency drift of 22.5 Hz/s is assumed. This is intended to model the frequency drift expected from a temperature controlled crystal oscillator—based frequency reference.

In terms of output power it is assumed that the base station is configured with 43 dBm output power per 200 kHz channel. This corresponds to a typical GSM macro deployment scenario. For the device side a 33 dBm power class is assumed, which again corresponds to a common GSM implementation. Also, a new lower power class of 23 dBm is investigated. This power class is of particular interest because it is commonly understood to facilitate reduced device complexity, as elaborated on in Section 4.7.4.

For the evaluation of the EC-PDTCH coverage the MCS-1 (modulation and coding scheme 1) is used. MCS-1 uses Gaussian Minimum Shift Keying (GMSK) modulation and a code rate of roughly 0.5, which makes it the most robust EC-GSM-IoT MCS. Each MSC-1 block carries 22 bytes of payload. An overview of the radio related simulations assumptions is given in Table 4.1.

Table 4.1 Simulation assumptions

Parameter	Value
Frequency band	900 MHz
Propagation condition	Typical Urban (TU)
Fading	Rayleigh, 1 Hz
Device initial oscillator inaccuracy	20 ppm (applied in FCCH/EC-SCH evaluations)
Device frequency drift	22.5 Hz/s
Device NF	5 dB
Base station NF	3 dB
Device power class	33 or 23 dBm
Base station power class	43 dBm
Modulation and coding scheme	MCS-1

4.2.2 COVERAGE PERFORMANCE

Table 4.2 presents the downlink (DL) coverage performance for each of the logical channels recorded when configuring a set of simulations according to the evaluations assumptions introduced in Section 4.2.1. The FCCH and EC-SCH synchronization performance is presented in terms of the time, at the 90th percentile, required to synchronize to a cell at the maximum coupling loss of 164 dB. The control and broadcast channels performances are presented as the maximum coupling loss at which 10% BLER is experienced.

Table 4.2 EC-GSM-IoT downlink maximum coupling loss performance [1,7]

#	Logical channel name	EC-PDTCH/D		EC-PACCH/D	EC-CCCH/D	EC-BCCH	EC-SCH	FCCH/EC-SCH
1	Performance	0.5 kbps[a]	2.3 kbps[b]	10% BLER	10% BLER	10% BLER	10% BLER	1.15 s
Transmitter								
2	Total BS Tx power [dBm]	43	43	43	43	43	43	43
Receiver								
3	Thermal noise [dBm/Hz]	−174	−174	−174	−174	−174	−174	−174
4	Receiver noise figure [dB]	5	5	5	5	5	5	5
5	Interference margin [dB]	0	0	0	0	0	0	0
6	Channel bandwidth [kHz]	271	271	271	271	271	271	271
7	Effective noise power [dBm] = (3) + (4) + (5) + 10 $\log_{10}$(6)	−114.7	−114.7	−114.7	−114.7	−114.7	−114.7	−114.7
8	Required DL SINR [dB]	−6.3	3.7	−6.4	−8.8	−6.5	−8.8	−6.3
9	Receiver sensitivity [dBm] = (7) + (8)	−121	−111	−121.1	−123.5	−121.2	−123.5	−121
10	Receiver processing gain [dB]	0	0	0	0	0	0	0
11	MCL [dB] = (2) − (9) + 10	164	154	164.1	166.5	164.2	166.5	164

[a]Assuming a 33 dBm device feedbacks the EC-PACCH/U.
[b]Assuming a 23 dBm device feedbacks the EC-PACCH/U.

Table 4.3 EC-GSM-IoT uplink maximum coupling loss performance [1,7,8]

#	Physical channel name	EC-PDTCH/U		EC-PACCH/U		EC-RACH	
1	Performance	0.5 kbps	0.6 kbps	10% BLER	20% BLER	20% BLER	20% BLER
Transmitter							
2	Total device Tx power [dBm]	33	23	33	23	33	23
Receiver							
3	Thermal noise [dBm/Hz]	−174	−174	−174	−174	−174	−174
4	Receiver noise figure [dB]	3	3	3	3	3	3
5	Interference margin [dB]	0	0	0	0	0	0
6	Channel bandwidth [kHz]	271	271	271	271	271	271
7	Effective noise power [dBm] = (3) + (4) + (5) + $10\log_{10}(6)$	−116.7	−116.7	−116.7	−116.7	−116.7	−116.7
8	Required UL SINR [dB]	−14.3	−14.3	−14.3	−14.3	−15	−15
9	Receiver sensitivity [dBm] = (7) + (8)	−131.0	−131.0	−131.0	−131.0	−131.7	−131.7
10	Receiver processing gain [dB]	0	0	0	0	0	0
11	MCL [dB] = (2) − (9) + 10	164.0	154.0	164.0	154.0	164.7	154.7

For the EC-PDTCH/D Table 4.2 presents a *physical layer data rate* of 0.5 kbps achieved at the 90th percentile of the throughput CDF generated from a simulation modelling the HARQ procedure depicted in Figure 4.1. In this case a 33 dBm device is assumed to feedback the EC-PACCH/U control information. The EC-PDTCH/D performance is also presented at 154 dB's coupling loss under the assumption of a 23 dBm device sending the EC-PACCH/U Ack/Nack feedback. In this case a physical layer data rate of 2.3 kbps is achievable. The 10 dB reduction in coverage is motivated by the 10 dB lower output power of the 23 dBm device. The presented data rates correspond to physical layer throughput over the access stratum where no special consideration is given to the SNDCP, LLC, RLC, and MAC overheads accumulated across the higher layers. Simplicity motivates the use of this metric, which is also used in the LTE-M and NB-IoT performance evaluations in Chapters 6 and 8.

The uplink (UL) performance is summarized in Table 4.3. Two device power classes are evaluated, i.e., 33 and 23 dBm. At 33 dBm output power, 164 dB Maximum Coupling Loss (MCL) is achievable while for 23 dBm, support for 154 dB coupling loss is accomplished. Although the 164 dB MCL target is not within reach for the 23 dBm case, it is still of interest as the lower power class reduces device complexity. A low output power is also beneficial because it lowers the requirement, in terms of supported power amplifier drain current, on the battery feeding the device with power. The presented EC-PDTCH/U physical layer data rates were derived based on the HARQ model depicted in Figure 4.1.

The results in Tables 4.2 and 4.3 clearly show that EC-GSM-IoT meets the targeted MCL requirement of 164 dB for a physical layer data rate of 0.5 kbps.

4.3 DATA RATE

Section 4.2 presents EC-PDTCH physical layer data rates in the range of 0.5—0.6 kbps and 0.5—2.3 kbps in the UL and DL, respectively. These data rates are applicable under extreme coverage conditions. To ensure a spectrally efficient network operation and a high end-user throughput, it is, equally relevant to consider the throughput achievable for radio conditions sufficiently good to guarantee no or a limited level of block errors. Under such conditions, the network can configure the use of the highest supported modulation and coding scheme, i.e., 8PSK, on the maximum number of supported time slots (TSs). Up to eight DL time slots can be supported by EC-GSM-IoT according to the 3GPP specifications, although it is expected that support for four or five time slots in practice will be a popular design choice.

In the DL, a device is dynamically scheduled on its assigned resources and a base station will in best case transmit eight MCS-9 blocks on the eight time slots during four consecutive TDMA frames. Each MCS-9 block contains a RLC/MAC header of 5 bytes and two RLC blocks, each of 74 bytes. The maximum supported EC-GSM-IoT RLC window size of 16 limits the number of RLC blocks that at any given time can be outstanding with a pending acknowledgment status. The base station uses the *RRBP* field in the RLC header of the EC-PDTCH/D block to poll the device for a Packet Downlink Ack/Nack (PDAN) report. The device responds earliest 40 ms after the end of the EC-PDTCH/D transmission time interval (TTI) as illustrated in Figure 4.2. Assuming that the base station needs 20 ms to process the PDAN report before resuming the EC-PDTCH/D transmission implies that eight MCS-9 blocks each of 153 bytes size can be transmitted every 100 ms. This limits the peak DL physical layer data rate of EC-GSM-IoT to 97.9 kbps.

This data rate can be compared with the often referred to *instantaneous peak physical layer data rate* of 489.6 kbps that can be reached across the EC-PDTCH/D 20 ms TTI. High data rates on link level can be translated into a high spectral efficiency, which is of importance for the system as a whole in terms of improved capacity. For the individual device the support of a flexible range of data rates in combination with a proper link adaptation equates to improved latency and battery life when radio conditions improve.

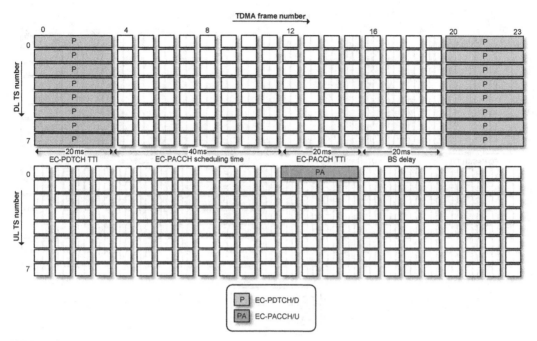

FIGURE 4.2

EC-GSM-IoT downlink scheduling cycle.

In the UL, EC-GSM-IoT uses the concept of Fixed Uplink Allocations (FUA) (see Section 3.3.2.1.2) to schedule traffic. For devices supporting 8PSK the best performance is achieved when eight MCS-9 blocks are scheduled on eight time slots. Again, the RLC window size of 16 sets a limitation on the number of scheduled blocks. After the EC-PACCH/D carrying the FUA information element has been transmitted, a minimum scheduling gap of 40 ms delays the EC-PDTCH/U transmission of the MCS-9 blocks as illustrated in Figure 4.3. After the end of the EC-PDTCH/U transmission the timer $T3326$ [9] needs to expire before the network can send the next EC-PACCH/D containing an Ack/Nack report as well as a new FUA. Just as for the DL, this implies that eight MCS-9 blocks can be transmitted every 100 ms. This limits the UL peak physical layer data rate of EC-GSM-IoT to 97.9 kbps.

The EC-PDTCH/U instantaneous peak physical layer data rate matches the EC-PDTCH/D 489.6 kbps across the 20 ms TTI. For devices only supporting GMSK modulation on the transmitter side, the highest modulation and coding scheme is MCS-4, which contains a RLC/MAC header of 4 octets and a single RLC block of 44 octets. In this case 16 MCS-4 RLC blocks can be scheduled during 40 ms every 120 ms leading to an UL peak physical layer data rate of 51.2 kbps.

The EC-PDTCH/U instantaneous peak physical layer data rate for the GMSK only devices is limited to 153.6 kbps over the 20 ms TTI.

Tables 4.4 and 4.5 summarizes the findings of this and the previous section in terms of physical layer data rates supported at 164 dB MCL and the peak physical layer data rates experienced under error-free conditions. In addition it presents the physical layer data rates simulated at coupling losses

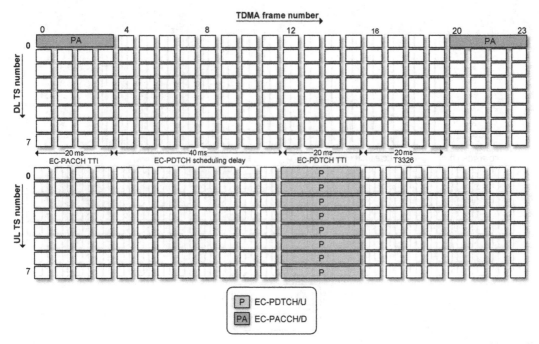

FIGURE 4.3

EC-GSM-IoT uplink scheduling cycle.

of 154 and 144 dB. For the 33 dBm device MCS-1 is providing the best performance at 164 and 154 dB coupling loss. At 144 dB coupling loss MCS-3 is the best choice in the UL even when 8PSK is supported, while MCS-4 provides the highest data rate for the DL. For the 23 dBm device MCS-1 is giving best performance at 144 and 154 dB. The evaluation assumptions used when deriving these performance figures are the same as presented in Section 4.2.1.

Table 4.4 EC-GSM-IoT physical layer data rates 33 dBm device [5]					
	Physical layer data rate			Peak physical layer data rate [kbps]	Instantaneous peak physical layer data rate [kbps]
	164 dB MCL [kbps]	154 dB CL [kbps]	144 dB CL [kbps]		
Downlink	0.5	3.7	45.6	97.9	489.6
Uplink, 8PSK supported	0.5	2.7	39.8	97.9	489.6
Uplink, GMSK supported	0.5	2.7	39.8	51.2	153.6

Table 4.5 EC-GSM-IoT physical layer data rates 23 dBm device [5]

	Physical layer data rate			Peak physical layer data rate [kbps]	Instantaneous peak physical layer data rate [kbps]
	164 dB MCL	154 dB CL [kbps]	144 dB CL [kbps]		
Downlink	–	2.3	7.4	97.9	489.6
Uplink, 8PSK supported	–	0.6	2.7	97.9	489.6
Uplink, GMSK supported	–	0.6	2.7	51.2	153.6

4.4 LATENCY

For large data transmissions the data rate is decisive for the user experience. For short data transfers expected in an IoT network the latency, including the time to establish a connection and transmitting the data, is a more relevant metric for characterizing the experienced quality of service. Hence, to guarantee a minimum level of service quality also under the most extreme conditions, EC-GSM-IoT agreed should be capable of delivering a so-called Exception report within 10 seconds after waking up from it most energy-efficient state.

4.4.1 EVALUATION ASSUMPTIONS

The Exception report is a concept specified in 3GPP Release 13 and corresponds to a message of high urgency that is prioritized by the radio and core networks. The EC-RACH channel request message contains a code point that indicates the transmission of an Exception report, which allows the network to prioritize the scheduling of the same. The 96-byte Exception report studied for EC-GSM-IoT includes 20 bytes of application layer data; 65 bytes of protocol overhead from the *COAP* application, *DTLS* security, *UDP* transport, and *IP* protocols; and 11 bytes of overhead from *SNDCP* and *LLC* layers, i.e., the GPRS core network (GPRS CN) protocols [5].

Table 4.6 EC-GSM-IoT exception report definitions including application, security, transport, IP, and GPRS CN protocol overhead [1]

Type	Size [Bytes]
Application data	20
COAP	4
DTLS	13
UDP	8
IP	40
SNDCP	4
LLC	7
Total	96

Table 4.6 summarizes the packet size definitions assumed in the evaluation of the EC-GSM-IoT latency performance. It should be noted that the 40-byte IP overhead can optionally be reduced to 4 bytes if Robust Header Compression is successfully applied in the core network. This would significantly reduce the message size and improve the latency performance.

Besides the time to transmit the 96 byte Exception report once a connection has been established, using the EC-PDTCH/U and EC-PACCH/D, the latency calculations include the time to synchronize to the network over the FCCH and EC-SCH and the time to perform the random access procedure using the EC-RACH and the EC-AGCH. In addition to the actual transmission times for the various channels, also processing delays in the device and base station as well as scheduling delays are accounted for when evaluating the EC-GSM-IoT service latency.

The acquisition of system information is not included in the latency calculations because its content can be assumed to be known by the device due to its semi-static characteristics. EC-GSM-IoT actually mandates reading of the system information not more often than once in every 24 h. It should also be remembered that the EC-SCH is demodulated as part of the synchronization procedure, and it contains the most crucial information concerning frame synchronization, access barring, and modification indication of the system information.

Figure 4.4 illustrates the signaling and packet transfers taken into account in the latency evaluation [10]. Three specific parts are identified, namely the time to acquire synchronization T_{SYNC}, the time to

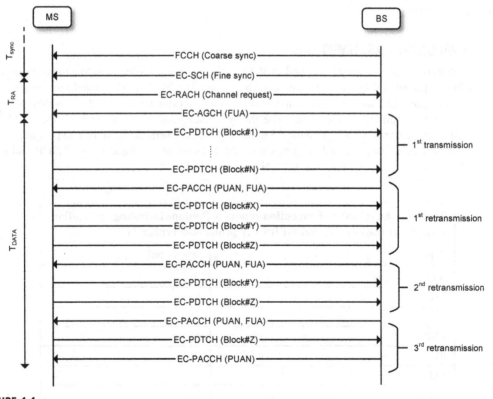

FIGURE 4.4

EC-GSM-IoT exception report procedure.

perform the Random Access procedure to access the system T_{RA}, and the time to transmit the data T_{DATA}. In the depicted example it is assumed that a first EC-PDTCH/U transmission of the report is followed by three HARQ retransmissions.

4.4.2 LATENCY PERFORMANCE

Figure 4.5 shows the time to detect a cell and perform time, frequency, and frame synchronization using the FCCH and EC-SCH under the assumption that a device wakes up from deep sleep with frequency error as large as 20 ppm corresponding to a frequency offset of 18 kHz in the 900 MHz band. This is a reasonable requirement on the frequency accuracy of the Real Time Clock (RTC) responsible for keeping track of time and scheduled events during periods of deep sleep, as discussed in Section 4.7.1. The synchronization time T_{SYNC} used in the latency calculations is derived from the 90th percentile in the synchronization time cumulative distribution function (CDF) depicted in Figure 4.5.

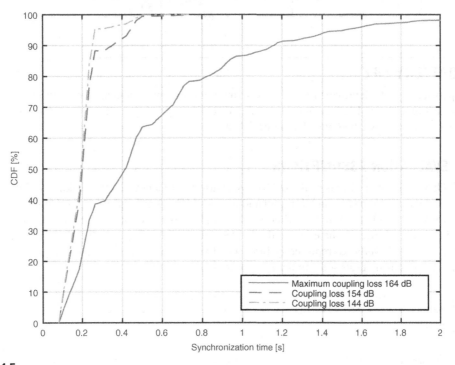

FIGURE 4.5

EC-GSM-IoT time to FCCH and EC-SCH synchronization.

T_{RA} corresponding to the time needed to perform random access and receive the Fixed Uplink Allocation (FUA) is dependent on the assumed coverage class, i.e., the number of blind transmissions, of the EC-RACH and EC-AGCH that guarantees 20% and 10% BLER respectively, for the applicable coupling loss of the studied scenario.

T_{DATA} is based on the HARQ transmission of six EC-MCS-1 blocks needed to deliver the 96 byte Exception report, following the procedure illustrated in Figure 4.1. The delay derived at the 90th percentile of the EC-PDTCH latency CDF described in Section 4.2.1.1.3 is used for determining T_{DATA}.

Table 4.7 summarizes the total time to deliver the Exception report at coupling losses of 144, 154 and 164 dB.

Table 4.7 EC-GSM-IoT exception report latencies for devices using 23 or 33 dBm output power [5,10,11]

Coupling loss [dB]	23 dBm device [s]	33 dBm device [s]
144	1.2	0.6
154	3.5	1.8
164	–	5.1

4.5 BATTERY LIFE

To support large-scale deployments of IoT systems with minimal maintenance requirements, it was required that an EC-GSM-IoT device for a range of traffic scenarios supports operation over at least 10 years on a pair of AA batteries delivering 5 Wh.

4.5.1 EVALUATION ASSUMPTIONS

The EC-GSM-IoT battery life is evaluated for a scenario where a device after waking up from it most energy-efficient state transmits an UL report and receives a DL *Application Acknowledgment*. The UL report size is set to 50 or 200 bytes, while the DL application layer acknowledgment packet size is assumed to equal 65 bytes at the entry of the GPRS core network. These packet sizes are assumed to contain the protocol overheads reported in Table 4.6. Two different report triggering intervals, of 2 and 24 hours, are assumed. Table 4.8 summarizes these assumptions.

Table 4.8 EC-GSM-IoT packet sizes at the entry of the GPRS CN for evaluation of battery life [1]

Message type	UL report		DL application acknowledgment
Size [byte]	200	50	65
Triggering interval	Once in every 2 h or once every day		

Before the higher layers in a device triggers the transmission of a report, the device is assumed to be in idle mode in which it may enter Power Save Mode (PSM) to suspend all its idle mode tasks and optimize energy consumption. Ideally it is only the RTC (see Section 4.7.1) that is active in this deep sleep state. When receiving and transmitting, the device baseband and radio frequency (RF) front end increase the power consumption. The transmit operation dominates the overall power consumption due to the high output power and the moderate power amplifier (PA) efficiency. This is especially the case for the 33 dBm power class where the transmitter side is expected to consume roughly 40 times more power than the receiver side. The different modes of device operation and their associated power consumption levels assumed during the evaluations of EC-GSM-IoT battery life are summarized in Table 4.9.

Table 4.9 EC-GSM-IoT power consumption [1]				
TX, 33 dBm	TX, 23 dBm	RX	Idle and connected mode, light sleep	Idle mode, deep sleep
4.051 W	0.503 W	99 mW	3 mW	15 µW

Figure 4.6 illustrates the UL and DL packet flows modeled in the battery life evaluation. Not illustrated is a one second period of light sleep between the end of the UL report and the start of the DL application acknowledgment message. A period of light sleep after the end of final EC-PACCH/U is also modeled. This period is assumed to be configured by the Ready timer to 20 seconds, during which the device uses a Discontinuous Reception (DRX) cycle that allows for two paging opportunities

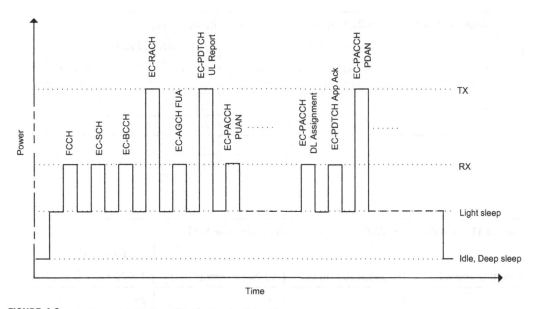

FIGURE 4.6

EC-GSM-IoT packet flow used in the evaluation of battery life.

to enable DL reachability. An important difference compared with the latency evaluation in Section 4.4 is that the number of EC-PDTCH blocks modeled in this evaluation corresponds to the average number of blocks needed to be sent, including retransmissions, to secure that the uplink report and downlink application acknowledgment are successfully received. For the synchronization time, the average FCCH and EC-SCH acquisition time was used and not the 90th percentile value used in the latency evaluations. This is a reasonable approach for the modeling of device power consumption for over more than 10 years of operation.

4.5.2 BATTERY LIFE PERFORMANCE

The resulting battery life for the investigated scenarios are presented in Tables 4.10 and 4.11. It is seen that a 10-year battery life is feasible for the reporting interval of 24 hours. It is also clear that the 2 hours reporting interval is a too aggressive target when the devices are at the MCL of 164 dB. Under these assumptions a battery life of a pair of years is achievable for the assumed pair of AA batteries. For devices with requirements on longer battery life than presented in Table 4.10, this can obviously be achieved by adding battery capacity.

In these evaluations, an ideal battery power source was assumed. It delivers 5 Wh without any losses or imperfections that typically can be associated with most battery types. EC-GSM-IoT requires a drain current in the order of 1 Ampere, which may require extra consideration when selecting the battery technology to support an EC-GSM-IoT device. A highly optimized RF front end is also assumed in these investigations with a PA efficiency of 50%. A lower efficiency will deteriorate the reported battery life.

Table 4.10 EC-GSM-IoT, 33 dBm device, battery life time [11]

Reporting Interval [h]	DL Packet Size [bytes]	UL Packet Size [bytes]	Battery Life [years]		
			144 dB CL	154 dB CL	164 dB CL
2	65	50	22.6	13.7	2.8
		200	18.4	8.5	1.2
24		50	36.0	33.2	18.8
		200	35.0	29.5	11.0

Table 4.11 EC-GSM-IoT, 23 dBm device, battery life time [11]

Reporting interval [h]	DL packet size [bytes]	UL packet size [bytes]	Battery life [years]	
			144 dB CL	154 dB CL
2	65	50	26.1	12.5
		200	22.7	7.4
24		50	36.6	32.5
		200	36.0	28.3

4.6 CAPACITY

As a carrier of IoT services, it is required of EC-GSM-IoT to support a large volume of devices. More specifically, a supported system capacity of atleast 60,680 devices per square kilometer (km^2) is expected to be supported by EC-GSM-IoT [1]. This objective is based on an assumed deployment in downtown London with a household density of 1517 homes/km^2 and 40 devices active per household. For a hexagonal cell deployment with an intersite distance of 1732 m, this results in 52,547 devices per cell. This is clearly an aggressive assumption that includes the underlying assumption that EC-GSM-IoT will serve all devices in every household, while in real life we use a multitude of different solutions to connect our devices. Table 4.12 summarizes the assumptions behind the targeted system capacity.

Table 4.12 Assumption on required system capacity [1]

Household density [homes/km²]	Devices per home	Intersite distance [m]	Devices per cell
1517	40	1732	52,547

4.6.1 EVALUATION ASSUMPTIONS

For the cellular IoT solutions, coverage is a key criterion, which makes the use of the sub-GHz frequency bands very attractive. Therefore in the evaluation of EC-GSM-IoT capacity it is assumed that the studied system is deployed in the 900 MHz frequency band, supported also by the current operation of GSM. To model coverage, a distance-dependent path loss model is assumed in combination with a large-scale shadow fading with a standard deviation of 8 dB and correlation distance of 110 m. This is intended to model an urban environment where buildings and infrastructure influence the received signal characteristics. All devices modeled in the network are assumed to be stationary and indoor. In addition to the distance-dependent path loss, a very aggressive outdoor to indoor penetration loss model with losses ranging up to 60 dB is assumed to achieve an overall maximum coupling loss of 164 dB in the studied system.

Just as for the coverage evaluations, a Rayleigh fading Typical Urban channel model with 1 Hz Doppler spread is assumed to model the expected signal variations due to items moving in the proximity of the assumed stationary EC-GSM-IoT devices.

The overall coupling loss distribution is presented in Figure 4.7 together with its path loss and outdoor to indoor loss components. The coupling loss is defined as the loss in signal power calculated as the difference in power measured at the transmitting and receiving antenna ports. Besides the path loss, outdoor to indoor loss, and shadow fading the coupling loss captures base station and device antenna gains. For the base station, a single transmit and two receive antennas with 18 dBi directional gain is assumed. This corresponds to a macro deployment with over-the-rooftop antennas. For the device side a −4 dBi antenna gain is assumed. This antenna loss is supposed to capture an antenna integrated in a device where form factor is prioritized over antenna efficiency.

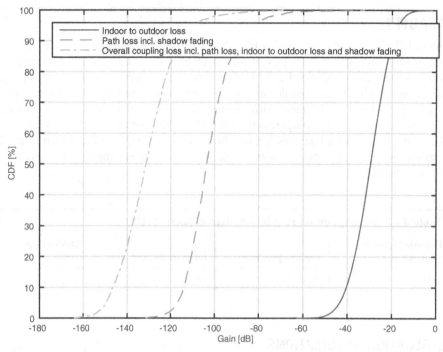

FIGURE 4.7

Distance-dependent path loss, outdoor to indoor penetration loss, and overall coupling loss distributions recorded in an EC-GSM-IoT system simulator.

The simulated system assumes that a single 200 kHz channel is configured per cell. The first time slot in every TDMA frame is configured with GSM synchronization, broadcast, and control channels, while the second time slot is used for the EC-GSM-IoT version of the same logical channels. The remaining six time slots are reserved for the EC-PDTCH and EC-PACCH. The EC-GSM-IoT channel mapping is in detailed covered in Section 3.2.5.

Table 4.13 captures a set of the most relevant EC-GSM-IoT system simulation parameter settings.

The system capacity is determined for two types of traffic scenarios as described in the next two sections.

Table 4.13 System level simulation assumptions [1]	
Parameter	**Model**
Cell structure	Hexagonal grid with three sectors per site
Cell intersite distance	1732 m
Frequency band	900 MHz
System bandwidth	2.4 MHz
Frequency reuse	12
Frequency channels (ARFCN) per cell	1
Base station transmit power	43 dBm

Table 4.13 System level simulation assumptions [1]—cont'd

Parameter	Model
Base station antenna gain	18 dBi
Channel mapping	TS0: FCCH, SCH, BCCH, CCCH TS1: EC-SCH, EC-BCCH, EC-CCCH TS2-7: EC-PACCH, EC-PDTCH
Device transmit power	33 or 23 dBm
Device antenna gain	−4 dBi
Device mobility	0 km/h
Path loss model	$120.9 + 37.6 \times \log_{10}(d)$, with d being the base station to device distance in km
Shadow fading standard deviation	8 dB
Shadow fading correlation distance	110 m

4.6.1.1 Autonomous Reporting and Network Command

In the first traffic scenario, corresponding to a deployment of smart utility meters, it is assumed that 80% of all devices autonomously triggers an UL report with a Pareto distributed payload size ranging between 20 and 200 bytes as illustrated in Figure 4.8.

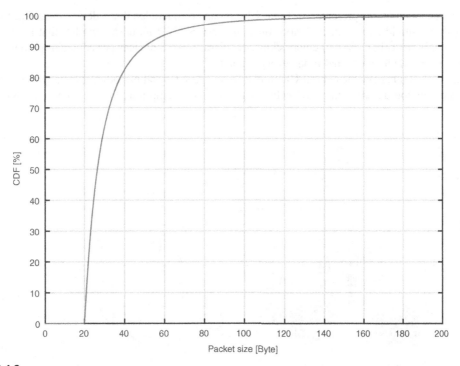

FIGURE 4.8

Pareto distributed payload ranging between 20 and 200 bytes and with a mean size of 33.6 bytes.

For the part of the meters sending the autonomous report, a set of different triggering intervals ranging from twice per hour to once per day, as captured in Table 4.14, is investigated. In 50% of the cases the device report is assumed to trigger an application-level acknowledgment resulting in a DL transmission following the UL report. The payload of the application level acknowledgment is for simplicity assumed to be zero bytes which means that the content of the DL transmission is defined by the 76 bytes protocol overhead defined in Table 4.6.

The network is assumed to send a 20-byte DL command to the 20% of the devices not transmitting an autonomous UL report. Also the network command follows the distribution and periodicity captured in Table 4.14. Every second device is expected to respond to the network command with an UL report. Also this report follows the Pareto distribution ranging between 20 and 200 bytes.

Table 4.14 Device autonomous reporting and network command periodicity and distribution [1]	
Device report and network command periodicity [h]	**Device distribution [%]**
24	40
2	40
1	15
0.5	5

Given the assumptions presented in Table 4.14, it can be concluded that a device on average makes the transition from idle to connected mode once in every ~ 128.5 min. This implies that for a load of 52,547 devices per cell, or 60,680 devices/km^2, on average ~ 6.8 devices per cell attempt to make the transition from idle to connected mode every second.

Figure 4.9 summarizes the overall UL and DL message sizes and periodicities taking the details of the device autonomous reporting and network command assumptions into account. The presented

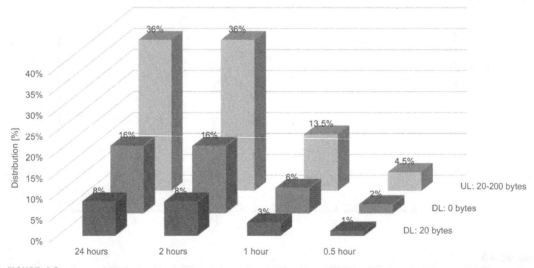

FIGURE 4.9

Mobile autonomous reporting and network command periodicity and packet size distribution [1].

packet sizes do not account for the protocol overheads of Table 4.6, which should be added to get a complete picture of the data volumes transferred over the access stratum.

It can be seen that the traffic model is UL heavy, which is a typical characteristic of IoT traffic. At the same time, it can be seen that DL traffic constitutes a substantial amount of the overall traffic generated.

4.6.1.2 Software Download

In the second traffic scenario, the system's ability to handle large DL transmission to all devices in a network is investigated. It is in general foreseen that UL traffic will dominate the load in LPWANs, but that one of the important exceptions is the download of software e.g., to handle firmware upgrades. A Pareto distributed download packet size between 200 and 2000 bytes is here considered to model a firmware upgrade targeting all devices in the network. Again the protocol overheads of Table 4.6 are added on top of this packet size. The devices are expected to receive a software upgrade on average once in every 180 days. These assumptions are presented in Table 4.15.

Table 4.15 Software download periodicity and distribution [1]	
Periodicity [days]	**Device distribution [%]**
180	100

4.6.2 CAPACITY PERFORMANCE

In the first scenario, the periodicity of the device autonomous reports and network commands in combination with the targeted load of 52,547 users per cell resulted in an overall 6.8 users per second and cell making the transition from idle to connected mode to access the system.

For this load, the consumed EC-GSM-IoT radio resources are presented in Figure 4.10. The EC-CCCH, EC-PDTCH, and EC-PACCH usage is presented in terms of average fraction of the available

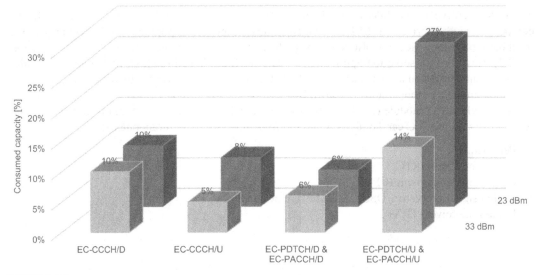

FIGURE 4.10

Average percentage of radio resources consumed to serve 52,547 users per cell [3].

resources consumed. The DL common control signaling to set up 6.8 connections per second on average occupies roughly 10% of the available EC-CCCH/D resources on TS1. For the EC-CCCH/U the load is even lower. For the case of the 33 dBm device power class the EC-PDTCH/U and EC-PACCH/U consume on average around 14% of the six configured time slots to deliver the UL data and Ack/Nack reports. For the 23 dBm power class roughly 27% of the six UL time slots are on average occupied. The DL resource consumption is significantly lower than the UL resource consumption because of the UL heavy traffic model. At this level of system load the percentage of failed connection attempts was kept below 0.1%. These results indicate that a significantly higher load than the targeted load of 52,547 users per cell can be supported by an EC-GSM-IoT system.

It should be noted that no paging load is considered on the EC-CCCH/D. In reality, the load on the paging channel will be dependent not only on the number of devices that the network needs to page, but also on the paging strategy taken by the network. That is, when the network tries to reach a device, it will not exactly know where the device is located and needs to send the paging message to multiple cells to increase the chance of getting a response. With a device that has negotiated a long eDRX cycle, it can take a very long time to reach the device. Hence, there is a clear trade-off between paging load and paging strategy that will have an impact on the overall mobile-terminated reach-ability performance. Any load caused by paging should be added to the resource usage presented in Figure 4.10, specifically to the EC-CCCH/D load. In case of a too significant paging load increase, the network can allocate up to four time slots, i.e., TS 1, 3, 5, and 7, for EC-CCCH/D, and by that increase the EC-CCCH capacity by well over 400%.

In the second scenario, where a software download is studied, it is assumed that the load can be evenly spread over time. The download periodicity of 180 days in combination with the load of 52,547 users per cell resulted in a single user making the transition from idle to connected mode on average once in every 5 min. With this assumption the consumed number of resources is close to being negligible despite the assumption of large packet sizes.

4.7 DEVICE COMPLEXITY

To be competitive in the Internet of Things market place, it is of high importance to offer a competitive module price. GSM/EDGE, which is the currently most popular cellular technology for machine-type communication, offers, for example, a module price in the area of USD 5 (see Section 3.1.2.4). However, for some IoT applications, this price-point is still too high to enable large-scale, cost-efficient implementations. For EC-GSM-IoT it is therefore a target to offer a significantly reduced complexity compared to GSM/EDGE.

An EC-GSM-IoT module can, to a large extent, be implemented as a system on chip (SoC). The functionality on the SoC can be divided into the following five major components:

- Peripherals
- Real Time Clock (RTC)
- Central processing unit (CPU)
- Digital signal processor (DSP) and hardware accelerators
- Radio transceiver (TRX)

In addition to these blocks, a number of parts may be located outside the SoC, as discrete components on a printed circuit board (PCB). The power amplifier (PA) defining the device power class and crystal oscillators (XO) providing the module frequency references are such components.

4.7.1 PERIPHERALS AND REAL TIME CLOCK

The peripherals block provides the module with external interfaces to, e.g., support a SIM, serial communication, graphics, and general purpose input and output. This is a generic block that can be found in most communication modules. The range of supported interfaces is more related to the envisioned usage of the device than to the radio access technology providing its connectivity.

The RTC is a low power block keeping track of time and scheduled tasks during periods of deep sleep. Ideally, it is the only component consuming power when a device is in deep sleep, e.g., during periods of eDRX or PSM. The RTC is typically fed by a low power XO running at 32 kHz, which may be located outside of the SoC as illustrated in Figure 4.11. For EC-GSM-IoT the accuracy of the XO during periods of sleep is assumed to equal 20 ppm.

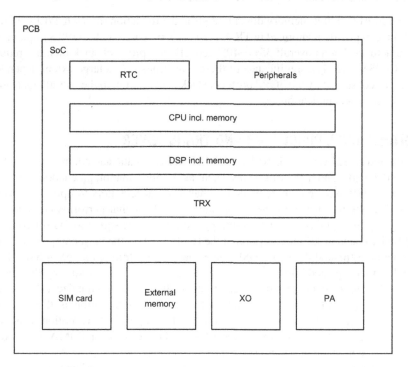

FIGURE 4.11

EC-GSM-IoT module architecture.

Both the Peripherals block and the RTC are generic components that can be expected to be found in all cellular devices regardless of the supported access technology. With this in mind it is important to understand that the cost associated with functionality related to the radio access technology is only a part of the total price on a communications module.

4.7.2 CENTRAL PROCESSING UNIT

The CPU is responsible for generic functions such as booting, running drivers, and applications. It also contains the supported protocol stacks including the GSM protocol stack, i.e., SNDCP, LLC, RLC, and MAC. It contains a controller as well as a memory. A reduction in the protocol stack reduces the CPU memory requirements. But a reduction in and simplifications of the applications supported by the module will also allow reduced computational load and memory requirements to facilitate a less advanced implementation.

The protocol stack in an EC-GSM-IoT device is favorably impacted by the following facts:

- Circuit switched (CS) voice is not supported.
- The only mandatory modulation and coding schemes are MCS-1 to MCS-4.
- The RLC window size is only 16 (compared to 64 for GPRS or 1024 for EGPRS).
- There is a significant reduction in the number of supported RLC/MAC messages and procedures compared with GPRS.
- Concurrent UL and DL data transfers are not supported.

A reduction in RLC buffer memory down to 2 kB and a reduction in SNDCP/LLC buffer memory down to 43 kB are explicitly mentioned in TR 45.820 [1]. Compared with a GPRS implementation, TR 45.820 also concludes that an overall 35%−40% reduction in protocol stack memory requirements is feasible for EC-GSM-IoT. The significance of these reductions is to a large extent dependent on how large the memory consumed by the 3GPP protocol stack is in relation to the overall controller memory for booting, drivers, and applications.

4.7.3 DIGITAL SIGNAL PROCESSOR AND TRANSCEIVER

The DSP feeds, and is fed by, the CPU with RLC/MAC headers, data, and control blocks. It handles the modem baseband parts and performs tasks such as symbol mapping, encoding, decoding, and equalization. The DSP may be complemented by hardware accelerators to optimize special purpose tasks such as FIR filtering. It passes the bit stream to the TRX that performs tasks such as GMSK modulation, analog to digital conversion, filtering, and mixing the signal to radio frequency (RF).

For the DSP baseband tasks, the reception of the EC-PDTCH is the most computational demanding task Consuming an estimated 88×10^3 DSP cycles per TDMA frame, i.e., per 4.6 ms. For coverage class 2, 3, and 4, four repeated bursts are mapped on consecutive TSs. Assuming that the four bursts can be combined on IQ-level (see Section 3.2.8.2) allows the device to equalize a single burst and not four as in the case of GPRS. Therefore although EC-PDTCH reception is the most demanding operation, it is significantly less demanding than GSM/EDGE PDTCH reception. Compared with a GPRS reference supporting four receive TSs the 88×10^3 DSP cycles per TDMA frame correspond to a 66% reduction in computational complexity [1].

The IQ-combination poses new requirements on the DSP memory. Four bursts, each of 160 samples, stored using 2×16 bit IQ representation will, e.g., consume $4 \times 160 \times 2 \times 16 = 2.56$ kB. This is, however, more than compensated for by the reduced requirements on soft buffer size stemming from the reduced RLC window and a reduced number of redundancy versions supported for EC-GSM-IoT, as explained in Section 3.3.2.2.

Based on the above calculations, the overall reduction in the DSP memory size compared to an EGPRS reference results in an estimated saving in ROM and RAM memory of 160 kB. This corresponds to a ROM memory savings of 48% and RAM memory savings in the range of 19%−33% [1].

For the TRX RF components, it is positive that EC-GSM-IoT supports only four global frequency bands. This minimizes the need to support frequency-specific variants of the same circuitry. Also the fact the EC-GSM-IoT operates in half duplex has a positive impact on the RF front end as it facilitates the use a RX-TX antenna switch instead of a duplexer.

4.7.4 OVERALL IMPACT ON DEVICE COMPLEXITY

Based on the above findings, in terms of reduction in higher and lower layers' memory requirements, procedures, and computational complexity, it has been concluded that a 20% reduction in the SoC size is within reach [1].

In addition to components on the chip, it is mainly the PA that is of interest to consider for further complexity reduction. For EC-GSM-IoT, in GSM commonly used 33 dBm power class is supported by the specification. However, because of its high power and drain current, it needs to be located outside of the chip. At 50% PA efficiency the PA would, e.g., generate 4 W power, of which 2 W will be dissipated as heat. The 23 dBm power class was therefore specified to allow the PA to be integrated on the chip. At 3.3 V supply voltage and an on-chip PA efficiency of 45%, the heat dissipation is reduced to 250 mW and the drain current is down at 135 mA, which is believed to facilitate a SoC including the PA. This will further reduce the overall module size and complexity. The potential cost/complexity benefit from the integration of the PA onto the chip has not been quantified for EC-GSM-IoT but is more in detailed investigated for LTE-M (see Chapter 6), which can at least give an indication of the potential complexity reduction also for other technologies.

4.8 OPERATION IN A NARROW FREQUENCY DEPLOYMENT

GSM is traditionally operating the BCCH frequency layer over at least 2.4 MHz by using a 12 frequency reuse. This is also the assumption used when evaluating EC-GSM-IoT capacity in Section 4.6. For a LPWAN network, it is, clearly an advantage to support operation in a smaller frequency allocation. For EC-GSM-IoT operation over 9 or 3 frequencies, i.e., using 1.8 MHz or 600 kHz, are investigated in this section. More specifically, the performance is evaluated in the areas of idle mode procedures, common control channels, and data traffic and dedicated control channels.

The results presented in Sections 4.8.1 and 4.8.2 clearly show that EC-GSM-IoT can be deployed in a frequency reuse as tight as 600 kHz, with limited impact on system performance.

4.8.1 IDLE MODE PROCEDURES

Reducing the frequency reuse for the BCCH layer may impact tasks such as synchronization to, identification of, and signal measurements on a cell via the FCCH and EC-SCH. Especially, the FCCH detection is vulnerable because the FCCH signal definition (see Section 3.2.6.1) is the same in all cells. A suboptimal acquisition of the FCCH may negatively influence tasks such as Public Land Mobile Network (PLMN) selection, cell selection, and cell reselection.

4.8.1.1 Public Land Mobile Network and Cell Selection

For initial PLMN or cell selection, a device may need to scan the full range of supported bands and Absolute Radio-Frequency Channel Numbers (ARFCNs) in the search for an EC-GSM-IoT deployment. A quad band device supporting GSM 850, 900, 1850, and 1900 frequency bands needs to search in total 971 ARFCNs. In worst case, a device requires 2 seconds to identify an EC-GSM-IoT cell when at 164 dB MCL, as depicted in Figure 4.5. This was proven to be the case regardless of the frequency reuse, since thermal noise dominates over external interference in deep coverage locations even in a tight frequency deployment. In a scenario where only a single base station is within coverage, and where this base station is configured using the last frequency being searched by a device, a sequential scan over the 971 ARFCNs would demand 971×2 seconds $= 32$ minutes of active RF reception and baseband processing. By means of an interleaved search method the search time can be reduced down to 10 minutes as presented in Table 4.16 [3]. In practice it is also expected that it is sufficient for an EC-GSM-IoT device to support the two sub-GHz bands for global coverage, which has the potential to further reduce the worst case full band search time.

Table 4.16 Worst case of full band search time for a quad band device at 164 dB coupling loss from the serving cell [3]			
System bandwidth	600 kHz	1.8 MHz	2.4 MHz
Time of PLMN selection	10 min		

After the initial full band scan, a serving cell needs to be selected. To improve performance in an interference limited network the signal strengths of a set of highly ranked cells are measured over the FCCH and EC-SCH (see Section 3.3.1.1) with the measurements excluding contributions from interference and noise. With this new approach of measuring for cell selection, the likelihood of selecting the strongest cell as serving cell is, as summarized in Table 4.17, close to being independent of the frequency reuse.

Table 4.17 The probability for an EC-GSM-IoT device to select the optimal serving cell [3]			
System bandwidth	600 kHz	1.8 MHz	2.4 MHz
Probability of selecting strongest cell as serving cell	89.3%	89.7%	90.1%

4.8.1.2 Cell Reselection

After the initial cell selection, EC-GSM-IoT mobility relies on the idle mode cell reselection procedure where significant importance is put on accurate evaluation of the quality of the serving cell (see Section 3.3.1.2). One important scenario for EC-GSM-IoT is that a device waking up after a long period of deep sleep can successfully synchronize to and reconfirm the identity of the serving cell. The reconfirmation of the serving cell is, as seen in Table 4.18, only slightly impacted by the tighter frequency reuse.

Table 4.18 The probability and time required for an EC-GSM-IoT device to successfully reconfirm the serving cell after a period of deep sleep [3]

System bandwidth	600 kHz	1.8 MHz	2.4 MHz
Probability of reconfirming serving cell	98.7%	99.9%	99.9%
Synchronization time, 99th percentile	0.32 s	0.12 s	0.09 s

4.8.2 DATA AND CONTROL CHANNEL PERFORMANCE

The data and control channel capacity is evaluated under the same assumptions as elaborated on in Section 4.6.1. In addition to the 12 frequency reuse, consuming 2.4 MHz, also 9 and 3 frequency reuse patterns are investigated. The relative radio resource consumption is summarized in Figure 4.12 in terms of average fraction of the available radio resources consumed. In all three deployment scenarios the percentage of failed connection attempts are kept below 0.1% at the investigated system load.

The impact on resource utilization is seen as negligible when going from 2.4 to 1.8 MHz, while it becomes noticeable when going down to a 600 kHz deployment. This is especially seen for the DL traffic channels that are more severely hit by the increased interference level.

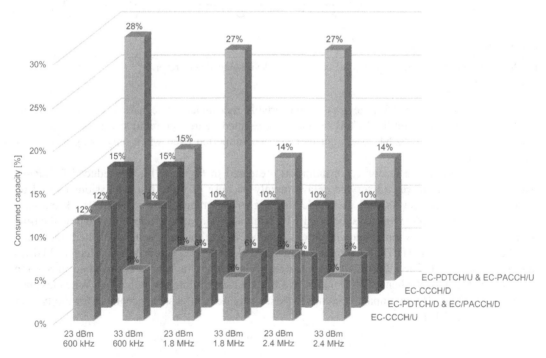

FIGURE 4.12

Radio resource utilization at a load of 52,547 users per cell [3].

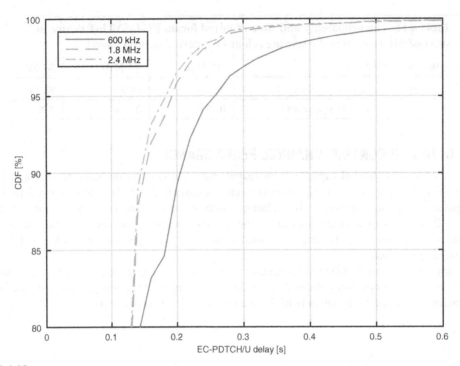

FIGURE 4.13

EC-PDTCH/U transmission delay during the delivery of a device autonomous report.

The targeted load of 52,547 devices is comfortably met under all scenarios. In relation to the available radio resources on the BCCH carrier, the presented figures are relatively modest indicating that a load well beyond 52,547 users per cell may be supported even when only using a 600 kHz frequency deployment.

Besides the increased resource consumption presented in Figure 4.12, the reduced frequency reuse also results in increased service delays mainly because of more retransmissions caused by increased interference levels, and users selecting higher coverage classes. Figures 4.13 and 4.14 illustrate the impact on the time to successfully transmit a device autonomous report and on the time to transmit a DL application acknowledgment once a connection has been established, including EC-PACCH and EC-PDTCH transmission times and thereto associated delays. Here a 33 dBm device is considered. The UL and DL packet sizes follows the characteristics specified in Section 4.6.1. Again the impact when going from 2.4 to 1.8 MHz is negligible for both cases. When taking a further step to 600 kHz the impact becomes more accentuated but is still acceptable for the type of services EC-GSM-IoT targets.

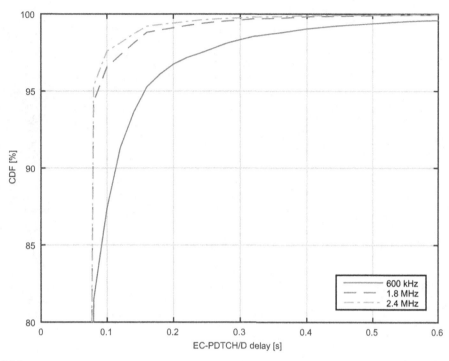

FIGURE 4.14

EC-PDTCH/D transmission delay during the delivery of an application acknowledgment.

REFERENCES

[1] Third Generation Partnership Project, Technical Report 45.820 v13.0.0, Cellular System Support for Ultra-Low Complexity and Low Throughput Internet of Things, 2016.

[2] Ericsson, et al., GP-151039 New Work Item on Extended Coverage GSM (EC-GSM) for Support of Cellular Internet of Things, 3GPP TSG GERAN Meeting #67, 2015.

[3] Third Generation Partnership Project, Technical Specification 45.050 v13.1.0 Background for Radio Frequency (RF) Requirements, 2016.

[4] Third Generation Partnership Project, Technical Specification 45.005 v14.0.0, Radio Transmission and Reception, 2016.

[5] Ericsson, et al., GP-150419 EC-GSM, Throughput, Delay and Resource Analysis, 3GPP TSG GERAN Meeting #66, 2015.

[6] Ericsson, et al., GP-140608 Radio Propagation Channel for the Cellular IoT Study, 3GPP TSG GERAN Meeting #63, 2014.

[7] Ericsson, et al., GP-160292 Introduction of Radio Frequency Colour Code, 3GPP TSG GERAN Meeting #70, 2016.

[8] Ericsson, et al., GP-160032 Additional EC-RACH Mapping, 3GPP TSG GERAN Meeting #69, 2016.

[9] Third Generation Partnership Project, Technical Specification 44.060 v14.0.0, General Packet Radio Service (GPRS); Mobile Station (MS) - Base Station System (BSS) Interface; Radio Link Control/Medium Access Control (RLC/MAC) Protocol, 2016.

[10] Ericsson, et al., GP-150449 EC-GSM − Exception Report Latency Performance Evaluation, 3GPP TSG GERAN Meeting #66, 2015.

[11] Ericsson, et al., GP-160351 EC-GSM -IoT Performance Summary, 3GPP TSG GERAN Meeting #70, 2015.

LTE-M

CHAPTER OUTLINE

Cellular Internet of Things. http://dx.doi.org/10.1016/B978-0-12-812458-1.00005-8

Abstract

In this chapter, we describe the Long-Term Evolution for Machine-Type Communications (LTE-M) physical layer design with an emphasis on how the physical channels are designed to fulfill the objectives that LTE-M targets, namely low device cost, deep coverage, and long battery lifetime while maintaining capacity for a large number of devices per cell, with performance and functionality suitable for both low-end and mid-range applications for the Internet of Things (IoT).

Section 5.1 describes the background behind the introduction of LTE-M in the Third Generation Partnership Project (3GPP) specifications and the design principles of the technology. Section 5.2 focuses on the physical channels with an emphasis on how these channels are designed to fulfill the objectives that LTE-M is intended to achieve, namely low device cost, deep coverage, and long battery lifetime, while maintaining capacity for a large number of devices per cell, with performance and functionality suitable for both low-end and mid-range applications for the IoT. Section 5.3 covers LTE-M procedures in idle and connected mode, including all activities from initial cell selection to completing a data transfer. The idle mode procedures include the initial cell selection, which is the procedure that a device has to go through when it is first switched on or is attempting to select a new cell to camp on. Idle mode activities also include acquisition of system information, paging, and random access. Descriptions of some fundamental connected mode procedures include scheduling, power control, and mobility. Finally, a summary of the most recent improvements accomplished in Release 14 of the 3GPP specifications are presented in Section 5.4.

5.1 BACKGROUND
5.1.1 3GPP STANDARDIZATION

LTE-M extends LTE with features for improved support for *Machine-Type Communications* (MTC) and IoT. The recently deployed low-cost device category M1 (Cat-M1) and coverage enhancement (CE) modes have their origin in the 3GPP study item *Study on provision of low-cost Machine-Type Communications (MTC) User Equipments (UEs) based on LTE* [1], in the following sections referred to as the *LTE-M study item*.

Since then, a number of related 3GPP work items have been carried out:

- Release 12 work item *Low cost and enhanced coverage MTC UE for LTE* [2], sometimes referred to as the *MTC work item*, which introduced device Cat-0
- Release 13 work item *Further LTE Physical Layer Enhancements for MTC* [3], sometimes referred to as the *eMTC work item*, which introduced device Cat-M1 and the CE modes A and B
- Release 14 work item *Further Enhanced MTC for LTE* [4], sometimes referred to as the *feMTC work item* which introduced device Cat-M2 and various other improvements

In this book we use the term LTE-M when we refer to the Cat-M device category series, the CE modes, and all functionality that can be supported by the Cat-M devices or the CE modes, such as the power consumption reduction techniques *Power Saving Mode* (PSM) and *Extended Discontinuous Reception* (eDRX).

The descriptions of the LTE-M physical layer and procedures in idle and connected mode in Sections 5.2 and 5.3 focus on Release 13. The further enhancements of LTE-M in Release 14 are described in Section 5.4.

5.1.2 RADIO ACCESS DESIGN PRINCIPLES
5.1.2.1 Low Device Complexity and Cost

During the LTE-M study item [1], various device cost reduction techniques were studied, with the objective to bring down the LTE device cost substantially to make LTE attractive for low-end MTC applications that have so far been adequately handled by GSM/GPRS. It was estimated that this would correspond to a device modem manufacturing cost in the order of 1/3 of that of the simplest LTE modem, which at that time was a single-band LTE device Cat-1 modem.

The study identified the following cost reduction techniques as most promising:

- Reduced peak rate
- Single receive antenna

- Half-duplex operation
- Reduced bandwidth
- Reduced maximum transmit power.

A first step was taken in Release 12 with LTE device Cat-0 that compared to Cat-1 supported a reduced peak rate for user data (1 Mbps in DL and UL instead of 10 Mbps in DL and 5 Mbps in UL), a single receive antenna (instead of at least two), and optionally *half-duplex frequency-division duplex* (HD-FDD) operation.

The next step was taken in Release 13 with LTE device Cat-M1 that includes all the cost reduction techniques of Cat-0 plus a reduced bandwidth (1.4 MHz instead of 20 MHz) and optionally a lower device power class (maximum transmit power of 20 dBm instead of 23 dBm).

With these cost reduction techniques, the *Bill of Material* cost for the Cat-M1 modem was estimated to reach that of an Enhanced GPRS (EGPRS) modem. For further information on LTE-M cost estimates, refer to Section 6.7.

5.1.2.2 Coverage Enhancement

The LTE-M study item [1] also studied CE techniques, with the objective to achieve a 20 dB better coverage than in then existing LTE networks to provide coverage for devices with challenging coverage conditions, for example stationary utility metering devices located in basements.

The study identified various forms of prolonged transmission time as the most promising CE techniques. The fact that many of the MTC applications of interest have very relaxed requirements on data rates and latency can be exploited to enhance the coverage through repetition or retransmission techniques. The study concluded that 20 dB CE can be achieved using the identified techniques, but after taking into account other aspects such as spectral efficiency and required standardization effort, 3GPP went on to pursue 15 dB as a CE target.

Release 13 standardized two CE modes: CE mode A, supporting up to 32 subframe repetitions of the data channel, and CE mode B, supporting up to 2048 repetitions. Recent evaluations indicate that the initial coverage target of 20 dB seems to be actually reachable using the repetitions available in CE mode B. For further information on LTE-M coverage and data rate estimates, refer to Sections 6.2 and 6.3.

In this book we refer to LTE devices with CE mode support as *LTE-M devices*. These devices may be low-cost Cat-M1 devices or they may higher LTE device categories configured in a CE mode. For more information on the CE modes refer to Section 5.2.3.3.

5.1.2.3 Long Device Battery Lifetime

Support for a device battery lifetime of many years, potentially decades, has been introduced in a first step in the form of the PSM in Release 12 and in a second step in the form of the eDRX in Release 13. These features are supported for LTE-M devices and also for other 3GPP radio access technologies.

Compared to ordinary LTE devices, LTE-M devices can have a reduced power consumption during their "on" time mainly thanks to the reduced transmit and receive bandwidths.

PSM and eDRX are described in Sections 2.2.3, 5.3.1.4, and 5.3.1.5, and the battery lifetime for LTE-M is evaluated in Section 6.5.

5.1.2.4 Support of Massive Number of Devices

The handling of massive numbers of devices in LTE was improved already in Releases 10 and 11, for example in the form of Access Class Barring (ACB) and overload control, as discussed in Section 2.2.1. Further improvements have been introduced later on, for example, in the form of the *Radio Resource Control (RRC) Suspend/Resume* mechanism described in Section 2.2.2, which helps reduce the required signaling when resuming an RRC connection after a period of inactivity as long as the device has not left the cell in the meanwhile.

For more information on LTE-M capacity estimates, refer to Section 6.6.

5.1.2.5 Deployment Flexibility

LTE-M supports the same system bandwidths at the network side as LTE (1.4, 3, 5, 10, 15, and 20 MHz). If an operator has a large spectrum allocation for LTE, then there is also a large bandwidth available for LTE-M traffic. The DL and UL resources on an LTE carrier can serve as a resource pool that can be fully dynamically shared between LTE traffic and LTE-M traffic. It may furthermore be possible to schedule delay-tolerant LTE-M traffic during periods when the ordinary LTE users are less active, thereby minimizing the performance impact from the LTE-M traffic on the LTE traffic.

5.2 PHYSICAL LAYER

In this section, we describe the LTE-M physical layer design with an emphasis on how these channels are designed to fulfill the objectives that LTE-M targets, namely low device cost, deep coverage, and long battery lifetime, while maintaining capacity for a large number of devices per cell, with performance and functionality suitable for both low-end and mid-range applications for the IoT.

5.2.1 GUIDING PRINCIPLES

LTE-M extends the LTE physical layer with features for improved support for MTC. The LTE-M design therefore builds on the solutions already available in LTE.

The fundamental DL and UL transmission schemes are the same as in LTE, meaning Orthogonal Frequency-Division Multiplexing (OFDM) in DL and Single-Carrier Frequency-Division Multiple-Access (SC-FDMA) in UL, with the same numerologies (channel raster, subcarrier spacing, cyclic prefix (CP) lengths, resource grid, frame structure, etc.). This means that LTE-M transmissions and LTE transmissions related to, for example, smartphones and mobile broadband modems can coexist in the same LTE cell on the same LTE carrier and the resources can be shared dynamically between LTE-M users and ordinary LTE users.

Most of the physical layer changes in LTE-M compared to LTE are motivated by the requirements on low device cost, deep coverage, and long battery lifetime. The low device cost is enabled by reduced transmit and receive bandwidths and other simplifications. The deep coverage is mainly achieved through repetition techniques. The long battery lifetime is made possible by the introduction of long sleeping cycles and efforts to keep the overhead from both higher and lower layer control signaling as small as possible. The physical layer design takes these new aspects into account.

5.2.2 PHYSICAL LAYER NUMEROLOGY

5.2.2.1 Channel Raster

LTE-M is reusing LTE's *Primary Synchronization Signal* (PSS), *Secondary Synchronization Signal* (SSS), and the core part of the *Physical Broadcast Channel* (PBCH) carrying the *Master Information Block* (MIB). These physical signals are located in the center of the LTE system bandwidth and this center frequency is aligned with a channel raster of 100 kHz. The absolute frequency that the LTE and LTE-M system is centered around can be deduced from the *E-UTRA Absolute Radio Frequency Channel Number* (EARFCN).

LTE-M specific physical channels and signals are transmitted within the so-called *narrowbands*, each mapped over six *Physical Resource Blocks* (PRBs). The PRB definition is given in Section 5.2.2.3 and the narrowband concept including its location within the LTE system bandwidth is described in Section 5.2.3.2.

5.2.2.2 Frame Structure

The overall time frame structure on the access stratum for LTE and LTE-M is illustrated in Figure 5.1. On the highest level one hyperframe cycle has 1024 hyperframes that each consists of 1024 frames. One frame consists of 10 subframes, each dividable into two slots of 0.5 ms as shown in the figure. Each slot is divided into 7 OFDM symbols in case of normal CP length and 6 OFDM symbols in case of *extended CP* length. The normal CP length is designed to support propagation conditions with a

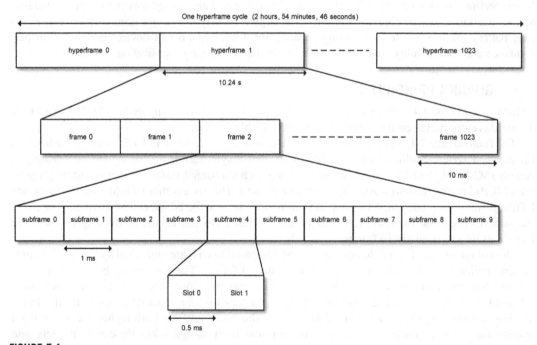

FIGURE 5.1

Frame structure for LTE and LTE-M.

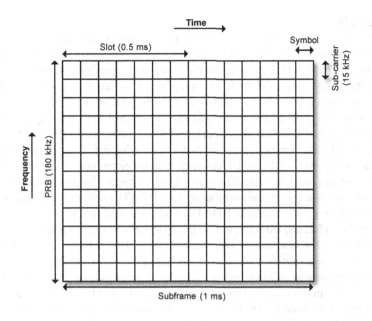

FIGURE 5.2

Physical resource block (PRB) pair in LTE and LTE-M.

delay spread up to 4.7 µs, while the extended CP is intended to support deployments where the delay spread is up to 16.7 µs. All illustrations in this book assume the normal CP length because it is much more commonly used than the extended CP length.

Each subframe can be uniquely identified by a *hyper system frame number* (H-SFN), a *system frame number* (SFN), and *subframe number* (SN). The ranges of H-SFN, SFN, and SN are 0−1024, 0−1024, and 0−9, respectively.

5.2.2.3 Resource Grid

One PRB spans 12 subcarriers, which with the 15-kHz subcarrier spacing correspond to 180 kHz. The smallest resource unit that can be scheduled to a device is one PRB pair mapped over two slots, which for the normal CP length case (with 7 OFDM symbols per slot) corresponds to 12 subcarriers over 14 OFDM symbols as illustrated in Figure 5.2. An even smaller resource unit used in the physical layer specifications is the Resource Element (RE) that refers to one subcarrier in one OFDM symbol.

5.2.3 TRANSMISSION SCHEMES

The fundamental DL and UL transmission schemes are the same as in LTE. This means that the DL uses OFDM and the UL uses SC-FDMA, with 15 kHz subcarrier spacing in both DL and UL [5]. In the DL, a *direct current (DC) subcarrier* is reserved at the center of the system bandwidth. Both normal and extended CP lengths are supported. DL *transmission modes* (TM) supporting beamforming from up to four antenna ports are supported (see Section 5.2.4.6 for more information on the DL TM).

5.2.3.1 Duplex Modes

LTE-M supports both *frequency-division duplex* (FDD) operation and *time-division duplex* (TDD) operation. In FDD operation, two different carrier frequencies are used for DL and UL. If the device supports *full-duplex FDD* (FD-FDD) operation, it can perform reception and transmission at the same time, whereas if the device only supports HD-FDD operation, it has to switch back and forth between reception and transmission. According to the basic LTE behavior for HD-FDD devices that is referred to as *HD-FDD operation type A*, a device that only supports HD-FDD is only expected to be able to do DL reception in subframes where it does not perform UL transmission. In HD-FDD operation type A, the switching back and forth between reception and transmission is fast but relies on the existence of two separate local oscillators for DL and UL carrier frequency generation. To facilitate implementation of low-cost devices employing just a single local oscillator for carrier frequency generation for both DL and UL, *HD-FDD operation type B* was introduced [5,6]. HD-FDD operation type B is used for LTE-M devices and also for LTE device Cat-0. In HD-FDD operation type B, a guard subframe is inserted at every switch from DL to UL and from UL to DL, giving the device time to retune its carrier frequency.

In TDD operation, where the same carrier frequency is used for DL and UL transmission, the division of so-called *normal subframes* within a frame into DL and UL subframes depends on the cell-specific *UL−DL configuration* as indicated in Table 5.1. The switching from DL to UL takes place during a guard period within a so-called *special subframe*, indicated by "S" in the table. The symbols before the guard period are used for DL transmission and the symbols after the guard period are used for UL transmission. The location and length of the guard period within the special subframe is given by a cell-specific *special subframe configuration*. Interested readers can refer to Reference [5] for more details.

LTE-M devices can be implemented with support for FD-FDD, HD-FDD operation type B, TDD, or any combination of these duplex modes. This means that LTE-M can be deployed both in paired FDD bands and unpaired TDD bands, and that both full-duplex and half-duplex device implementations are possible, allowing for trade-off between device complexity and performance.

5.2.3.2 Narrowband Operation

The supported LTE system bandwidths are {1.4, 3, 5, 10, 15, 20} MHz including guard bands. Discounting the guard bands, the maximum bandwidth that can be scheduled in the largest system

Table 5.1 UL−DL configurations for TDD operation in LTE and LTE-M

UL−DL configuration	Subframe number									
	0	1	2	3	4	5	6	7	8	9
0	DL	S	UL	UL	UL	DL	S	UL	UL	UL
1	DL	S	UL	UL	DL	DL	S	UL	UL	DL
2	DL	S	UL	DL	DL	DL	S	UL	DL	DL
3	DL	S	UL	UL	UL	DL	DL	DL	DL	DL
4	DL	S	UL	UL	DL	DL	DL	DL	DL	DL
5	DL	S	UL	DL	DL	DL	DL	DL	DL	DL
6	DL	S	UL	UL	UL	DL	S	UL	UL	DL

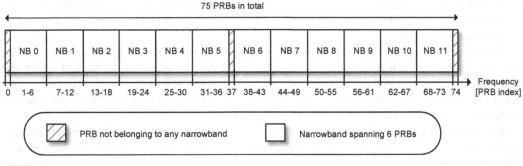

FIGURE 5.3

LTE-M narrowbands in 15 MHz LTE system bandwidth.

bandwidth is 100 PRBs or 18 MHz. Ordinary LTE devices support transmission and reception spanning the full system bandwidth.

LTE-M introduces low-cost devices that are only required to support a reduced bandwidth for transmission and reception. These low-cost devices are sometimes referred to as *Bandwidth-reduced Low-complexity* (BL) devices in the standard specifications. The simplest LTE-M device supports a maximum channel bandwidth of 6 PRBs [6]. Resource allocation for LTE-M devices is based on nonoverlapping *narrowbands* of size 6 PRBs as illustrated in Figure 5.3 for the 15-MHz system bandwidth case.

For all system bandwidths except for the smallest one, the system bandwidth cannot be evenly divided into narrowbands which means that there are some PRBs that are not part of any narrowband. For the system bandwidths which have an odd total number of PRBs, the PRB at the center is not included in any narrowband, and if there are any remaining PRBs not included in any narrowband, they are evenly distributed at the edges of the system bandwidth, i.e., with the lowest and highest PRB indices, respectively [5]. The number of narrowbands and the PRBs not belonging to any narrowband are listed in Table 5.2. The PRBs not belonging to any narrowband cannot be used for LTE-M related transmissions on the physical channels *MTC Physical Downlink Control Channel* (MPDCCH),

Table 5.2 LTE-M narrowbands

LTE system bandwidth including guard bands (MHz)	Total number of PRBs in system bandwidth	Number of narrowbands	PRBs not belonging to any narrowband
1.4	6	1	None
3	15	2	1 on each edge + 1 at the center (=3)
5	25	4	1 at the center
10	50	8	1 on each edge (=2)
15	75	12	1 on each edge + 1 at the center (=3)
20	100	16	2 on each edge (=4)

Physical Downlink Shared Channel (PDSCH), and *Physical Uplink Shared Channel* (PUSCH) but can be used for LTE-M related transmissions on other physical channels/signals and for any ordinary LTE transmissions in the cell.

The center frequency of a narrowband is not necessarily aligned with the 100-kHz channel raster of LTE. However, as explained in Section 5.2.2, the signals and channels essential for cell search and basic system information (SI) acquisition, i.e., PSS, SSS, and PBCH (see Section 5.2.4), are common with LTE, and therefore still located at the center of the LTE system bandwidth (around the DC subcarrier) and aligned with the 100-kHz channel raster [7].

To ensure good frequency diversity even for devices with reduced bandwidth, frequency hopping is supported for many of the physical signals and channels (see Section 5.3.3.2 for more information on frequency hopping).

5.2.3.3 Coverage Enhancement Modes

LTE-M implements a number of coverage enhancement (CE) techniques, the most significant one being the support of repetition of most physical signals and channels. The motivation for the CE is twofold.

First, low-cost LTE-M devices may implement various simplifications to drive down the device complexity, for example, a single-antenna receiver and a lower maximum transmission power. These simplifications are associated with some performance degradation that would result in a coverage loss compared to LTE unless it is compensated for through some CE techniques.

Second, it is expected that some LTE-M devices will experience very challenging coverage conditions. Stationary utility metering devices mounted in basements serve as an illustrative example. This means that it may not be sufficient that LTE-M provides the same coverage as LTE, but in fact the LTE-M coverage needs to be substantially improved compared to LTE.

To address these aspects, LTE-M introduces two CE modes. *CE mode A* provides sufficient CE to compensate for all the simplifications that can be implemented by low-cost LTE-M devices and then some additional CE beyond normal LTE coverage. *CE mode B* goes a step further and provides the deep coverage that may be needed in more challenging coverage conditions. CE mode A is optimized for moderate CE achieved through a small amount of repetition, whereas CE mode B is optimized for substantial CE achieved through a large amount of repetition. If a device supports CE mode B, then it also supports CE mode A.

The low-cost LTE-M device categories (Cat-M1 and so on) have mandatory support for CE mode A and can optionally also support CE mode B. These low-cost devices always operate in one of the two CE modes. The CE modes support efficient operation of low-cost LTE-M devices, which, for example, means that resource allocation in CE mode is based on narrowbands introduced in Section 5.2.3.2.

Higher LTE device categories (Cat-0, Cat-1, and so on) can optionally support the CE modes— either just CE mode A or both CE mode A and B. These more capable devices will typically only operate in CE mode if this is needed in order to stay in coverage, i.e., when they are outside the normal LTE coverage. When these devices are in normal LTE coverage they will typically use normal LTE operation rather than CE mode and enjoy the higher performance available in normal LTE operation in terms of, e.g., data rates and latency.

5.2.4 DOWNLINK PHYSICAL CHANNELS AND SIGNALS

LTE-M supports the set of DL channels and signals depicted in Figure 5.4. The physical layer provides data transport services to higher layers through the use of *transport channels* via the *Medium Access Control* (MAC) layer [8]. The *Downlink Control Information* (DCI) is strictly speaking not a transport channel, which is indicated by the dashed line. The MAC layer in turn provides data transport services through the use of *logical channels* that are also shown in the figure for completeness [9]. For more information on the higher layers, refer to Section 5.3.

In this section we focus on the DL physical channels and signals. PSS, SSS, and PBCH are transmitted periodically in the center of the LTE carrier. MPDCCH and PDSCH are transmitted in a narrowband (see Section 5.2.3.2). DL *Reference Signals* (RS) are transmitted in all PRBs.

5.2.4.1 Downlink Subframes

A cell-specific *subframe bitmap* can be broadcasted in the SI (see Section 5.3.1.2) to indicate which DL subframes are valid for LTE-M transmission. The bitmap length is 10 or 40 bits corresponding to the subframes within 1 or 4 frames. A network can, for example, choose to indicate subframes that are used as *Positioning Reference Signal* (PRS) or *Multimedia Broadcast Multicast Service Single Frequency Network* (MBSFN) subframes as invalid for LTE-M, but this is up to the network implementation.

Figure 5.5 shows an example with a 10-bit LTE-M subframe bitmap indicating that subframes #5 and #7 are invalid. Assume that the DL (MPDCCH or PDSCH) transmission that starts in subframe #4 in the first frame should use subframe repetition factor 4. If all subframes were valid, the repetitions denoted R1, R2, R3, and R4 in the figure would be mapped to subframes #4, #5, #6, and #7, respectively, but due to the invalid subframes, the repetitions are instead mapped to valid subframes #4, #6, #8, and #9, respectively.

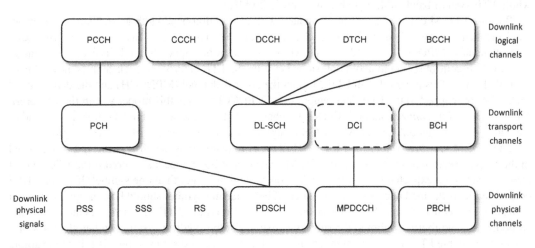

FIGURE 5.4

Downlink channels and signals used in LTE-M.

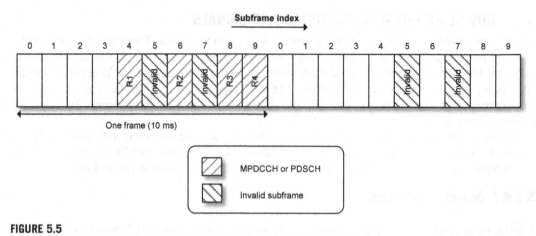

FIGURE 5.5

LTE-M subframe bitmap example.

The DL subframe structure in LTE-M only uses a part of the DL subframe REs in LTE. As shown in Figure 5.6, the DL subframe structure in LTE consists of an *LTE control region* and an *LTE data region*. The LTE control region consists of one or more OFDM symbols in the beginning of the subframe and the LTE data region consists of the remaining OFDM symbols in the subframe. In LTE, data transmissions on PDSCH are mapped to the LTE data region, whereas a number of control channels (*Physical Control Format Indicator Channel* (PCFICH), *Physical Downlink Control Channel* (PDCCH), and *Physical Hybrid Automatic Repeat Request Indicator Channel* (PHICH)) are mapped to the LTE control region. These control channels are all wideband channels spanning almost the whole LTE system bandwidth, which can be up to 20 MHz.

Because LTE-M devices can be implemented with a reception bandwidth as small as one narrowband, the mentioned wideband LTE control channels are not used for LTE-M. Instead, a new narrowband control channel (MPDCCH) is used for LTE-M devices and it is mapped to the LTE data region rather than the LTE control region to avoid collisions between the LTE control channels and the new LTE-M control channel. This means that in LTE-M, both the control channel (MPDCCH) and the data channel (PDSCH) are mapped to the LTE data region. (The MPDCCH shares this property with the *Enhanced Physical Downlink Control Channel* (EPDCCH) channel that was introduced in LTE Release 11, and as we will see in Section 5.2.4.5, the MPDCCH design is in fact based on the EPDCCH design).

The LTE-M *starting symbol* for MPDCCH/PDSCH transmissions is cell-specific and broadcasted in the SI (see Section 5.3.1.2). An early LTE-M starting symbol can be configured if the LTE control channel load is not expected to require an LTE control region longer than one symbol. If a larger LTE control region is deemed necessary, then a later LTE-M starting symbol should be configured to avoid collisions between LTE and LTE-M transmissions. The possible LTE-M starting symbols are the second, third, and fourth symbol in the subframe, except for the smallest system bandwidth (1.4 MHz) where the possible LTE-M starting symbols are the third, fourth, and fifth symbol [10]. In the example in Figure 5.6, the starting symbol is the fourth symbol. In the TDD case, in subframes #1 and #6, the LTE-M starting symbol is no later than the third symbol because of the position of PSS/SSS (see Section 5.2.4.2).

When an LTE-M device needs to retune from one DL narrowband in a first subframe to another DL narrowband in a second subframe (or from an UL narrowband to a DL narrowband with a different

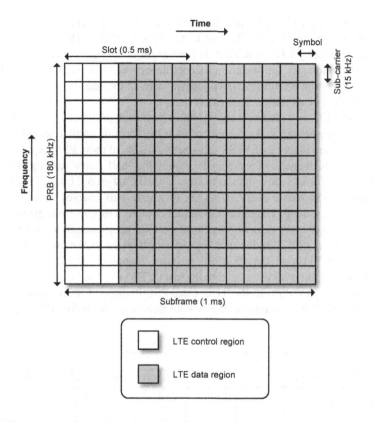

FIGURE 5.6

Downlink subframe structure in LTE.

center frequency in case of TDD), the device is allowed to create a *guard period for narrowband retuning* by not receiving the first two OFDM symbols in the second subframe [5]. This means that the guard period falls partly or completely within the LTE control region and that the impact on the LTE-M transmission can be expected to be minimal.

5.2.4.2 Synchronization Signals

Subframes in FDD	#0 and #5 for both PSS and SSS
Subframes in TDD	#1 and #6 for PSS, #0 and #5 for SSS
Subframe periodicity	5 ms for both PSS and SSS
Sequence pattern periodicity	5 ms for PSS, 10 ms for SSS
Subcarrier spacing	15 kHz
Bandwidth	62 subcarriers (not counting the DC subcarrier)
Frequency location	At the center of the LTE system bandwidth

LTE-M devices rely on LTE's PSS and SSS for acquisition of a cell's carrier frequency, frame timing, CP length, duplex mode, and *Physical Cell Identity* (PCID). The LTE signals can be used without modification even by LTE-M devices in challenging coverage conditions. Because PSS and SSS are transmitted periodically, the device can accumulate the received signal over multiple frames to achieve sufficient acquisition performance, without the need to introduce additional repetitions on the transmit side (at the cost of increased acquisition delay).

LTE supports 504 PCIDs divided into 168 groups where each group contains 3 identities. In many cases the 3 identities correspond to 3 adjacent cells in the form of sectors served by the same base station. The 3 identities are mapped to 3 PSS sequences and one of these PSS sequences is transmitted every 5 ms in the cell, which enables the device to acquire the "half-frame" timing of the cell. For each PSS sequence there are 168 SSS sequences indicative of the PCID group. Like PSS, SSS is transmitted every 5 ms, but the 2 SSSs within every 10 ms are different. This enables the device to acquire the PCID as well as the frame timing. The same SSS sequence pattern repeats itself every 10 ms. The same PCID can be used in two or more cells as long as they are far apart enough to avoid ambiguity due to overhearing, so the number of PCIDs does not impose a limit on the total number of cells in a network.

Figures 5.7 and 5.8 illustrate the PSS/SSS resource mapping for FDD cells and TDD cells, respectively. In the FDD case, PSS is mapped to the last OFDM symbol in slots #0 and #10 and SSS is mapped to the symbol before PSS. In the TDD case, PSS is mapped to the third OFDM symbol in subframes #1 and #6 and SSS is mapped to the symbol three symbols before PSS [5]. This means that

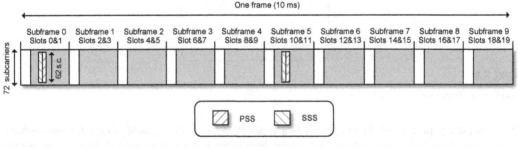

FIGURE 5.7

Synchronization signals in LTE FDD.

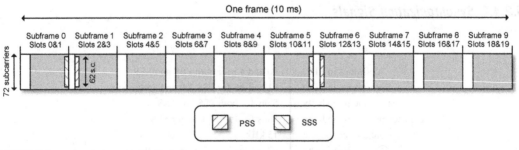

FIGURE 5.8

Synchronization signals in LTE TDD.

the duplex mode (FDD or TDD) can be detected from the synchronization signals, although this is normally not needed because a given frequency band typically only supports one of the duplex modes [7]. The exact PSS/SSS symbol positions vary slightly depending on the CP length, which means that the device can also detect whether it should assume normal or extended CP length based on the detection of the synchronization signals.

As shown in the figures, PSS and SSS are mapped to the center 62 subcarriers (around the DC subcarrier) of the LTE carrier. This means that the signal fits within the smallest LTE-M device bandwidth that corresponds to 72 subcarriers. The PSS/SSS region is not aligned with any of the narrowbands (see Section 5.2.3.2) except when the smallest system bandwidth (1.4 MHz) is used, which means that the LTE-M device may need to do frequency retuning (see Section 5.2.4.1) whenever it needs to receive PSS/SSS.

5.2.4.3 Downlink Reference Signals

Subframe	Any
Subcarrier spacing	15 kHz
Cell-specific Reference Signal (CRS) bandwidth	Full system bandwidth
Demodulation Reference Signal (DMRS) bandwidth	Same as associated MPDCCH/PDSCH
CRS frequency location	According to Figure 5.9 in every PRB
DMRS frequency location	According to Figure 5.10 in affected PRBs

DL RS are predefined signals transmitted by the base station to allow the device to estimate the DL propagation channel to be able to demodulate the DL physical channels [5] and perform DL quality measurements [11].

The CRS can be used for demodulation of PBCH or PDSCH and is transmitted from one, two, or four logical antenna ports numbered 0−3, where in the typical case each logical antenna port corresponds to a physical antenna. The CRS for different antenna ports is mapped to REs in every PRB and in every (non-MBSFN) subframe in the cell as shown in Figure 5.9. The CRS mapping shown in Figure 5.9 is one example and may be frequency shifted up by one or two subcarriers depending on the PCID value.

The DMRS can be used for demodulation of PDSCH or MPDCCH and is configured per device and is not PCID dependent, except in case of *MPDCCH Common Search Space* (see Section 5.3.3.1) where the *DMRS sequence initialization* is based on PCID. The DMRS is transmitted on the same logical antenna port as the associated PDSCH or MPDCCH. If the logical antenna port is mapped to multiple physical antennas, the coverage and capacity can be improved through antenna techniques such as beamforming. DMRS can be transmitted to different devices from up to 4 logical antenna ports, numbered 7−10 for PDSCH and 107−110 for MPDCCH, mapped to REs as shown in Figure 5.10. CRS is also transmitted but not shown in the Figure 5.10. As can be seen from the figure, DMRS for antenna ports 7 and 8 is mapped to the same set of REs but separated by an *Orthogonal Cover Code* (OCC), and the same holds for DMRS for antenna ports 9 and 10. The DMRS for the four different antenna ports can thus be distinguished by the device.

The LTE-M physical layer also supports the PRS needed for the *Observed Time Difference of Arrival* (OTDOA) multilateration positioning method (see Section 5.4.3).

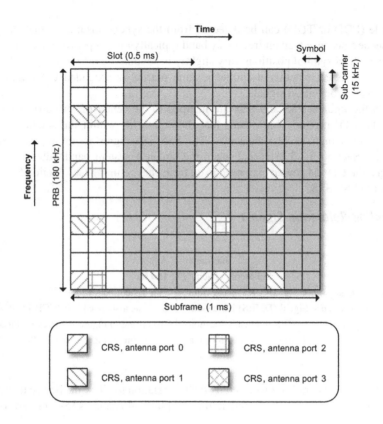

FIGURE 5.9

Cell-specific reference signal in LTE and LTE-M.

5.2.4.4 PBCH

Subframes in FDD	#0 for core part, #9 for repetitions
Subframes in TDD	#0 for core part, #5 for repetitions
Basic transmission time interval (TTI)	40 ms
Repetitions	Core part plus 0 or 4 repetitions
Subcarrier spacing	15 kHz
Bandwidth	72 subcarriers (not counting the DC subcarrier)
Frequency location	At the center of the LTE system bandwidth

The PBCH is used to deliver the MIB that provides essential information for the device to operate in the network (see Section 5.3.1.1.3 for more details on MIB).

The PBCH is mapped to the center 72 subcarriers in the LTE system bandwidth. The PBCH of LTE serves as the *PBCH core part* in LTE-M and the LTE-M specification adds additional

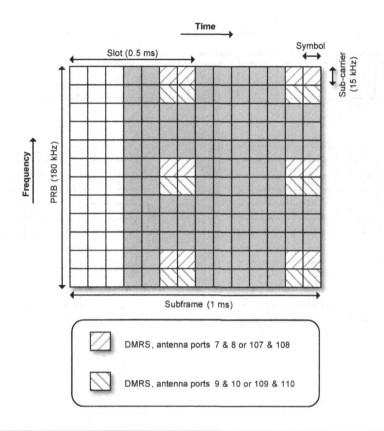

FIGURE 5.10

Device-specific demodulation reference signal in LTE and LTE-M.

PBCH repetitions for improved coverage. It is up to the network whether to enable the PBCH repetitions in a cell. Enabling the repetitions is only motivated in cells that need to support deep coverage.

The TTI for PBCH is 40 ms and the *transport block size* (TBS) is 24 bits. A 16-bit *cyclic redundancy check* (CRC) is attached to the transport block. The CRC is masked with a bit sequence that depends on the number of CRS transmit antenna ports on the base station (see Section 5.2.4.3), which means that the device learns the number of CRS transmit antenna ports as a by-product in the process of decoding PBCH [12].

Together, the 40 bits from the 24-bit transport block and the 16-bit CRC are encoded using the LTE tail-biting convolutional code (TBCC), and rate matched to generate 1920 encoded bits. The encoded bits are scrambled with a cell-specific sequence (for randomization of inter-cell interference) and segmented into four segments distributed to four consecutive frames. Each segment is 480 bits long and mapped to 240 quadrature phase shift keying (QPSK) symbols distributed over the 72 subcarriers. Transmit diversity is applied for PBCH based on *Space-Frequency Block Coding* (SFBC) in case of

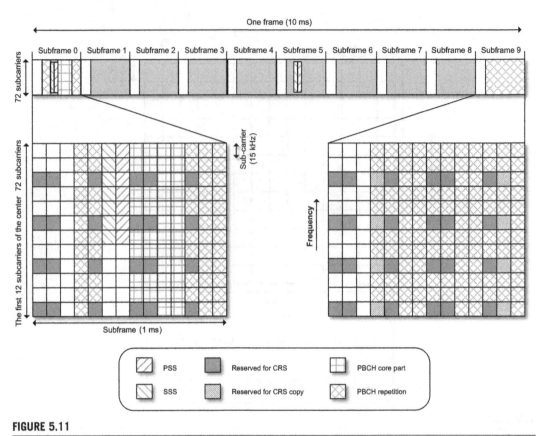

FIGURE 5.11

PBCH core part and PBCH repetition in LTE FDD.

two antenna ports and on a combination of SFBC and *Frequency-Switched Transmit Diversity* (FSTD) in case of four antenna ports [5].

The PBCH core part is always transmitted as four OFDM symbols in subframe #0 in every frame. When the PBCH repetitions are enabled, they are transmitted in subframes #0 and #9 in the FDD case and in subframes #0 and #5 in the TDD case, as illustrated in Figures 5.11 and 5.12. Note that the zoomed-in parts only show the first 12 of 72 subcarriers. The PBCH repetitions part contains four copies of each one of the four OFDM symbols in the PBCH core part, resulting in a repetition factor of five for each OFDM symbol. If the copied OFDM symbol contains CRS, the CRS is also copied. In the FDD case, the fact that subframes 0 and 9 are adjacent facilitates coherent combination across PBCH repetitions, for example for frequency estimation purposes. In the TDD case, the fact that subframes 0 and 5 have been selected means that PBCH repetition can be supported in all UL−DL configurations because these subframes are DL subframes in all UL−DL configurations (see Table 5.1).

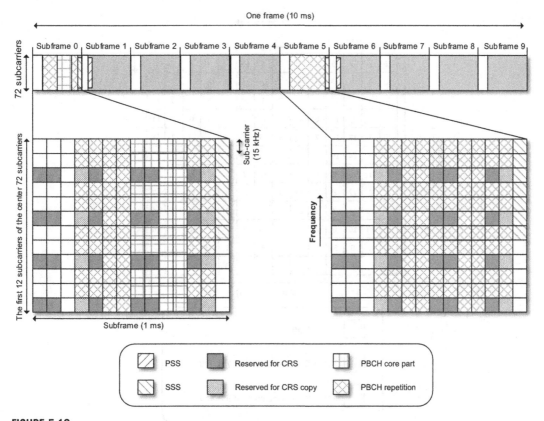

FIGURE 5.12

PBCH core part and PBCH repetition in LTE TDD.

5.2.4.5 MPDCCH

Subframe	Any
Basic TTI	1 ms
Repetitions	1, 2, 4, 8, 16, 32, 64, 128, 256
Subcarrier spacing	15 kHz
Bandwidth	2, 4 or 6 PRBs
Frequency location	Within a narrowband
Frequency hopping	Between 2 and 4 narrowbands if configured

The MPDCCH is used to carry Downlink Control Information (DCI). An LTE-M device needs to monitor MPDCCH for the following types of information [12].

- UL power control command (DCI Format 3/3A)
- UL grant information (DCI Format 6-0A/6-0B in CE mode A/B)

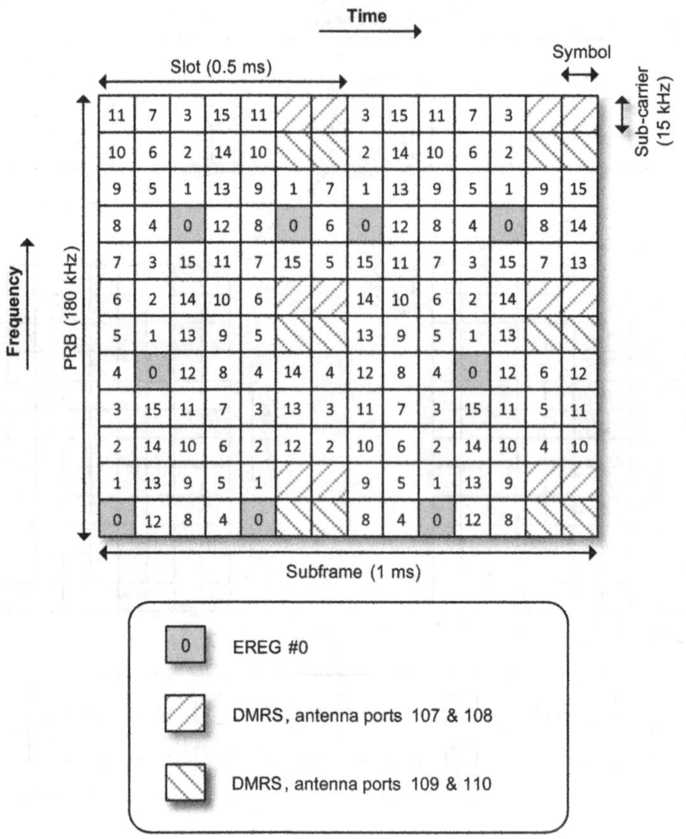

FIGURE 5.13

Enhanced resource-element groups (EREGs) for MPDCCH.

- DL scheduling information (DCI Format 6-1A/6-1B in CE mode A/B)
- Indicator of paging or SI update (DCI Format 6-2)

The MPDCCH design is based on the EPDCCH of LTE, which was introduced in LTE in Release 11. This means that the REs in one PRB pair are divided into 16 *enhanced resource-element groups* (EREGs) with each EREG containing 9 REs, as illustrated in Figure 5.13, where EREG #0 is highlighted. Furthermore, EREGs can be further combined into *enhanced control channel elements* (ECCEs). In normal subframes with normal CP length, each ECCE is composed of 4 EREGs and thus 36 REs.

An MPDCCH can span 2, 4, or 6 PRBs, and within these PRBs the transmission can either be *localized* or *distributed*. Localized transmission means that each ECCE is composed of EREGs from the same PRB, and distributed transmission means that each ECCE is composed of EREGs from different PRBs [5]. To achieve sufficient coverage, multiple ECCEs can furthermore be aggregated in an MPDCCH, according to the *ECCE aggregation level* of the MPDCCH. In normal subframes with normal CP length, aggregation of 2, 4, 8, 16, or 24 ECCEs is supported, where the highest aggregation level corresponds to aggregation of all REs in 6 PRBs. The device attempts to decode multiple MPDCCH candidates according to the *MPDCCH search space* as discussed in Section 5.3.3.1.

The MPDCCH carries the DCI and a 16-bit CRC is attached to the DCI. The CRC is masked with a sequence determined by the Radio Network Temporary Identifier (RNTI). The RNTI is an identifier used for addressing one or more devices and the RNTIs that can be monitored by a device are listed in Table 5.18. After the CRC attachment and RNTI masking, TBCC encoding and rate matching is used to generate a code word with a length matched to the number of encoded bits available for MPDCCH transmission. The determination of the number of available bits takes into account the MPDCCH aggregation level, modulation scheme (QPSK) and the REs not available for MPDCCH, i.e., the REs before the LTE-M starting symbol (see Section 5.2.4.1) in the subframe and the REs occupied by CRS (see Section 5.2.4.3). The MPDCCH transmission and its associated DMRS are masked with a scrambling sequence which is cell- or device-specific depending on whether it addresses a common or dedicated RNTI (see Table 5.18).

Further MPDCCH CE beyond what can be achieved with the highest ECCE aggregation level can be provided by repeating the subframe up to 256 times. For CE mode B, to simplify combination of the repetitions in receiver implementations that use combining on I/Q sample level, the scrambling sequence is repeated over multiple subframes (4 subframes in FDD and 10 subframes in TDD). Furthermore, the device can assume that any potential precoding matrix (for beamforming) stays the same over a number of subframes indicated in the SI (see Section 5.3.3.2).

5.2.4.6 PDSCH

Subframe	Any
Basic TTI	1 ms
Repetitions	Maximum 32 in CE mode A, maximum 2048 in CE mode B
Subcarrier spacing	15 kHz
Bandwidth	1–6 PRBs in CE mode A, 4 or 6 PRBs in CE mode B
Frequency location	Within a narrowband
Frequency hopping	Between 2 or 4 narrowbands if configured

The PDSCH is primarily used to transmit unicast data. The data packet from higher layers is segmented into one or more *transport blocks* (TB), and PDSCH transmits one TB at a time. PDSCH is also used to broadcast information such as SI (see Section 5.3.1.2), paging messages (see Section 5.3.1.4), and random access related messages (see Section 5.3.1.6).

Table 5.3 shows the *modulation and coding schemes* (MCS) and TBS for PDSCH in CE mode A and B. However, the low-cost LTE-M device Cat-M1 is restricted to a maximum TBS of 1000 bits, so the TBS values larger than 1000 bits do not apply to Cat-M1, only to higher device categories configured with CE mode A (see Section 5.2.3.3 for more information on the CE modes).

Further restrictions apply when PDSCH is used for broadcast. The modulation scheme is then restricted to QPSK and special TBS may apply (see Section 5.3.1).

A 24-bit CRC is attached to the TB. The channel coding is the standard LTE turbo coding with 1/3 code rate, 4 redundancy versions, rate matching, and interleaving [12]. PDSCH is not mapped to REs

Table 5.3 PDSCH modulation and coding schemes and TBS in LTE-M

MCS index	Modulation scheme	TBS index	CE mode A # PRBs						CE mode B # PRBs	
			1	2	3	4	5	6	4	6
0	QPSK	0	16	32	56	88	120	152	88	152
1	QPSK	1	24	56	88	144	176	208	144	208
2	QPSK	2	32	72	144	176	208	256	176	256
3	QPSK	3	40	104	176	208	256	328	208	328
4	QPSK	4	56	120	208	256	328	408	256	408
5	QPSK	5	72	144	224	328	424	504	328	504
6	QPSK	6	328	176	256	392	504	600	392	600
7	QPSK	7	104	224	328	472	584	712	472	712
8	QPSK	8	120	256	392	536	680	808	536	808
9	QPSK	9	136	296	456	616	776	936	616	936
10	16QAM	9	136	296	456	616	776	936		
11	16QAM	10	144	328	504	680	872	1032		
12	16QAM	11	176	376	584	776	1000	1192	Unused	
13	16QAM	12	208	440	680	904	1128	1352		
14	16QAM	13	224	488	744	1000	1256	1544		
15	16QAM	14	256	552	840	1128	1416	1736		

before the LTE-M starting symbol (see Section 5.2.4.1) and not to REs occupied by RS (see Section 5.2.4.3). In CE mode A, the PDSCH is modulated with QPSK or 16QAM and mapped to between 1 and 6 PRBs anywhere within a narrowband. In CE mode B, the PDSCH is modulated with QPSK and mapped to 4 or 6 PRBs within a narrowband. The modulation scheme restrictions facilitate low-cost LTE-M device receiver implementations with relaxed requirements on demodulation accuracy compared to ordinary LTE devices which support at least up to 64QAM. For more information on PDSCH scheduling, refer to Section 5.3.2.1.1.

LTE-M supports the following PDSCH transmission modes (TM) inherited from LTE:

- **TM1**: Single-antenna transmission (supported in both CE mode A and B)
- **TM2**: Transmit diversity (supported in both CE mode A and B)
- **TM6**: Closed-loop codebook-based precoding (supported in CE mode A only)
- **TM9**: Non-codebook-based precoding (supported in both CE mode A and B)

TM2 is based on SFBC in case of two antenna ports and on a combination of SFBC and FSTD in case of four antenna ports [5]. Feedback of precoding matrix recommendations for TM6 and TM9 and other feedback (DL channel quality indicator and Hybrid Automatic Repeat Request (HARQ) feedback) are discussed in Section 5.2.5.5. Because most LTE-M devices are expected to be low-cost devices with a single receive antenna, *multiple-input multiple-output* operation is not supported.

Table 5.4 PDSCH/PUSCH repetition factors in CE mode A	
Broadcasted maximum number of PDSCH/PUSCH repetitions for CE mode A	PDSCH/PUSCH repetition factors that can be selected from the DCI on MPDCCH
No broadcasted value (default)	1, 2, 4, 8
16	1, 4, 8, 16
32	1, 4, 16, 32

CE can be provided by repeating the subframe up to 2048 times. The maximum numbers of repetitions in CE modes A and B, respectively, are configurable per cell according to Tables 5.4 and 5.5.

For PDSCH demodulation, the device uses CRS for TM1/TM2/TM6 and DMRS for TM9 (see Section 5.2.4.3). The PDSCH is masked with a scrambling sequence generated based on the PCID and the RNTI. For CE mode B, to simplify combination of the repetitions in receiver implementations that use combining on I/Q sample level, the same scrambling sequence and the same redundancy version (RV) are repeated over multiple subframes (4 subframes in FDD and 10 subframes in TDD). Furthermore, the device can assume that any potential precoding matrix (for beamforming) stays the same over a number of subframes indicated in the SI (see Section 5.3.3.2).

5.2.5 UPLINK PHYSICAL CHANNELS AND SIGNALS

LTE-M supports the set of UL channels depicted in Figure 5.14. The physical layer provides data transport services to higher layers through the use of transport channels via the MAC layer [8]. The *Uplink Control Information* (UCI) is strictly speaking not a transport channel, which is indicated by the dashed line. The MAC layer in turn provides data transport services through the use of logical channels, which are also shown in the figure for completeness [9]. For more information on the higher layers, refer to Section 5.3.

Table 5.5 PDSCH/PUSCH repetition factors in CE mode B	
Broadcasted maximum number of PDSCH/PUSCH repetitions for CE mode B	PDSCH/PUSCH repetition factors that can be selected from the DCI on MPDCCH
No broadcasted value (default)	4, 8, 16, 32, 64, 128, 256, 512
192	1, 4, 8, 16, 32, 64, 128, 192
256	4, 8, 16, 32, 64, 128, 192, 256
384	4, 16, 32, 64, 128, 192, 256, 384
512	4, 16, 64, 128, 192, 256, 384, 512
768	8, 32, 128, 192, 256, 384, 512, 768
1024	4, 8, 16, 64, 128, 256, 512, 1024
1536	4, 16, 64, 256, 512, 768, 1024, 1536
2048	4, 16, 64, 128, 256, 512, 1024, 2048

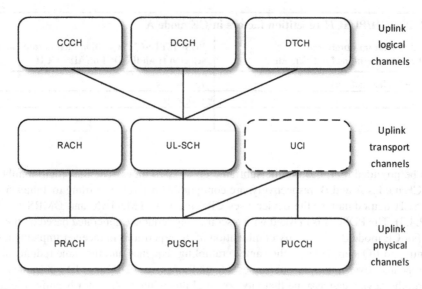

FIGURE 5.14

Uplink channels used in LTE-M.

In this section we focus on the UL physical channels. Due to the adopted UL transmission scheme in LTE (i.e., SC-FDMA), the transmission from a device needs to be contiguous in the frequency domain. To maximize the chances that large contiguous allocations are available for UL data transmission on the PUSCH for LTE and LTE-M users, it is often considered beneficial to allocate the resources for *Physical Random Access Channel* (PRACH) and *Physical Uplink Control Channel* (PUCCH) near the edges of the system bandwidth. The UL RS are not shown in Figure 5.14 but are transmitted together with PUSCH or PUCCH or separately for sounding of the radio channel.

5.2.5.1 Uplink Subframes

A cell-specific *subframe bitmap* can be broadcasted in the SI (see Section 5.3.1.2) to indicate which subframes are valid for LTE-M transmission. For FDD UL, the bitmap length is 10 bits corresponding to the UL subframes within 1 frame. For TDD, the bitmap length is 10 or 40 bits corresponding to the subframes within 1 or 4 frames. This bitmap could, for example, facilitate so-called dynamic TDD operation within the LTE cell. Typically, all UL subframes are configured as valid.

When an LTE-M device needs to retune from one UL narrowband in a first subframe to another UL narrowband in a second subframe, the device creates a *guard period for narrowband retuning* by not transmitting two of the SC-FDMA symbols [5]. If the two subframes both carry PUSCH or they both carry PUCCH, the guard period is created by truncating the last symbol in the first subframe and the first symbol in the second subframe. If one of the subframes carries PUSCH and the other one carries PUCCH, truncation of PUCCH is avoided by instead truncating up to two symbols of PUSCH. The rationale for this rule is that PUSCH is protected by more robust channel coding and retransmission scheme compared to PUCCH.

A *shortened format* may be used for PUSCH/PUCCH to make room for *Sounding Reference Signal* (SRS) transmission in the last SC-FDMA symbol in an UL subframe (see Section 5.2.5.3).

5.2.5.2 PRACH

Subframe	Any
Basic TTI	1, 2, or 3 ms
Repetitions	1, 2, 4, 8, 16, 32, 64, 128
Subcarrier spacing	1.25 kHz
Bandwidth	839 subcarriers (ca. 1.05 MHz)
Frequency location	Any
Frequency hopping	Between two frequency locations if configured

The PRACH is used by the device to initialize connection and allows the serving base station to estimate the time of arrival of UL transmission. The time of arrival of the received PRACH signal reflects the round-trip propagation delay between the base station and device. Figure 5.15 shows the structure of the LTE PRACH preamble.

LTE-M reuses the LTE PRACH formats listed in Table 5.6, where T_s is the *basic LTE time unit* $1/(15,000 \times 2048)$ s.

In LTE, the *PRACH configuration* is cell-specific and there are many possible configurations in terms of mapping the signal on the subframe structure [5]. A configuration can be sparse or dense in

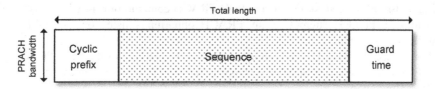

FIGURE 5.15

LTE PRACH preamble structure.

Table 5.6 PRACH formats in LTE-M						
PRACH format	CP length (T_s)	Sequence length (T_s)	Total length (ms)	Cell range from guard time (km)	FDD PRACH configurations	TDD PRACH configurations
0	3,168	24,576	1	15	0–15	0–19
1	21,024	24,576	2	78	16–31	20–29
2	6,240	2 × 24,576	2	30	32–47	30–39
3	21,024	2 × 24,576	3	108	48–63	40–47

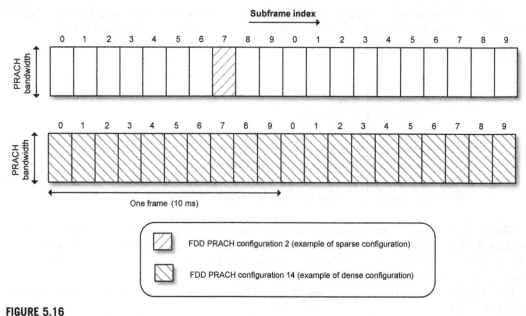

FIGURE 5.16

Example PRACH configurations for FDD PRACH Format 0 in LTE and LTE-M.

time, as illustrated by the examples for PRACH Format 0 in Figure 5.16, where FDD PRACH configuration 2 uses every 20th subframe and FDD PRACH configuration 14 uses every subframe. A device can make a PRACH attempt in any PRACH opportunity using one out of the (max 64) configured PRACH preamble sequences.

LTE-M introduces PRACH CE through up to 128 times repetition of the basic PRACH preamble structure in Figure 5.15. The repetitions are mapped onto the PRACH subframes that are included in the PRACH configuration.

In a cell supporting CE mode B, up to 4 *PRACH CE levels* can be defined. If the cell only supports CE mode A, up to 2 PRACH CE levels can be defined. The network has several options for separating the PRACH resources that correspond to different PRACH CE levels.

- **Frequency domain**: The network can separate the PRACH resources of the PRACH CE levels by configuring different *PRACH frequencies* for different PRACH CE levels.
- **Time domain**: The network can separate the PRACH resources of the PRACH CE levels by configuring different *PRACH configurations* and *PRACH starting subframe periodicities* for different PRACH CE levels.
- **Sequence domain**: The network can separate the PRACH resources of the PRACH CE levels by configuring nonoverlapping *PRACH preamble sequence groups* for different PRACH CE levels.

For more information on the random access procedure and on frequency hopping, see Sections 5.3.1.6, 5.3.2.2, and 5.3.3.2.

5.2.5.3 Uplink Reference Signals

Subframe	Any
Subcarrier spacing	15 kHz
DMRS bandwidth	Same as associated PUSCH/PUCCH
SRS bandwidth	4 PRBs
DMRS frequency location	Same as associated PUSCH/PUCCH
SRS frequency location	Configurable

UL RS [5] are predefined signals transmitted by the device to allow the base station to estimate the UL propagation channel to be able to demodulate UL physical channels, perform UL quality measurements, and issue timing advance commands. Figure 5.17 depicts the UL RS for LTE-M.

The DMRS for PUSCH and PUCCH are transmitted in the SC-FDMA symbols indicated in Table 5.7 in each slot in the transmitted UL subframe. The DMRS bandwidth is equal to the channel bandwidth, i.e., 1–6 PRBs for PUSCH and 1 PRB for PUCCH. Multiplexing of multiple PUCCH in

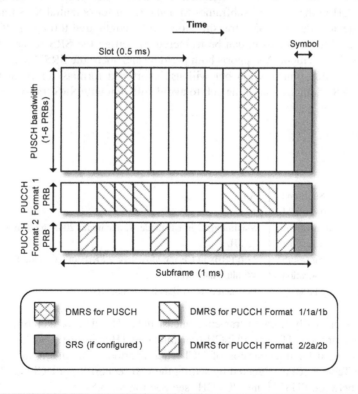

FIGURE 5.17

Uplink reference signals for LTE-M.

Table 5.7 DMRS locations in LTE-M

Physical channel	DMRS position within each slot (SC-FDMA symbol indices starting with 0)	
	Normal CP length (7 symbols per slot)	Extended CP length (6 symbols per slot)
PUSCH	3	2
PUCCH Format 1/1a/1b	2, 3, 4	2, 3
PUCCH Format 2	1, 5	3
PUCCH Format 2a/2b	1, 5	N/A

the same time-frequency resource is enabled by the possibility to generate multiple orthogonal DMRS sequences by applying a *Cyclic Shift*, and in case of PUCCH Format 1/1a/1b also by applying an *OCC* on top of the cyclic time shift.

The network can reserve the last SC-FDMA symbol of some UL subframes in a cell for SRS transmission for sounding of the radio channel. The device will then use shortened format for PUSCH and PUCCH in the affected subframes to make room for potential SRS transmission from itself or another device. Periodic SRS transmission can be configured through RRC configuration, whereas aperiodic SRS transmission can be triggered by setting the SRS request bit in DCI (see Tables 5.16 and 5.17). CE mode A supports both periodic and aperiodic SRS transmission. CE mode B does not support SRS transmission but will use shortened formats for PUSCH and PUCCH according to the SRS configuration in the cell to avoid collision with SRS transmissions from other devices.

5.2.5.4 PUSCH

Subframe	Any
Basic TTI	1 ms
Repetitions	Maximum 32 in CE mode A, maximum 2048 in CE mode
Subcarrier spacing	15 kHz
Bandwidth	1−6 PRBs in CE mode A, 1 or 2 PRBs in CE mode B
Frequency location	Within a narrowband
Frequency hopping	Between 2 narrowbands if configured

The PUSCH is primarily used to transmit unicast data. The data packet from higher layers is segmented into one or more TB, and PUSCH transmits one TB at a time.

PUSCH is also used for transmission of UCI when aperiodic *Channel State Information* (CSI) transmission (see Table 5.10) is triggered by setting the CSI request bit in DCI (see Table 5.17) or in case of collision between PUSCH and PUCCH (see Section 5.2.5.5).

Table 5.8 shows the MCS and TBS for PUSCH in CE mode A and B. However, the low-cost LTE-M device Cat-M1 is restricted to a maximum TBS of 1000 bits, so the TBS values larger than 1000 bits

Table 5.8 PUSCH modulation and coding schemes and transport block sizes in LTE-M

MCS index	Modulation scheme	TBS index	CE mode A # PRBs						CE mode B # PRBs	
			1	2	3	4	5	6	1	2
0	QPSK	0	16	32	56	88	120	152	56	152
1	QPSK	1	24	56	88	144	176	208	88	208
2	QPSK	2	32	72	144	176	208	256	144	256
3	QPSK	3	40	104	176	208	256	328	176	328
4	QPSK	4	56	120	208	256	328	408	208	408
5	QPSK	5	72	144	224	328	424	504	224	504
6	QPSK	6	328	176	256	392	504	600	256	600
7	QPSK	7	104	224	328	472	584	712	328	712
8	QPSK	8	120	256	392	536	680	808	392	808
9	QPSK	9	136	296	456	616	776	936	456	936
10	QPSK	10	144	328	504	680	872	1032	504	1032
11	16QAM	10	144	328	504	680	872	1032	Unused	
12	16QAM	11	176	376	584	776	1000	1192		
13	16QAM	12	208	440	680	904	1128	1352		
14	16QAM	13	224	488	744	1000	1256	1544		
15	16QAM	14	256	552	840	1128	1416	1736		

do not apply to Cat-M1, only to higher device categories configured with CE mode A (see Section 5.2.3.3 for more information on the CE modes).

A 24-bit CRC is attached to the TB. The channel coding is the standard LTE turbo coding with 1/3 coding rate, 4 redundancy versions, rate matching, and interleaving [12]. PUSCH is mapped to the SC-FDMA symbols that are not used by DMRS (see Section 5.2.5.3). In CE mode A, the PUSCH is modulated with QPSK or 16QAM and mapped to between 1 and 6 PRBs anywhere within a narrowband. In CE mode B, the PUSCH is modulated with QPSK and mapped to 1 or 2 PRBs within a narrowband. For more information on PUSCH scheduling, refer to Section 5.3.2.1.2.

CE can be provided by repeating the subframe up to 2048 times. The maximum numbers of repetitions in CE modes A and B, respectively, are configurable per cell according to Tables 5.4 and 5.5. HD-FDD devices supporting CE mode B can indicate to the network that they need to insert periodic *uplink transmission gaps* in case of long PUSCH transmissions in which case the device will insert a 40-ms gap every 256 ms [5]. This gap is used by the device to correct the frequency error by measuring the DL RS.

PUSCH is masked with a scrambling sequence generated based on the PCID and the RNTI. For CE mode B, to simplify combination of the repetitions in receiver implementations that use combining on I/Q sample level, the same scrambling sequence and the RV are repeated over multiple subframes (4 subframes in FDD, 5 subframes in TDD).

5.2.5.5 PUCCH

Subframe	Any
Basic TTI	1 ms
Repetitions in CE mode A	1, 2, 4, 8
Repetitions in CE mode B	4, 8, 16, 32 (Release 14 also supports 64 and 128)
Subcarrier spacing	15 kHz
Bandwidth	1 PRB
Frequency location	Any PRB
Frequency hopping	Between 2 PRB locations

The PUCCH is used to carry the following types of Uplink Control Information (UCI).

- UL scheduling request (SR)
- DL HARQ feedback (ACK or NACK)
- DL CSI

A PUCCH transmission is mapped to a configurable *PUCCH region* that consists of two PRB locations with equal distance to the center frequency of the LTE system bandwidth, typically chosen to be close to the edges of the system bandwidth. Inter-subframe (not intra-subframe) frequency hopping takes place between the two PRB locations in the PUCCH region (see Figures 5.21 and 5.22 and Section 5.3.3.2). PUCCH is mapped to the SC-FDMA symbols that are not used by DMRS (see Section 5.2.5.3). CE can be provided by repeating the subframe up to 32 times in Release 13 (and up to 128 times in Release 14).

The PUCCH formats supported by LTE-M are listed in Table 5.9. If a device in connected mode has a valid PUCCH resource for SR, it can use it to request an UL grant when needed, otherwise it has to rely on the random access procedure for this. A PUCCH resource for HARQ feedback is allocated when a DL HARQ transmission is scheduled (see Section 5.3.2.1.1). By applying different cyclic time shifts and OCCs (see Section 5.2.5.3), it is in principle possible to multiplex up to 36 PUCCH Format 1/1a/1b in the same time-frequency resource.

The CSI modes supported by LTE-M are listed in Table 5.10 [10]. The *Channel Quality Information* (CQI) report reflects the device's recommendation regarding what PDSCH MCS to use when

Table 5.9 PUCCH formats in LTE-M

PUCCH format	Description	Modulation scheme	Comment
1	Scheduling request	On-off keying (OOK)	Supported in CE mode A and B
1a	1-bit HARQ feedback	BPSK	Supported in CE mode A and B
1b	2-bit HARQ feedback for TDD	QPSK	Only supported in CE mode A
2	20-bit CSI report	QPSK	Only supported in CE mode A
2a	20-bit CSI report + 1-bit HARQ feedback	QPSK + BPSK	Only supported in CE mode A
2b	20-bit CSI report + 2-bit HARQ feedback in TDD	QPSK + QPSK	Only supported in CE mode A

Table 5.10 Downlink CSI modes in LTE-M

CSI mode	Description	Triggering	Physical channel	Comment
1-0	Wideband CQI in TM1/TM2/TM9	Periodic	PUCCH Format 2/2a/2b	Only supported in CE mode A
1-1	Wideband CQI and PMI in TM6/TM9	Periodic	PUCCH Format 2/2a/2b	Only supported in CE mode A
2-0	Subband CQI in TM1/TM2/TM9	Aperiodic	PUSCH	Only supported in CE mode A

targeting 10% BLock Error Rate (BLER) for the first HARQ transmission. The *Precoding Matrix Indicator* (PMI) report is the device's recommendation regarding what precoding matrix to use in PDSCH TM6 and TM9 (see Section 5.2.4.6). For PDSCH TM9, either closed-loop or open-loop beamforming may be used, and in the latter case no PMI reporting is needed. In LTE-M, the CQI and PMI reports are based on CRS measurements in the narrowbands monitored by the device for MPDCCH monitoring. The wideband CQI report reflects the quality when all monitored narrowbands are used for a transmission, whereas the subband CQI report contains one separate CQI report per monitored narrowband. By applying different cyclic time shifts (see Section 5.2.5.3) it is in principle possible to multiplex up to 12 PUCCH Format 2/2a/2b in the same time-frequency resource.

The number of PUCCH subframe repetitions to use in connected mode is configured per device. In CE mode A, the possible PUCCH repetition numbers are {1, 2, 4, 8} and different repetition numbers can be configured for PUCCH Format 1/1a/1b and PUCCH Format 2/2a. In CE mode B, the possible repetition numbers for PUCCH Format 1/1a/1b are {4, 8, 16, 32} in Release 13 and {4, 8, 16, 32, 64, 128} in Release 14. Similar ranges apply for the PUCCH carrying HARQ feedback for the PDSCH carrying Message 4 during the random access procedure described in Section 5.3.1.6, but the values are broadcasted in System Information Block 2 (SIB2).

Simultaneous transmission of more than one PUSCH or PUCCH transmission from the same device is not supported. If a device is scheduled to transmit both PUSCH and UCI in a subframe, and both PUSCH and PUCCH are without repetition, then the UCI is multiplexed into the PUSCH according to ordinary LTE behavior, but if PUSCH or PUCCH is with repetition then PUSCH is dropped in that subframe. If a device is scheduled to transmit two or more of HARQ feedback, SR, and periodic CSI in a subframe, and PUCCH is without repetition, then ordinary LTE behavior applies, but if PUCCH is with repetition then only the highest priority UCI is transmitted, where HARQ feedback has the highest priority and periodic CSI has the lowest priority.

5.3 IDLE AND CONNECTED MODE PROCEDURES

In this section, we describe LTE-M physical layer procedures and higher layer protocols, including all activities from initial cell selection to setting up and controlling a connection. This section uses physical layer related terminology introduced in Section 5.2.

The idle mode procedures include the initial cell selection, SI acquisition, cell reselection, and paging procedures. The transition from idle to connected mode involves the procedures for random

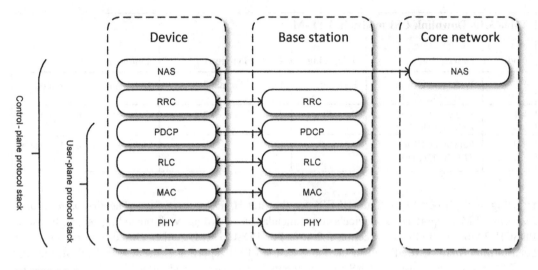

FIGURE 5.18

Protocol stack for LTE-M.

access and access control. The connected mode operation includes procedures for scheduling, retransmission, power control, and mobility support. Idle mode procedures and connected mode procedures are treated in Sections 5.3.1 and 5.3.2, respectively. Additional physical layer procedures common for idle and connected mode are treated in Section 5.3.3.

The LTE-M radio protocol stack is inherited from LTE [8] and is illustrated in Figure 5.18. The main changes are in the physical layer, but there are also changes to the higher layers. Changes to the control plane are mainly covered in Section 5.3.1 and changes to the user plane are mainly covered in Section 5.3.2. The mappings between physical channels, transport channels, and logical channels are illustrated in Figures 5.4 and 5.14.

5.3.1 IDLE MODE PROCEDURES

The first idle mode procedure that the device needs to carry out is cell selection. Once a cell has been selected, most of the interaction between the device and the base station relies on transmissions addressed by the base station to the device using a 16-bit RNTI [9]. The RNTIs monitored by LTE-M devices in idle mode are listed in Table 5.18 together with references to the relevant book sections.

5.3.1.1 Cell Selection

The main purpose of cell selection is to identify, synchronize to, and determine the suitability of a cell. The general steps in the LTE-M cell selection procedure (which to a large extent follows the LTE cell selection procedure) are as follows:

1. Search for the PSS to identify the presence of an LTE cell and to synchronize in time and frequency to the LTE carrier frequency and half-frame timing.

2. Synchronize to the SSS to identify the frame timing, PCID, CP length, and duplex mode (FDD or TDD).
3. Acquire the MIB to identify the SFN, the DL system bandwidth, and the scheduling information for the LTE-M-specific SIB1.
4. Acquire the SIB1 to identify, for example, the H-SFN, public land mobile network (PLMN) identity, tracking area, unique cell identity, UL−DL configuration (in case of TDD), and scheduling information for other SI messages and to prepare for verification of the cell suitability.

These procedures are in detail described in the next few sections.

5.3.1.1.1 Time and Frequency Synchronization

The *initial cell selection* procedure aims to time-synchronize to PSS and to obtain a *carrier frequency error* (CFO) estimation. As shown in Figures 5.7 and 5.8, PSS is transmitted every 5 ms at the center 62 subcarriers of the DL carrier. The device can assume that the allowed carrier frequencies are aligned with a 100-kHz channel raster [7], i.e., the carrier frequencies to search for are multiples of 100 kHz. The initial oscillator inaccuracy for a low-cost device may be as high as 20 ppm (parts per million), corresponding to, for example, 18 kHz initial CFO for a 900-MHz band. This means that there is relatively large uncertainty both regarding time and frequency during initial cell selection, which means that time and frequency synchronization at initial cell selection can take significantly longer than at *non-initial cell selection* or *cell reselection* (cell reselection is described in Section 5.3.1.3).

By time synchronizing to PSS the device detects the 5-ms (half-frame) timing. PSS synchronization can be achieved by correlating the received signal with the three predefined PSS sequences. Time and frequency synchronization can be performed in a joint step using subsequent PSS transmissions. For further details on the synchronization signals, refer to Section 5.2.4.2.

5.3.1.1.2 Cell Identification and Initial Frame Synchronization

Like the PSS, the SSS is transmitted every 5 ms at the center 62 subcarriers of the DL carrier. As discussed in Section 5.2.4.2, the SSS can be used to acquire the frame timing, the PSS and SSS sequences together can be used to identify the cell's PCID, and the relative positions of the PSS and SSS transmissions within a frame can be used to detect the duplex mode (FDD or TDD) and the CP length (normal or extended).

5.3.1.1.3 Master Information Block Acquisition

After acquiring the PCID, the device knows the CRS placement within a resource block as the subcarriers that CRS REs are mapped to are determined by PCID. It can thus demodulate and decode PBCH, which carries the MIB. For further details on PBCH, refer to Section 5.2.4.4. One of the information elements carried in the MIB is the 8 most significant bits (MSBs) of the SFN. Because the SFN is 10 bits long, the 8 MSBs of SFN change every 4 frames, i.e., every 40 ms. As a result, the TTI of PBCH is 40 ms. A MIB is encoded to a PBCH code block, consisting of 4 code subblocks. PBCH is transmitted in subframe 0 in every frame, and each PBCH subframe carries a code subblock. A code subblock can be repeated as explained in Section 5.2.4.4 for enhanced coverage. Initially the device does not know which subblock is transmitted in a specific frame. Therefore, the device needs to form four hypotheses to decode a MIB during the initial cell selection process. This is referred to as blind decoding. In addition, to correctly decode the MIB CRC the device needs to hypothesize whether 1, 2, or 4 antenna ports are used for CRS transmission at the base station. A successful MIB decoding is indicated by having a correct CRC. At that point, the device has acquired the information listed below:

- Number of antenna ports for CRS transmission
- SFN
- DL system bandwidth
- Scheduling information for the LTE-M-specific SIB1

Typically, the system bandwidth is the same in DL and UL but in principle they can be different. The DL system bandwidth is indicated in MIB, whereas the UL system bandwidth is indicated in SIB2, which is described in Section 5.3.1.2.2. Table 5.2 lists the supported system bandwidths.

The presence of the scheduling information for the LTE-M-specific SIB1 is an indication that LTE-M is supported by the cell. Hereafter the LTE-M-specific SIB1 will be referred to as SIB1 for short.

5.3.1.1.4 Cell Identity and Hyper System Frame Number Acquisition

After acquiring the MIB including the scheduling information about SIB1 the device is able to locate and decode SIB1. We will describe more about how device acquires SIB1 in Section 5.3.1.2.1. From a cell search perspective, it is important to know that the SIB1 carries the H-SFN, the PLMN identity, tracking area, and cell identity. Unlike the PCID, the cell identity is unique within the PLMN. After acquiring SIB1, the device has achieved complete synchronization to the framing structure shown in Figure 5.1. Based on the information provided in SIB1, the device will be able to determine whether the cell is suitable for camping, and whether the device can attempt to attach to the network. Somewhat simplified, a cell is considered suitable if the PLMN is available, the cell is not barred, and the cell's signal strength exceeds a minimum requirement.

Figure 5.19 illustrates how the device acquires complete framing information during the initial cell search procedure.

After completing the initial cell search, the device is expected to have a time and frequency accuracy that offers robust operation in subsequent transmission and reception during connected and idle mode operations.

5.3.1.2 *System Information Acquisition*

After selecting a suitable cell to camp on a device needs to acquire the full set of SI. Table 5.11 summarizes the supported SIB types [8]. SIB1 and SIB2 contain the most critical SI, required by the device to be able to select the cell for camping and for accessing the cell. The other SIBs may or may not be transmitted in a given cell depending on the network configuration. LTE-M inherits the SIBs from LTE. LTE SIBs related to functionality not supported by LTE-M are not included in the table. Interested readers can refer to Reference [13] for additional details regarding SIBs.

Although LTE-M reuses the SIB definitions from LTE, it should be noted that the SI for LTE-M devices is transmitted separately from the SI for ordinary LTE devices. The main reason for this is that the SI transmissions in LTE are scheduled using LTE's PDCCH, and may be scheduled to use any number of PRBs for PDSCH, which together means SI transmission occupies in general a too large channel bandwidth to be received by LTE-M devices with reduced receive bandwidth, and as a

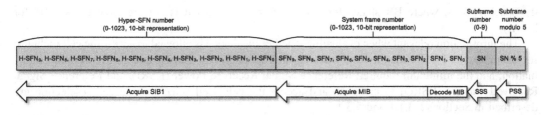

FIGURE 5.19

Illustration of how the LTE-M device acquires complete timing information during the initial cell search.

consequence LTE-M devices are not able to receive LTE's ordinary SI transmissions. SI transmissions for LTE-M devices are transmitted without an associated physical DL control channel on the PDSCH described in Section 5.2.4.6, masked with a scrambling sequence generated using the SI-RNTI defined in the standard [9].

5.3.1.2.1 System Information Block 1

SIB1 carries information relevant when evaluating if the device is allowed to camp on and access the cell, as well as scheduling information for the other SIBs. SIB1 for LTE-M is referred to as SIB1-BR in

Table 5.11 SIB types in LTE-M	
SIB	**Content**
SIB1	Information relevant when evaluating if a device is allowed to access a cell and scheduling information for other SIBs
SIB2	Configuration information for common and shared channels
SIB3	Cell reselection information, mainly related to the serving cell
SIB4	Information about the serving frequency and intrafrequency neighboring cells relevant for cell reselection
SIB5	Information about other frequencies and interfrequency neighboring cells relevant for cell reselection
SIB6	Information about UMTS 3G frequencies and neighboring cells relevant for cell reselection
SIB7	Information about GSM frequencies cells relevant for cell reselection
SIB8	Information about CDMA2000 3G frequencies and neighboring cells relevant for cell reselection
SIB9	Name of the base station in case it is a small so-called home base station (femto base station)
SIB10	Earthquake and Tsunami Warning System (ETWS) primary notification
SIB11	ETWS secondary notification
SIB12	Commercial Mobile Alert System warning notification
SIB14	Information about Extended Access Barring for access control
SIB16	Information related to GPS time and Coordinated Universal Time
SIB17	Information relevant for traffic steering between E-UTRAN and WLAN
SIB20	Information related to Single-Cell Point-to-Multipoint multicast (supported in Release 14)

the specifications, where BR stands for *bandwidth-reduced*, but it is here referred to as SIB1 for brevity.

The information relevant when evaluating whether the device is allowed to access the cell includes, for example, the PLMN identity, the tracking area identity, cell barring and cell reservation information, and the minimum required Reference Signal Received Power (RSRP) and Reference Signal Received Quality (RSRQ) to camp on and access the cell. Cell selection and cell reselection are described in Sections 5.3.1.1 and 5.3.1.3.

The scheduling information for the other SIBs is described in Section 5.3.1.2.2. In addition, SIB1 contains information critical for scheduling of DL transmissions in general. This information includes the H-SFN, the UL−DL configuration (in case of TDD), the LTE-M DL subframe bitmap, and the LTE-M starting symbol. For further information on these timing aspects, refer to Sections 5.2.2.2, 5.2.3.1 and 5.2.4.1.

The scheduling information for SIB1 itself is signaled using 5 bits in MIB [10]. If the value is zero then the cell is an LTE cell that does not support LTE-M devices, otherwise the TBS and number of repetitions for SIB1 are derived according to Table 5.12.

A SIB1 transport block is transmitted on PDSCH according to an 80 ms long pattern that starts in frames with SFN exactly divisible by 8 [5]. As indicated in Table 5.12, the PDSCH is repeated in 4, 8, or 16 subframes during this 80-ms period (except for system bandwidths smaller than 5 MHz, which only support 4 repetitions). Exactly what subframes are used for SIB1 transmission depends on the PCID and duplex mode according to Table 5.13. The PCID helps randomize the interference between SIB1 transmissions from different cells.

The LTE-M DL subframe bitmap signaled in SIB1 has no impact on the SIB1 transmission itself. Similarly, the LTE-M starting symbol signaled in SIB1 does not apply to the SIB1 transmission. Instead, the starting symbol for the PDSCH carrying SIB1 is always the 4th OFDM symbol in the subframe except for the smallest system bandwidth (1.4 MHz) where it is always the 5th OFDM symbol in the subframe [10].

SIB1 is transmitted on PDSCH with QPSK modulation using all 6 PRBs in a narrowband and RV cycling across the repetitions. The frequency locations and frequency hopping for SIB1 are fixed in the standard as described in Section 5.3.3.2. In case the scheduling causes collision between SIB1 and other MPDCCH/PDSCH transmission in a narrowband in a subframe, the SIB1 transmission takes precedence and the other transmission is dropped in that narrowband in that subframe.

The SIB1 information is unchanged at least during a *modification period* of 5.12 s, except in the rare cases of *Earthquake and Tsunami Warning System* or *Commercial Mobile Alert System* notifications, or when the ACB information changes (see Section 5.3.1.7 for more information). In practice the time between SI updates is typically much longer than 5.12 s.

5.3.1.2.2 System Information Blocks 2−20

SIB1 contains scheduling information for the other SIBs listed in Table 5.11. SIBs other than SIB1 are carried in *SI messages*, where each SI message can contain one or more SIBs [13]. Each SI message is configured with an *SI periodicity*, a TBS and a starting narrowband. The possible periodicities are {8, 16, 32, 64, 128, 256, 512} frames, and the possible TBS are {152, 208, 256, 328, 408, 504, 600, 712, 808, 936} bits. Each SI message can also be configured with its own *SI value tag* as described in Section 5.3.1.2.3.

Table 5.12 Transport block size and number of PDSCH repetitions for SIB1 in LTE-M

SIB1 scheduling information signaled in MIB	SIB1 transport block size in bits	Number of PDSCH repetitions in an 80-ms period
0	LTE-M not supported in the cell	
1	208	4
2		8
3		16
4	256	4
5		8
6		16
7	328	4
8		8
9		16
10	504	4
11		8
12		16
13	712	4
14		8
15		16
16	936	4
17		8
18		16
19–31	Reserved values	

Table 5.13 Subframes for SIB1 transmission

System bandwidth (MHz)	Number of PDSCH repetitions in an 80-ms period	PCID	Subframes with SIB1 transmission in FDD case	Subframes with SIB1 transmission in TDD case
<5	4	Even	Subframe #4 in even frames	Subframe #5 in odd frames
		Odd	Subframe #4 in odd frames	Subframe #5 in odd frames
≥5	4	Even	Subframe #4 in even frames	Subframe #5 in odd frames
		Odd	Subframe #4 in odd frames	Subframe #0 in odd frames
	8	Even	Subframe #4 in every frame	Subframe #5 in every frame
		Odd	Subframe #9 in every frame	Subframe #0 in every frame
	16	Even	Subframes #4 and #9 in every frame	Subframes #0 and #5 in every frame
		Odd	Subframes #0 and #9 every frame	Subframes #0 and #5 in every frame

The SI messages are periodically broadcasted during specific, periodic, and nonoverlapping time domain windows known as *SI windows* of configurable length. Possible SI window lengths are {1, 2, 5, 10, 15, 20, 40, 60, 80, 120, 160, 200} ms, where the smallest values are inherited from LTE and probably not that relevant for LTE-M. If the periodicity for the nth SI message is T_n frames, then that SI message is transmitted in the nth SI window after every frame that has an SFN evenly divisible by T_n. The intention with the specified behavior is to map different SI messages to different SI windows even if they have the same periodicity.

Furthermore, to support operation in extended coverage, SI messages can be repeated within their respective SI windows. Possible repetition patterns are {every frame, every 2nd frame, every 4th frame, and every 8th frame} throughout the SI window. All SI messages have the same repetition pattern.

Each SI message is transmitted on PDSCH with QPSK modulation using all 6 PRBs in its narrowband and RV cycling across the repetitions. Frequency hopping is supported as described in Section 5.3.3.2. In case the scheduling causes collision between SI messages and other MPDCCH/PDSCH transmission than SIB1 in a narrowband in a subframe, the SI message transmission takes precedence and the other transmission is dropped in that narrowband in that subframe [10].

The SI message content is unchanged during a configurable modification period where the possible values are {2, 4, 8, 16, 64} times the cell paging cycle. In practice the time between SI updates is typically much longer than this.

5.3.1.2.3 System Information Update

When the network modifies the SI in a cell, it can indicate this to the devices through the SI value tag [13]. The SI value tag is a 5-bit field in SIB1 that is changed every time the SI content has changed. It is also possible for the network to signal an SI value tag per SI message, which enables the device to just reacquire the SI messages that have actually changed instead of having to reacquire them all, which can be power consuming for the device. These SI value tags are 2-bit fields.

When the SI has changed, the network can also explicitly notify the devices through so-called *direct indication* in the DCI used for paging [12]. The meaning of the 8-bit direct indication field is described in Table 5.14 [13]. Paging and eDRX are discussed in Section 5.3.1.4.

Table 5.14 SI update notification through direct indication in DCI format for paging in LTE-M

Bit	Meaning
1	General SI update notification to devices not configured with eDRX
2	SIB10/11 update notification
3	SIB12 update notification
4	SIB14 update notification
5	General SI update notification to devices configured with eDRX
6	Reserved
7	Reserved
8	Reserved

Note that some SI updates, for example regularly changing parameters such as time information and access barring information, may not result in any SI value tag changes or explicit SI update notifications from the network. Also, when a device is configured with an eDRX cycle that is longer than the SI modification period, the device verifies that the stored SI is valid before trying to establish a connection or receive paging.

An LTE-M device considers its stored SI to be valid for several hours from the moment it was acquired, where the number of hours is either 3 or 24 h depending on network configuration [13]. When the SI has become invalid, the device should reacquire the SI before it accesses the network.

5.3.1.3 Cell Reselection

After selecting a cell, a device is mandated to monitor multiple neighbor cells. In simple words, in case the device detects that a neighbor cell has become stronger in terms of the RSRP than the currently serving cell, then the cell reselection procedure is triggered. A hysteresis value provided in SIB3 helps prevent ping-pong reselection [13]. Devices that are in good coverage, i.e., experiences a sufficiently high RSRP level in the serving cell, can be excluded from measuring for cell reselection. This helps improve the battery life of these devices. Besides securing that a device camps on the best cell, the cell reselection procedure is the main mechanism for idle mode mobility.

5.3.1.4 Paging and eDRX

In idle mode, the device monitors periodic *paging occasions* in a *paging narrowband* in the DL for potential attempts from the network to contact the device [14]. The paging transmission can contain multiple *paging records* intended for different devices in the cell. When a device receives a paging transmission at its paging occasion in its paging narrowband, it checks whether any of the paging records matches its device identity, and if there is a match, the device responds by initiating a connection to the cell using the random access procedure described in Section 5.3.1.6. The device can be identified either using a local *SAE Temporary Mobile Subscriber Identity* (S-TMSI) or the more rarely used global *International Mobile Subscriber Identity* (IMSI).

The monitoring of paging has implications on device battery lifetime and the latency of DL data delivery to the device. A compromise is achieved by configuring a *discontinuous reception* (DRX) cycle and/or an eDRX cycle. The maximum DRX cycle is 256 frames (2.56 s) in both idle and connected mode, and the maximum eDRX cycle is 256 hyperframes (about 44 min) in idle mode and 1024 frames (10.24 s) in connected mode. After each eDRX cycle, a *paging time window* occurs, configurable up to 2048 subframes (20.48 s) during which DL reachability is achieved through the configured DRX cycle. Figure 7.40 can serve as an illustration of these concepts, although it should be noted that the maximum values of the parameters are different in LTE-M and NB-IoT. The paging occasions for a device are determined by the DRX/eDRX configuration and the device identity (IMSI for DRX, S-TMSI for eDRX).

The total number of narrowbands that are used for paging is configurable per cell, and among these narrowbands the device determines its paging narrowband based on its device identity (IMSI). Frequency hopping is supported as described in Section 5.3.3.2.

The monitored physical channel is MPDCCH. MPDCCH repetition is supported using the *Type-1 MPDCCH Common Search Space* (CSS) described in Section 5.3.3.1. The MPDCCH carries DCI

Table 5.15 DCI Format 6-2 for paging and direct indication for LTE-M

Information	Size [bits]	Possible settings
Flag for paging/direct indication	1	Paging or direct indication (If this flag bit indicates direct indication then the remaining DCI content is according to Table 5.14)
PDSCH narrowband	1–4	Any narrowband in the system bandwidth
PDSCH TBS	3	{40, 56, 72, 120, 136, 144, 176, 208} bits
Number of PDSCH repetitions	3	One of the following ranges, depending on the setting of the DCI field "Number of MPDCCH repetitions": 00: {1, 2, 4, 8, 16, 32, 64, 128} 01: {4, 8, 16, 32, 64, 128, 192, 256} 10: {32, 64, 128, 192, 256, 384, 512, 768} 11: {192, 256, 384, 512, 768, 1024, 1536, 2048}
Number of MPDCCH repetitions	2	One of the following ranges, depending on the setting of the SIB2 parameter for max number of repetitions R_{max}: $R_{max} = 1$: {1} $R_{max} = 2$: {1, 2} $R_{max} = 4$: {1, 2, 4} $R_{max} = 8$: {1, 2, 4, 8} $R_{max} = 16$: {1, 4, 8, 16} $R_{max} = 32$: {1, 4, 16, 32} $R_{max} = 64$: {2, 8, 32, 64} $R_{max} = 128$: {2, 16, 64, 128} $R_{max} = 256$: {2, 16, 64, 256}

using DCI Format 6-2 [12]. This DCI format can either be used for carrying a direct indication field according to Table 5.14 or for scheduling a PDSCH carrying paging record(s) according to Table 5.15. The DCI CRC is masked with the *Paging RNTI* (P-RNTI), which is defined in the standard [9].

When the DCI schedules a PDSCH, the PDSCH is transmitted with QPSK modulation using all 6 PRBs in the narrowband [10]. PDSCH repetition is supported in a similar way as for DL unicast transmission as described in Section 5.3.2.1.1. The number of paging records that can be carried in one transport block depends on the size of each device identity. The size of each paging record can vary between 25 and 61 bits, meaning that the largest TBS (208 bits) can carry between 3 and 8 paging records.

When the *Mobility Management Entity* (MME) in the *core network* needs to page an LTE-M device, it will inform the involved base station(s) that the device is an LTE-M device so that the paging can be transmitted using the right format (with DCI Format 6-2, MPDCCH, and so on). The MME can also optionally provide a *Paging Coverage Enhancement Level*, which is an estimate between 1 and 256 of the required number of repetitions for MPDCCH [15]. In that case, the MME has been keeping the value as device history information since an earlier session in the same cell [13]. If the device is, for example, a stationary metering device that needs large CE because it is in a basement, it might be a good *paging strategy* to page the device right away with many MPDCCH repetitions in the cell where the device last accessed the network. However, if the device moves around when in idle mode there may not be any adequate history information in the MME because an LTE-M device in idle mode does not inform the network when the coverage situation changes [8]. For potential IoT use cases where the device is mobile but anyway frequently experiences bad coverage conditions, it may be difficult to find a suitable paging strategy, because it may not be acceptable from overhead point of view to page the device with many repetitions in multiple cells. In this case, some level of *Mobile Originated* (MO) traffic may be used to assist the network in keeping track of the device and thereby improve the DL reachability for the device. One example of this is device-triggered tracking area updates.

5.3.1.5 Power Saving Mode

The DRX and eDRX mechanisms described in the previous section provide minimum device power consumption for IoT applications where it needs to be possible to reach the device for *Mobile Terminated (MT)* traffic through paging within seconds or minutes. For applications where it is not required to be able to reach the device through paging faster than within half an hour or more, the PSM may be able to provide further power saving. The device will still be able to perform MO transmission in UL without delay. PSM is a standalone feature applicable to all 3GPP radio access technologies. For information about PSM, refer to Sections 2.2.3 and 7.3.1.5.

5.3.1.6 Random Access in Idle Mode

The random access procedure in LTE-M follows the same steps as in LTE [9]. After synchronizing to the network, confirming that access is not barred (as described in Section 5.3.1.7) and reading the PRACH configuration information in SIB2, the device can send a PRACH preamble to access the network (as described in Section 5.2.5.2). The random access procedure is also used when the device responds to a paging message. Use of random access in connected mode is described in Section 5.3.2.2.

If the base station detects a PRACH preamble, it sends back a *Random Access Response* (RAR), also known as *Message 2*. The RAR contains an UL *timing advance* (TA) command. The RAR further contains scheduling information pointing to the radio resources that the device can use to transmit a request to connect, also known as *Message 3*. In Message 3, the device will include its identity as part of an RRC message. The device can in some cases include its buffer status in Message 3 to facilitate the scheduling for subsequent UL transmissions. In *Message 4*, the network transmits a connection setup/resume message and contention resolution data that resolves any contention due to multiple devices transmitting the same preamble in the first step. The device finally replies with a connection setup/resume complete message to terminate the random access procedure and complete the transition to connected state. The device may also append UL data in the MAC layer of this message to optimize the latency of the data transfer. LTE-M supports both the ordinary RRC connection setup procedure and the RRC resume procedure described in Section 2.2.2. The message transfer is illustrated in Figure 2.2 and the latency is evaluated in Section 6.4.

The device needs to determine an appropriate PRACH resource configuration according to its coverage level estimation. The cell can configure up to three RSRP thresholds that are used by the device to select the PRACH resource configuration appropriate for its level of coverage. The PRACH resource configurations are signaled in SIB2. An example is given in Figure 5.20, in which three RSRP thresholds are configured and therefore there are four PRACH resources for three *PRACH CE levels*, respectively. The device performs a CRS-based RSRP measurement and selects a PRACH CE level in line with the measurement result. The higher the PRACH CE level, the larger the number of PRACH repetitions. If the device does not receive a RAR message in response to a PRACH attempt, it will

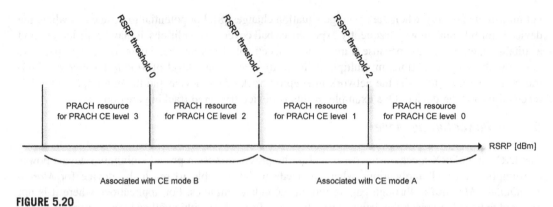

FIGURE 5.20

PRACH configurations and RSRP thresholds for LTE-M.

make further attempts until it receives RAR or until it has reached the maximum allowed number of attempts. As in ordinary LTE operation, PRACH preamble power ramping is applied, but in LTE-M the device will also be able to do PRACH CE level ramping, meaning that the device moves to the next PRACH CE level (increasing the number of PRACH repetitions per attempt) after a few unsuccessful attempts on one PRACH CE level.

After the PRACH preamble transmission, the device monitors a *RAR window* in the DL for a potential MPDCCH transmission that schedules a PDSCH transport block containing 56-bit RAR messages for one or more devices. The MPDCCH uses the *Type-2 MPDCCH CSS* described in Section 5.3.3.1. The DCI CRC is masked with the *Random Access RNTI* (RA-RNTI), which is determined from the PRACH transmission time according to a predefined rule in the standard [9]. The RAR message contains a *Temporary Cell-RNTI* (TC-RNTI) and a *RAR grant* and these are used to schedule the initial Message 3 transmission on PUSCH. Potential PUSCH HARQ retransmissions for Message 3 and all PDSCH HARQ (re)transmissions for Message 4 are scheduled using MPDCCH in Type-2 MPDCCH CSS with TC-RNTI. The HARQ-ACK feedback for Message 4 is transmitted on PUCCH, so the random access procedure makes use of five physical channels (PRACH for Message 1, MPDCCH+PDSCH for Message 2, PUSCH+MPDCCH for Message 3, and MPDCCH+PDSCH+PUCCH for Message 4).

The device is strictly speaking not configured in CE mode A or B (as described in Section 5.2.3.3) until it has entered connected mode. However, the PRACH CE levels are associated with the CE modes as illustrated in Figure 5.20. This means that if a PRACH preamble is successfully received on PRACH CE level 0 or 1 then the following messages (RAR, Message 3, and Message 4) will use DCI formats and various parameters signaled in SIB (e.g., maximum numbers of repetition and frequency hopping intervals) intended for CE mode A, and similarly if the PRACH preamble is successfully received on PRACH CE level 2 or 3 then the following messages will use DCI formats and SIB parameters intended for CE mode B. This means that the description of DL scheduling in connected mode in Section 5.3.2.1.1 is to a large extent valid for RAR and Message 4, and the description of UL scheduling in Section 5.3.2.1.2 is to a large extent valid for Message 3.

However, there are some differences compared to unicast. RAR is restricted to QPSK modulation and does not support HARQ retransmission. And the initial Message 3 transmission is not scheduled from MPDCCH but from a grant field in the RAR message. The RAR grant contains a PUSCH grant, which includes narrowband index, resource allocation within the narrowband and TBS, and an MPDCCH narrowband for scheduling of both potential HARQ retransmission(s) of Message 3 and HARQ (re)transmission(s) of Message 4. For further details on the RAR grant, refer to Section 6.2 in Reference [10].

5.3.1.7 Access Control

LTE-M supports access class barring (ACB) and extended access barring (EAB) as described in Section 2.2.1. SIB1 contains the scheduling information for SIB14 that carries the EAB information. Absence of SIB14 scheduling information in SIB1 implies that barring is not activated, whereas presence of SIB14 scheduling information informs the devices that a barring is activated. A change of the barring parameters can occur at any time, triggering the base station to also change the SIB14 scheduling information.

5.3.2 CONNECTED MODE PROCEDURES

Most of the interaction between the device and the base station relies on transmissions addressed by the base station to the device using a 16-bit RNTI [9]. The RNTIs monitored by LTE-M devices in connected mode are listed in Table 5.18 together with references to the relevant book sections.

5.3.2.1 Scheduling

In this section, we describe how scheduling for DL and UL transmissions works. When the base station needs to schedule a device dynamically, it sends a DCI which includes the resource allocation (in both time and frequency domains), modulation and coding scheme, and information needed for supporting the HARQ retransmission scheme. The DCI is carried on an MPDCCH, which is transmitted in an MPDCCH search space that the device is known to be monitoring (see Section 5.3.3.1) and the DCI has a CRC attached, which is masked with a device-specific Cell RNTI (C-RNTI) so that only the device for which the DCI is intended will decode it successfully, whereas other devices monitoring the same MPDCCH search space will discard the DCI because the CRC does not pass for them when they try to unmask the CRC using their C-RNTIs.

This section describes this *dynamic scheduling* of DL and UL as well as *semipersistent scheduling*.

5.3.2.1.1 Dynamic Downlink Scheduling

To allow low-complexity device implementation, LTE-M adopts the following scheduling principles:

- Cross-subframe scheduling (i.e., DCI and the scheduled data transmission do not occur in the same subframe) with relaxed processing time requirements.
- Optionally, HD-FDD operation at the device (i.e., no simultaneous transmission and reception at the device) allows time for the device to switch between transmission and reception modes.

Figure 5.21 shows an LTE-M DL scheduling example without repetition, with MPDCCH and PDSCH transmissions scheduled in parallel to a device in the same DL narrowband, and with the HARQ-ACK feedback transmitted on PUCCH in the UL. In ordinary LTE, the PDCCH or EPDCCH

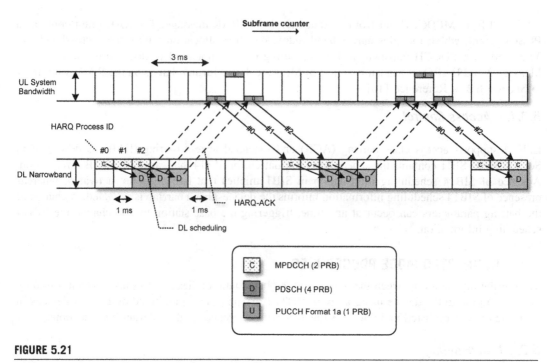

FIGURE 5.21

LTE-M downlink scheduling example with MPDCCH and PDSCH transmitted without repetition in the same downlink narrowband.

carrying the DCI and the PDSCH carrying the data are transmitted in the same DL subframe, but LTE-M has cross-subframe scheduling that shows in the figure as a 2-ms delay between the MPDCCH carrying the DCI and the scheduled PDSCH. Other than that, the timing relationships are similar to ordinary LTE, with a 4-ms delay between the PDSCH and the PUCCH carrying the associated HARQ-ACK feedback, and another 4-ms delay before a potential HARQ retransmission of the same data is scheduled. Due to the extra 2-ms scheduling delay, the DL HARQ *round-trip time* (RTT) is increased from $4 + 4 = 8$ to $2 + 4 + 4 = 10$ ms. The maximum number of DL HARQ processes depends on the duplex mode and CE mode as indicated in Table 5.16. HD-FDD devices cannot be scheduled more frequently than in this example because they cannot transmit and receive simultaneously and furthermore need a guard subframe at every switching between DL and UL (as discussed in Section 5.2.3.1). For a discussion on the impact of the 10-ms RTT on the DL peak rate in (half-duplex and full-duplex) FDD, see Sections 5.4.1.4 and 6.3.

The base station schedules DL transmission on PDSCH dynamically using DCI Format 6-1A and 6-1B in CE mode A and B, respectively. Table 5.16 shows the information carried in these DCI formats [12]. Some of the fields are specific to LTE-M and some of them are basically inherited from the ordinary LTE DCI formats. An effort has been made to make the DCI format for CE mode B as compact as possible because it is intended for situations where the device experiences a weak DL signal implying large numbers of repetitions, and every extra bit is expensive in terms of coverage and/or resource consumption. The most important fields are present in both DCI formats, such as the

Table 5.16 DCI Formats 6-1A and 6-1B used for scheduling PDSCH in CE modes A and B

Information	DCI Format 6-1A		DCI Format 6-1B	
	Size [bits]	Possible settings	Size [bits]	Possible settings
Format 6-0/6-1 differentiation	1	1	1	1
Frequency hopping flag	1	See Section 5.3.3.2	–	–
MCS	–	–	4	See Table 5.3
Resource block assignment	0–4	Narrowband index	0–4	Narrowband index
	5	0–20: Allocation of 1–6 PRBs	1	0: 4 PRBs (#0, ..., #3)
		21–31: Unused in Release 13		1: 6 PRBs (#0, ..., #5)
MCS	4	See Table 5.3	–	–
Number of PDSCH repetitions	2	See Table 5.4	3	See Table 5.5
HARQ process number	3–4	FDD: 0–7, TDD: 0–15	1	0–1
New data indicator	1	Toggle bit for new data	1	Toggle bit for new data
Redundancy version	2	0–3	–	–
PUCCH power control	2	See Section 5.3.2.3	–	–
Downlink assignment index	2	TDD-specific field	–	–
Antenna port and scrambling ID	2	TM9-specific field	–	–
SRS request	1	See Section 5.2.5.3	–	–
Precoding information	2 or 4	TM6-specific field	–	–
PMI confirmation	1	TM6-specific field	–	–
HARQ-ACK resource offset	2	PUCCH resource index offset	2	PUCCH resource index offset
Number of MPDCCH repetitions	2	See Table 5.19	2	See Table 5.19

PDSCH MCS, PDSCH resource block assignment, number of PDSCH repetitions, and number of repetitions of the MPDCCH carrying the DCI itself. The information about the number of MPDCCH repetitions is needed when calculating the starting subframe for the PDSCH transmission. When these DCI formats are used to schedule RAR as described in Section 5.3.1.6, some fields are reserved or repurposed (refer to Reference [12] for the detailed definition).

Figure 5.22 shows an LTE-M DL scheduling example where some repetition has been applied to enhance the coverage of the transmissions. The MPDCCH carrying the DCI is repeated in four subframes, the PDSCH carrying the data in eight subframes, and the PUCCH carrying the HARQ-ACK feedback in four subframes. All subframes are assumed to be configured as valid for transmission (see

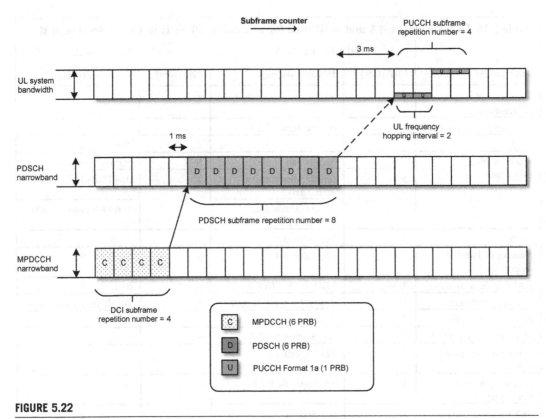

FIGURE 5.22

LTE-M downlink scheduling example with MPDCCH and PDSCH transmitted with repetition in different downlink narrowbands.

Sections 5.2.4.1 and 5.2.5.1). Note that the example makes use of the possibility to schedule PDSCH in a different narrowband than the MPDCCH narrowband. Furthermore, both MPDCCH and PDSCH are transmitted with the maximum channel bandwidth supported in the CE modes in Release 13, which is 1 narrowband (6 PRBs), which is in general beneficial for the device from coverage point of view. The reason for this is that because the total DL transmit power on the cell-carrier is typically (more or less) evenly distributed over all the PRBs in the system bandwidth, a large DL channel bandwidth usually also means a large chunk of the available transmit power.

The DCI field for *Number of PDSCH repetitions* contains a 2- or 3-bit value which is interpreted according to Tables 5.4 and 5.5. The DCI field for *Number of MPDCCH repetitions* is interpreted according to Table 5.19. Configuration of repetitions for PUCCH is discussed in Section 5.2.5.5. In the example in Figure 5.22, a 2-ms UL frequency hopping interval is assumed. Frequency hopping is discussed in Section 5.3.3.2. In CE mode A, the base station's scheduling decisions for PDSCH can be assisted by periodic or aperiodic CSI reports from the device, which is described in Section 5.2.5.5.

5.3.2.1.2 Dynamic Uplink Scheduling

Figure 5.23 shows an LTE-M UL scheduling example without repetition. Similar to LTE, a DCI carried on MPDCCH schedules a PUSCH transmission 4 ms later. A difference compared to LTE is that the UL HARQ scheme is asynchronous in LTE-M, whereas it is synchronous in LTE with a *Physical HARQ Indicator Channel* (PHICH) for HARQ feedback. This means that HARQ retransmissions in LTE-M are always explicitly scheduled using a DCI on an MPDCCH, i.e., there is no PHICH in LTE-M. Other than that, the HARQ operation in LTE-M is similar to the HARQ operation in LTE (this can be said for both UL HARQ and DL HARQ). The maximum number of UL HARQ processes depends on the CE mode as indicated in Table 5.17.

The base station schedules UL transmission on PUSCH dynamically using DCI Format 6-0A and 6-0B in CE mode A and B, respectively. Table 5.17 shows the information carried in these DCI formats [12]. Many aspects are similar to the PDSCH case described in the previous section. For example, there are fields indicating the number of repetitions for the PUSCH and for the MPDCCH itself, respectively, and a field for indicating the PUSCH narrowband. One notable difference compared to the DL case described in the previous section is that PUSCH transmission in CE mode B is always scheduled on very few PRBs (1 or 2 PRBs), whereas PDSCH transmission in CE mode B is always scheduled on a large portion of the allocated narrowband (4 or 6 PRBs). Unlike in DL, in UL an increase in channel bandwidth may not enable the transmitter to allocate higher power to the transmission—the device is probably already transmitting with maximum power and a larger bandwidth may simply be a waste of bandwidth.

If a device in connected mode has data to transmit but no PUSCH resource, it can request a PUSCH resource by transmitting a SR on PUCCH as described in Section 5.2.5.5. If the device has no valid PUCCH resource either, it will use the random access procedure instead as described in Section 5.3.2.2.

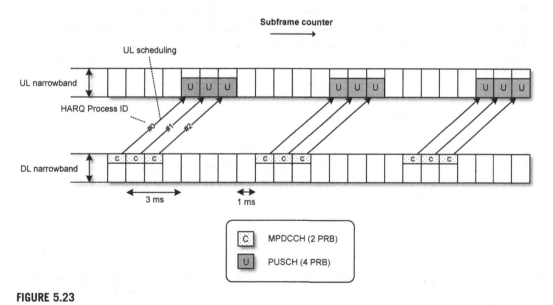

FIGURE 5.23

LTE-M uplink scheduling example without repetition.

Table 5.17 DCI Formats 6-0A and 6-0B used for scheduling PUSCH in CE modes A and B

Information	DCI Format 6-0A		DCI Format 6-0B	
	Size [bits]	Possible settings	Size [bits]	Possible settings
Format 6-0/6-1 differentiation	1	0	1	0
Frequency hopping flag	1	See Section 5.3.3.2	–	–
Resource block assignment	0–4	Narrowband index	0–4	Narrowband index
	5	0–20: Allocation of 1–6 PRBs	3	0–5: PRB index for 1 PRB
				6: 2 PRBs (#0 and #1)
		21–31: Unused in Release 13		7: 2 PRBs (#2 and #3)
MCS	4	See Table 5.8	4	See Table 5.8
Number of PUSCH repetitions	2	See Table 5.4	3	See Table 5.5
HARQ process number	3	0–7	1	0–1
New data indicator	1	Toggle bit for new data	1	Toggle bit for new data
Redundancy version	2	0–3	–	–
PUSCH power control	2	See Section 5.3.2.3	–	–
UL index	2	TDD-specific field	–	–
Downlink assignment index	2	TDD-specific field	–	–
CSI request	1	See Section 5.2.5.4	–	–
SRS request	1	See Section 5.2.5.3	–	–
Number of MPDCCH repetitions	2	See Section 5.3.3.1	2	See Section 5.3.3.1

5.3.2.1.3 Semipersistent Scheduling

Beside the dynamic scheduling described in the previous sections, LTE-M supports *semipersistent scheduling* (SPS) in CE mode A (but not in CE mode B) for downlink and uplink in a similar manner as LTE [9]. In LTE, SPS is mainly motivated by *Voice over Internet Protocol* (VoIP) services where periodic speech frames need to be scheduled and it is desired to avoid the physical control channel overhead associated with dynamic scheduling. For LTE-M devices, periodic sensor reporting could be a potential use case for SPS beside VoIP.

When SPS is configured, the device is configured by higher layers with a SPS-C-RNTI and a time interval. The SPS operation can then be activated or deactivated by a DCI addressed to the SPS-C-RNTI of the device. The activation DCI indicates what frequency resources, MCS, number of repetitions, etc. that should be used at the periodic persistent resources. The SPS-C-RNTI is also used for scheduling potential HARQ retransmissions. Note that SPS can be overridden by dynamic scheduling at any time if needed.

5.3.2.2 Random Access in Connected Mode

The device can initiate the random access procedure in connected mode when it needs to request an UL TA command and/or an UL grant. A *contention-based random access* is then performed, with similar RAR and random access message 3 transmissions as in the idle mode random access procedure described in Section 5.3.1.6. However, unlike in idle mode, message 3 does not include an RRC message, and because the device has already been assigned with a C-RNTI, which the device includes in message 3, the contention resolution in the fourth step is in this case performed using C-RNTI rather than TC-RNTI.

The base station can also order a device in CE mode A or B to initiate a random access procedure by sending a so-called *PDCCH order* to the device. This is useful during handover to another cell or when DL data transmission is resumed after a period of inactivity and the base station wants the device to reacquire UL time alignment for the expected UL responses to the DL data transmission. A modified version of DCI Format 6-1A or 6-1B is used to transmit the order. A starting PRACH CE level can be indicated in the order. As in LTE, a dedicated PRACH preamble index can be indicated already in the PDCCH order to allow *contention-free random access*, and in this case no explicit contention resolution phase is needed and the random access procedure ends already with the reception of the RAR. If no PRACH preamble index is indicated in the order, a contention-based random access is performed in the same way as for device-triggered connected mode random access.

5.3.2.3 Power Control

UL closed-loop *transmit power control* (TPC) commands for PUSCH/PUCCH can be sent to LTE-M devices in CE mode A using the TPC field in the DCI Format 6-0A/6-1A described in Sections 5.3.2.1.1 and 5.3.2.1.2, or using DCI Format 3/3A addressed with TPC-PUSCH-RNTI or TPC-PUCCH-RNTI in *Type-0 MPDCCH CSS* (see Section 5.3.3.1). A single DCI with DCI Format 3/3A can carry power control commands to multiple devices. This is similar to the power control behavior in LTE.

However, LTE-M devices in CE mode B are expected to be in bad coverage and will therefore always transmit using the configured maximum transmission power. Similarly, in the random access procedure, when a device reaches the highest PRACH CE level (PRACH CE level 3), it will always transmit at maximum power during PRACH transmission.

5.3.2.4 Mobility Support

Beside the cell selection and cell reselection mechanisms in idle mode described in Sections 5.3.1.1 and 5.3.1.3, LTE-M devices support connected mode mobility mechanisms such as handover, RRC connection release with redirection, RRC reestablishment, measurement reporting, etc., similarly as LTE devices [8].

However, low-cost LTE-M devices with reduced bandwidth support need *measurement gaps* for *intrafrequency* measurements in connected mode because they may need to retune their narrowband receiver to the center of the system bandwidth to receive the center 72 subcarriers because that is where the PSS/SSS signals are transmitted. The device may also perform *Radio Resource Management* measurements such as RSRP in the center while it has its receiver retuned. Support for *interfrequency* measurements for LTE-M devices is introduced in Release 14 (see Section 5.4).

A device in connected mode performs *Radio Link Monitoring* to determine whether it is *in sync* or *out of sync* with respect to the serving cell by comparing CRS-based measurements with thresholds known as *Qin* and *Qout* that correspond to 2% and 10% BLER of hypothetical MPDCCH transmissions, respectively [16]. The comparison is done over a period known as the *evaluation period*. If the evaluation results in out of sync for more than a certain number of times (which is a configurable parameter *N310*) over a certain period of time (upon expiry of the *T310* timer), the device declares *Radio Link Failure* and turns off its transmitter to avoid causing unwanted interference. The device may be able to find a better cell through cell selection and reestablish the connection.

5.3.3 PROCEDURES COMMON FOR IDLE AND CONNECTED MODE

5.3.3.1 MPDCCH Search Spaces

The transmission opportunities for MPDCCH are defined in the form of *search spaces*. Each device monitors an MPDCCH search space for potential DCI transmissions addressed to one of the RNTIs monitored by the device. An MPDCCH search space typically contains *blind decoding candidates* with different numbers of MPDCCH repetitions.

Release 13 supports the following MPDCCH search spaces [10]:

1. *Type-1 common search space* (Type1-CSS) is monitored by the device at its paging occasions in idle mode.
2. *Type-2 common search space* (Type2-CSS) is monitored by the device during the random access procedure in idle and connected mode.
3. *UE-specific search space* (USS) is a device-specific search space monitored by the device in connected mode. This is where scheduling of DL and UL data transmissions usually take place.
4. *Type-0 common search space* (Type0-CSS) is monitored by the device when it is configured with CE mode A in connected mode. It can be used for power control commands and as fallback if the device-specific search space fails.

An MPDCCH search space is defined by the following key parameters [13]:

1. *The MPDCCH narrowband index* indicates one of the narrowbands within the system bandwidth. The total number of narrowbands for each system bandwidth is shown in Table 5.2.
2. *The number of MPDCCH PRB pairs* is in the range {2, 4, 6} PRB pairs. For Type1-CSS and Type2-CSS, the number of MPDCCH PRB pairs is fixed to 6 PRB pairs.
3. *The MPDCCH resource block assignment* indicates the positions of the PRB pairs. If the number of PRB pairs is six (as is always the case for Type1-CSS and Type2-CSS) then this parameter is not needed because all PRB pairs within the narrowband are then included in the resource block assignment.
4. *The maximum MPDCCH repetition factor* (R_{max}) indicates the repetition factor for the candidate with the largest repetition factor in the search space. The range for this parameter is {1, 2, 4, 8, 16, 32, 64, 128, 256}.
5. *The relative MPDCCH starting subframe periodicity* (*G*) is used to determine the starting subframe periodicity for the search space. The range for this parameter is {1, 1.5, 2, 2.5, 4, 5, 8, 10} in FDD and {1, 2, 4, 5, 8, 10, 20} in TDD. The *absolute MPDCCH starting subframe periodicity* (*T*) in terms of subframes is calculated as $T = R_{max}G$.

The configuration parameters for Type1-CSS and Type2-CSS are signaled in SIB2, whereas the configuration parameters for USS and Type0-CSS are signaled in device-specific RRC signaling.

Table 5.18 MPDCCH search spaces, RNTIs, and DCI formats monitored by LTE-M devices

Mode	MPDCCH search space	RNTI	Usage	DCI format	Sections
Idle	–	SI-RNTI	Broadcast of system information	–	5.3.1.2
	Type-1 common	P-RNTI	Paging and SI update notification	6-2	5.3.1.4
	Type-2 common	RA-RNTI	Random access response	6-1A, 6-1B	5.3.1.6
		TC-RNTI	HARQ retransmission of random access message 3	6-0A, 6-0B	5.3.1.6
		TC-RNTI	Random access contention resolution with message 4	6-1A, 6-1B	5.3.1.6
Connected	UE-specific	C-RNTI	Random access order	6-1A, 6-1B	5.3.2.2
		C-RNTI	Dynamic DL scheduling	6-1A, 6-1B	5.3.2.1.1
		C-RNTI	Dynamic UL scheduling	6-0A, 6-0B	5.3.2.1.2
		SPS-C-RNTI	Semipersistent DL scheduling	6-1A	5.3.2.1.3
		SPS-C-RNTI	Semipersistent UL scheduling	6-0A	5.3.2.1.3
	Type-0 common (only in CE mode A)	C-RNTI	Random access order	6-1A	5.3.2.2
		C-RNTI	Dynamic DL scheduling	6-1A	5.3.2.1.1
		C-RNTI	Dynamic UL scheduling	6-0A	5.3.2.1.2
		SPS-C-RNTI	Semipersistent UL scheduling	6-0A	5.3.2.1.3
		TPC-PUCCH-RNTI	PUCCH power control	3, 3A	5.3.2.3
		TPC-PUSCH-RNTI	PUSCH power control	3, 3A	5.3.2.3
	Type-2 common	RA-RNTI	Random access response	6-1A, 6-1B	5.3.2.2
		TC-RNTI	HARQ retransmission of random access message 3	6-0A, 6-0B	5.3.2.2
		C-RNTI	Random access contention resolution	6-0A, 6-0B, 6-1A, 6-1B	5.3.2.2

A device in CE mode A monitors both USS and Type0-CSS, but this is facilitated in the device by the fact that these two search spaces share the same properties listed above. LTE-M devices in CE mode B or in idle mode never need to monitor more than a single search space at a time.

Further details about the search spaces are listed in Table 5.18 [10].

Within an MPDCCH search space, different candidates can have different ECCE aggregation levels and different repetition factors (R). As explained in Section 5.2.4.5, in normal subframes with normal CP length, aggregation of 2, 4, 8, 16 or 24 ECCEs is supported. With the smallest ECCE aggregation level (i.e., 2), half of the ECCEs available in a PRB pair are aggregated, and with the

Table 5.19 Repetition levels for MPDCCH search spaces

R_{max}	Repetition levels for Type1-CSS				Repetition levels for USS, Type0-CSS and Type2-CSS			
	R_1	R_2	R_3	R_4	R_1	R_2	R_3	R_4
256	2	16	64	256	32	64	128	256
128	2	16	64	128	16	32	64	128
64	2	8	32	64	8	16	32	64
32	1	4	16	32	4	8	16	32
16	1	4	8	16	2	4	8	16
8	1	2	4	8	1	2	4	8
4	1	2	4	–	1	2	4	–
2	1	2	–	–	1	2	–	–
1	1	–	–	–	1	–	–	–

highest ECCE aggregation level (i.e., 24), all the ECCEs available in a narrowband are aggregated. The available repetition levels R_1, R_2, R_3, and R_4 depend on R_{max} according to Table 5.19. For more details on search space definitions, see Section 9.1.5 in Reference [10].

Figure 5.24 shows an MPDCCH search space example with $R_{max} = 4$ and $G = 1.5$. In this search space, an MPDCCH can be scheduled without repetition ($R = 1$) in the subframes marked A, B, C, and D, or with two times repetition ($R = 2$) in the subframes marked AB and CD, or with four times

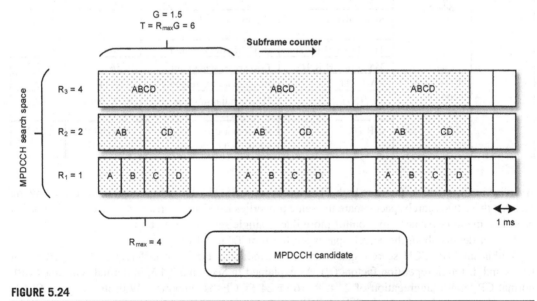

FIGURE 5.24

MPDCCH search space example for USS, Type0-CSS and Type2-CSS.

repetition ($R = 4$) in the subframes marked *ABCD*. If an MPDCCH is transmitted according to candidate *A* then candidates *AB* and *ABCD* are blocked in that *T* period but candidates *B, C, D*, and *CD* can still be used in the same *T* period. The $T - R_{max} = 2$ subframes between consecutive search spaces are not included in the search space (which means that the device can go to sleep during these subframes unless it has some other reason to stay awake during these subframes).

Figure 5.25 shows an MPDCCH search space example for Type1-CSS, the CSS used for paging. As can be seen, all candidates in the search space start in the same subframe. This allows devices in good coverage to go to sleep after they have detected that there is no transmission intended for them in the first subframe of the search space (i.e., in candidate *A*). If a search space such as the one in Figure 5.24 was used also for paging, the device would have to stay awake longer since it would have to be prepared for the eventuality that the base station choses to page the device in candidates *B, C* or *D*.

5.3.3.2 Frequency Hopping

LTE-M transmissions are restricted to a narrowband, but LTE-M provides means for frequency diversity through frequency hopping for all physical channels and signals except PSS/SSS and PBCH. As described in Sections 5.2.4.2 and 5.2.4.4, PSS/SSS and PBCH are always located at the center of the system bandwidth, similarly as in ordinary LTE.

Table 5.20 lists cell-specific configuration parameters for the time intervals and frequency offsets for frequency hopping in LTE-M. The time intervals indicate when the hops should take place in the time domain and the offsets indicate how large the hops should be in the frequency domain. The time intervals are synchronized so that the frequency hops for transmissions to/from different devices can take place at the same time. The frequency hops take place during the guard periods for frequency retuning described in Sections 5.2.4.1 and 5.2.5.1. The parameters for frequency hopping intervals for MPDCCH/PDSCH serve a double purpose since they also indicate the interval during which the device

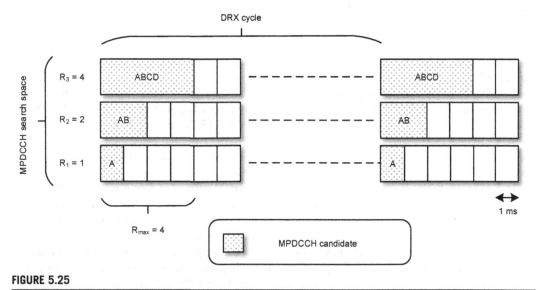

FIGURE 5.25

MPDCCH search space example for Type1-CSS.

Table 5.20 Cell-specific time intervals and frequency offsets for frequency hopping in LTE-M

Parameter	Possible values in FDD	Possible values in TDD	Signaled in
Number of narrowbands for frequency hopping for MPDCCH/PDSCH	2, 4	2, 4	SIB1
Frequency hopping interval [ms] for MPDCCH/PDSCH in CE mode A and during random access procedure for PRACH CE levels 0 and 1	1, 2, 4, 8	1, 5, 10, 20	SIB1
Frequency hopping interval [ms] for MPDCCH/PDSCH in CE mode B and during random access procedure for PRACH CE levels 2 and 3	2, 4, 8, 16	5, 10, 20, 40	SIB1
Frequency hopping interval [ms] for PUCCH/PUSCH in CE mode A and during random access procedure for PRACH CE levels 0 and 1	1, 2, 4, 8	1, 5, 10, 20	SIB2
Frequency hopping interval [ms] for PUCCH/PUSCH in CE mode B and during random access procedure for PRACH CE levels 2 and 3	2, 4, 8, 16	5, 10, 20, 40	SIB2
Frequency hopping offset for MPDCCH/PDSCH [narrowbands]	1−16	1−16	SIB1
Frequency hopping offset for PUSCH [narrowbands]	1−16	1−16	SIB2
Frequency hopping offset for PRACH [PRBs]	0−94	0−94	SIB2

can assume that the MPDCCH/PDSCH precoding remains the same (as mentioned in Sections 5.2.4.5 and 5.2.4.6).

For SIB1 and PUCCH, the frequency hopping is fixed in the LTE-M standard but for all other types of transmission it is up to the network whether to use frequency hopping or not. Table 5.21 lists the frequency hopping activation methods for different types of LTE-M transmission.

When frequency hopping is used in DL, a parameter in SIB1 (listed in Table 5.20) controls whether the hopping is between 2 and 4 narrowbands. The number of narrowbands used for frequency hopping for the PDSCH transmission that carries the SIB1 itself is fixed in the standard (as described in Table 5.21). The hopping pattern for the SIB1 transmission starts in a frame with an SFN evenly divisible by 8, in a narrowband that depends on the PCID, and hops between 2 and 4 narrowbands that have been selected so that they avoid the two center narrowbands in the system bandwidth to avoid collision with the center 72 subcarriers (the PSS/SSS/PBCH region).

When frequency hopping is used in UL, the hopping is always between two frequency locations. For PUCCH, frequency hopping is always active and the hopping occurs between two PRB locations that are symmetric with respect to the center frequency of the LTE system bandwidth. Figure 5.26

Table 5.21 Frequency hopping activation methods available in an LTE-M cell		
Type of transmission	Physical channel(s)	Frequency hopping activation method(s)
SIB1	PDSCH	The number of narrowbands for SIB1 transmission depends strictly on the DL system bandwidth signaled in MIB: For 1.4−3 MHz: no frequency hopping. For 5−10 MHz: hopping between 2 narrowbands. For 15−20 MHz: hopping between 4 narrowbands.
SI message	MPDCCH, PDSCH	Frequency hopping for SI messages and paging messages is activated by a common activation bit in SIB1.
Paging	MPDCCH, PDSCH	
Random access preamble	PRACH	Frequency hopping for PRACH is activated by an activation bit per PRACH CE level in SIB2.
Random access response and random access messages 3 and 4	MPDCCH, PDSCH, PUSCH, PUCCH	Frequency hopping for RAR, message 3, and message 4 is activated by an activation bit per PRACH CE level in SIB2.
Unicast DL data transmission	MPDCCH, PDSCH	Frequency hopping for unicast DL data transmission is activated by an activation bit in device-specific RRC signaling.
		In CE mode A, the frequency hopping can furthermore be deactivated by the frequency hopping bit in DCI Format 6-1A.
Unicast UL data transmission	MPDCCH, PUSCH	Frequency hopping for unicast UL data transmission is activated by an activation bit in device-specific RRC signaling.
		In CE mode A, the frequency hopping can furthermore be deactivated by the frequency hopping bit in DCI Format 6-0A.
HARQ-ACK, SR, CSI	PUCCH	Frequency hopping for PUCCH is always activated.

shows an example of UL frequency hopping with an UL frequency hopping interval of 2 ms, a PUCCH transmission with repetition factor 4, a PUSCH transmission with repetition factor 8, and a PUSCH frequency hopping offset of two narrowbands.

5.4 RELEASE 14 IMPROVEMENTS

The previous sections focused on describing the basic functionality specified for LTE-M in 3GPP Release 13. The work has continued in Release 14 with specification of the following features [4], which are described in this section.

1. Higher data rate support
2. Multicast transmission
3. Improved positioning
4. Voice enhancements
5. Mobility enhancements

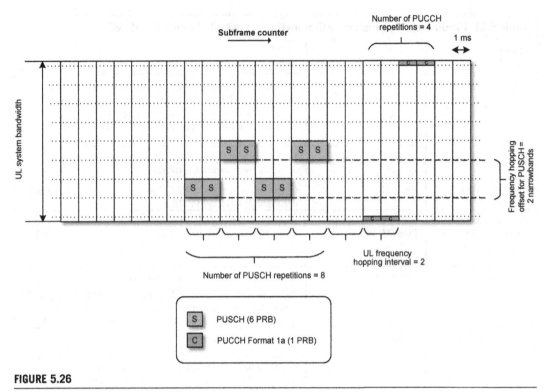

FIGURE 5.26

Uplink frequency hopping in LTE-M.

It is optional for LTE-M devices and LTE-M networks to support these new features. If a device supports a feature, it indicates its *capability* to the network, and then it is up to the network whether and how to take the capability into account when *configuring* the device.

5.4.1 HIGHER DATA RATE SUPPORT

The LTE-M device category introduced in Release 13 (Cat-M1) is suitable for MTC applications with low data rate requirements. Many utility metering applications would fall into this category. For these applications, the data rates supported by GSM/GPRS are fully adequate, and the design goal for Cat-M1 was to achieve similar device complexity and cost as an EGPRS device. Cat-M1 devices have instantaneous peak physical layer data rates in DL and UL of 1 Mbps. Taking scheduling delays into account, Cat-M1 devices supporting FD-FDD have peak data rates of 800 kbps in DL and 1 Mbps in UL, and Cat-M1 devices supporting HD-FDD have peak data rates of 300 kbps in DL and 375 kbps in UL.

However, there is a range of IoT applications with requirements on low device cost and long battery lifetime where LTE-M would be an attractive solution if the supported data rates would be a bit higher, closer to that of 3G devices or LTE Cat-1 devices. An important class of such applications is wearables such as smart watches. For this reason, the following features for higher data rate support for LTE-M devices are introduced in Release 14.

1. New device category Cat-M2
2. Wider bandwidth in CE modes
3. Larger UL TBS for Cat-M1
4. Ten DL HARQ processes in FDD
5. HARQ-ACK bundling in HD-FDD
6. Faster frequency retuning

These features are described in the following sections.

5.4.1.1 New Device Category Cat-M2

The most important data rate improvement is a new LTE-M device category, Cat-M2, with 5 MHz transmit and receive bandwidths instead of the 1.4 MHz supported in Release 13. The larger bandwidth allows data transmission in DL (on PDSCH) and UL (on PUSCH) with a maximum channel bandwidth of 24 PRBs (a so-called *wideband*) instead of just 6 PRBs (a *narrowband*).

A maximum TBS of 4008 bits in DL and 6968 bits in UL gives Cat-M2 instantaneous peak physical layer data rates of ~4 Mbps in DL and ~7 Mbps in UL. The reason for the larger maximum TBS in UL compared to DL is that it helps balance the DL and UL peak data rates in some DL-heavy TDD configurations (see e.g., UL–DL configuration #2 in Table 5.1). Increasing the UL TBS typically has a relatively small impact on the device complexity compared to increasing the DL TBS. Furthermore, the decoder complexity increase (in terms of number of *soft channel bits* that need to be stored) caused by the larger DL TBS is rather moderate thanks to the use of *Limited Buffer Rate Matching* [17].

The maximum channel bandwidth for control channels (MPDCCH, SIBs, etc.) is still 6 PRBs, because there is no strong need to increase the data rates for the control channels. This means that the implementation efforts required to upgrade existing LTE-M networks to support the higher data rates of Cat-M2 will be relatively small because most physical channels and procedures are the same as in Release 13. Note that a Cat-M2 device can operate as a Cat-M1 device in an LTE-M network that has not been upgraded because Cat-M2 is fully backward compatible with Cat-M1, and a Cat-M2 device only activates the advanced features when configured to do so by a base station.

In CE mode B, the maximum UL channel bandwidth is limited to 1.4 MHz, as shown in Table 5.22, since devices configured with CE mode B are expected to be so power limited that they cannot exploit a larger UL bandwidth.

Table 5.22 Maximum data channel bandwidths for LTE-M in Release 14

CE mode data channel bandwidth capability (MHz)	Introduced in	Associated Cat-M device	CE mode A		CE mode B	
			Downlink (MHz)	Uplink (MHz)	Downlink (MHz)	Uplink (MHz)
1.4	Release 13	Cat-M1	1.4	1.4	1.4	1.4
5	Release 14	Cat-M2	5	5	5	1.4
20	Release 14	–	20	5	20	1.4

5.4.1.2 Wider Bandwidth in CE Modes

As described in Section 5.2.3.3, the CE modes can be supported not only by the low-cost LTE-M device categories (Cat-M1, Cat-M2) but also by ordinary LTE device categories. The maximum data channel bandwidths supported by the CE modes have been increased in Release 14 as shown in Table 5.22 to increase the achievable data rates.

An ordinary LTE device can indicate support for CE mode A, or A and B. Furthermore, the device can now indicate support for a maximum data channel bandwidth (1.4, 5, 20 MHz) in CE mode. A device indicating support for a particular bandwidth must also support bandwidths smaller than the indicated bandwidth.

The base station decides what maximum data channel bandwidth to configure for a device. Typically, an ordinary LTE device would only be configured in CE mode if it needs the coverage enhancement provided by the CE modes. However, even for a device in good coverage, it may be beneficial to be configured in CE mode with a relatively small bandwidth to save power. Therefore, Release 14 introduces assistance signaling that allows the device to indicate to the base station that the device would prefer to be configured in CE mode with a particular maximum bandwidth, and then the base station may choose to take this information into account when configuring the device.

5.4.1.3 Larger Uplink TBS for Cat-M1

As already mentioned, in DL-heavy TDD configurations, support of a larger TBS in UL than in DL will help balance the DL and UL peak data rates, and it may be possible to do this without increasing the device complexity significantly.

For this reason, Release 14 introduces the possibility to support a larger maximum UL TBS of 2984 bits instead of 1000 bits when operating in CE mode A with 1.4 MHz maximum UL data channel bandwidth. The larger UL TBS is an optional device capability that can be supported in any duplex mode. If the new maximum TBS is used in FD-FDD, it gives Cat-M1 an UL peak data rate of around 3 Mbps instead of 1 Mbps.

5.4.1.4 Ten Downlink HARQ Processes in FDD

As discussed in Sections 5.3.2.1.1 and 6.3, there is a relationship between the HARQ RTT and number of HARQ processes. LTE-M has inherited the number of DL HARQ processes for FDD (eight processes) from LTE even though LTE-M has a somewhat larger RTT than LTE (10 ms for LTE-M vs. 8 ms for LTE). This means that in CE mode A in FD-FDD in Release 13, DL data can be scheduled in eight consecutive subframes, but then there will be a 2-ms gap before the DL data transmission can continue.

Therefore, Release 14 introduces the possibility to support up to 10 DL HARQ processes in FDD. FD-FDD devices supporting 10 DL processes will be able to receive the maximum DL TBS in every subframe. This can be done without the need to increase the number of soft channel bits that need to be stored in the decoder in the device (because the maximum number of processes is only expected to be utilized under good channel conditions when relatively high code rate is used), and it increases the DL peak data rate for Cat-M1 from 800 kbps to 1 Mbps in FD-FDD.

5.4.1.5 HARQ-ACK Bundling in HD-FDD

In Release 13, every DL data transmission is associated with a HARQ-ACK feedback transmission in UL, as described in Section 5.2.5.5. Because the LTE-M UL in Release 13 does not support transmission of more than one HARQ-ACK feedback per subframe, a device in HD-FDD operation will (in the nonrepetition case) spend as long transmitting HARQ-ACK feedback in UL as it spends on the actual DL data transmission. As can be seen in Figure 5.21, an HD-FDD device will not be able to spend more than 3 of 10 subframes on actual DL data transmission.

Release 14 improves the situation by introducing the possibility to support HARQ-ACK bundling in HD-FDD. Instead of just transmitting three consecutive DL data subframes, a device supporting HARQ-ACK bundling will be able to transmit up to three consecutive so-called *HARQ-ACK bundles* where each bundle contains a number of DL data subframes. If the device supports 10 DL HARQ processes (described in previous section), then the 3 bundles can contain in total up to 10 consecutive DL data subframes. The bundles are sent back-to-back and then the corresponding HARQ-ACK feedbacks are sent back-to-back. There will only be a single HARQ-ACK feedback per bundle—the device sends ACK if all DL TBs within a bundle were successfully decoded, otherwise it sends NACK (Negative Acknowledgment) which means that the whole bundle will be retransmitted by the base station. This means that use of HARQ-ACK bundling is mainly useful for device that experience rather good channel conditions (which is often the case for many devices in a typical cell).

The result is that a Cat-M1 device supporting HARQ-ACK bundling and 10 DL processes can receive data in 10 of 17 subframes in HD-FDD and thereby almost double its DL peak rate from 300 to 588 kbps.

5.4.1.6 *Faster Frequency Retuning*

Due to the reduced radio frequency bandwidth of low-cost LTE-M devices, guard periods for frequency retuning are needed, as described in Sections 5.2.4.1 and 5.2.5.1. In Release 13, the device creates a guard period of two OFDM/SC-FDMA symbols. In Release 14, it is possible for the device to indicate that it can do faster frequency retuning so that the guard period can be smaller than two symbols. The device can indicate that it needs one symbol or even zero symbols—the latter value is mainly intended for ordinary LTE devices, which may have no need to do retuning to move between different narrowbands when operating in CE mode (because ordinary LTE devices can receive and transmit the full LTE system bandwidth rather than just a narrowband or wideband). Faster retuning means somewhat less truncation of the transmitted signal and therefore somewhat better link performance.

5.4.2 MULTICAST TRANSMISSION

Support for multicast transmission is seen as beneficial for some MTC applications where there is, for example, a need for efficient distribution of software/firmware upgrades to a large number of devices.

Due to the inherent narrowband property of LTE-M, it does not support LTE's MBSFN functionality because that is based on a channel bandwidth equal to the full LTE system bandwidth.

However, Release 14 does introduce support for multicast transmission based on the *Multimedia Broadcast Multicast Service* framework in the form of *Single-Cell Point-to-Multipoint* (SC-PTM) transmission. The solution is based on the SC-PTM feature introduced for ordinary LTE devices in Release 13. SC-PTM support was introduced in Release 14 for both LTE-M and NB-IoT, and the standardization work was partly carried out as a joint effort because the objectives were the same (to

Table 5.23 MPDCCH search spaces, RNTIs, and DCI formats for SC-PTM for LTE-M

Mode	MPDCCH search space	RNTI	Usage	DCI format
Idle	–	SI-RNTI	Broadcast of system information	–
	Type-1A common	SC-RNTI	Scheduling of SC-MCCH	6-2
	Type-2A common	G-RNTI	Scheduling of SC-MTCH	6-1A, 6-1B

extend Release 13 SC-PTM to support narrowband operation and coverage enhancement), so the SC-PTM solutions are very similar, although they use different physical channels in the physical layer.

SC-PTM for LTE-M (and NB-IoT) is only supported in idle mode. The new SIB20 can contain scheduling information for one Single Cell Multicast Control Channel (SC-MCCH) per cell, and SC-MCCH can contain scheduling information for one Single Cell Multicast Traffic Channel (SC-MTCH) per multicast service. The maximum channel bandwidth of an SC-MTCH can be either 1.4 or 5 MHz, with a maximum TBS of 1000 and 4008 bits, respectively (matching the capabilities of Cat-M1 and Cat-M2). Both SC-MCCH and SC-MTCH are transmitted on PDSCH transmissions scheduled by MPDCCH. Both MPDCCH and PDSCH can be repeated to achieve the required coverage. Table 5.23 lists the MPDCCH search spaces, RNTIs, and DCI formats used for SC-PTM.

5.4.3 IMPROVED POSITIONING

In principle, LTE-M already in Release 13 inherits the LTE positioning techniques *Enhanced Cell Identity* (E-CID) and OTDOA beside the basic CID based positioning. However, measurement performance requirements for E-CID and OTDOA are not introduced for LTE-M devices until Release 14.

Furthermore, Release 14 introduces OTDOA enhancements with respect to the PRS configurations the in time and frequency domains. Because the low-cost LTE-M devices have limited receive bandwidth (1.4 MHz for Cat-M1, 5 MHz for Cat-M2), they will benefit from a PRS that is mapped over a longer duration in the time domain rather than over a wide bandwidth in the frequency domain. Therefore, Release 14 introduces the possibility to configure PRS that are transmitted with a longer duration at every PRS occasion and/or at more frequent PRS occasions. Furthermore, optional PRS frequency hopping is introduced to provide frequency diversity gains also to narrowband LTE-M devices. The new PRS configurations in Release 14 allow LTE-M devices to achieve similar positioning accuracy as ordinary LTE devices. The exact PRS configuration in a cell will depend on the desired trade-off between PRS overhead and positioning accuracy.

5.4.4 VOICE ENHANCEMENTS

LTE-M already supports *Voice over LTE* (VoLTE) in CE mode A in Release 13. However, LTE-M in Release 13 was optimized for delay-tolerant MTC applications rather than for delay sensitive real-time applications such as VoLTE. For delay sensitive applications, any repetitions will need to be transmitted within the delay budget of the application. Release 14 introduces the following features which are intended to improve the VoLTE coverage especially for HD-FDD and TDD. However, there is nothing preventing the network from using these features also for other applications than VoLTE.

- New PUSCH repetition factors
- Modulation scheme restriction
- Dynamic HARQ-ACK delays
- SRS coverage enhancement

These features are described in the following sections.

5.4.4.1 New PUSCH Repetition Factors

During a VoLTE call, a *speech frame* is produced every 20 ms. If no *speech frame bundling* is applied, it needs to be possible to transmit a transport block each direction (UL and DL) every 20 ms. If speech frame bundling is applied, each transport block can carry multiple (e.g., 2 or 3) speech frames, meaning that the TBs are larger but less frequent (e.g., every 40 or 60 ms).

The LTE-M UL data channel (PUSCH) supports repetition factors which are powers of two, as can be seen in Table 5.4. The set of the available repetition factors in Release 13 is not considered optimal with respect to the VoLTE traffic pattern. Therefore, Release 14 introduces the possibility for a device to support a new range of PUSCH repetition factors. The new range is {1, 2, 4, 8, 12, 16, 24, 32}, where 12 and 24 are new values compared to Release 13. Furthermore, the whole range is immediately available through a 3-bit field in the DCI, i.e., the interpretation of the DCI field for number of PUSCH repetitions no longer depends on any higher layer parameter (the one referred to as *Broadcasted maximum number of PUSCH repetitions for CE mode A* in Table 5.4). This feature is intended to facilitate efficient scheduling of VoLTE traffic especially in HD-FDD and thereby improve VoLTE coverage.

5.4.4.2 Modulation Scheme Restriction

It has been observed that the standardized combinations of TBS and modulation scheme (shown in Tables 5.3 and 5.8) are not always optimal when repetition as applied. The link performance can sometimes be better if the modulation scheme is restricted to QPSK when 16QAM should be selected according to the tables. In UL, QPSK may also have an additional benefit in terms of lower *Peak-to-Average Power Ratio* compared to 16QAM, allowing for higher transmit power.

Therefore, Release 14 introduces the possibility to restrict the modulation scheme to QPSK. When a device is configured with the feature, PDSCH transmission will be limited to QPSK whenever PDSCH is scheduled with repetition, and PUSCH transmission will be limited to QPSK when a new 1-bit field in the DCI indicates that QPSK should be used instead of the default modulation scheme for the indicated MCS.

Support of modulation scheme restriction and support of the new PUSCH repetition factors mentioned above are bundled into a single capability, meaning that a device either supports both of them or neither of them.

5.4.4.3 Dynamic HARQ-ACK Delays

In LTE-M in Release 13, the delay between the end of a PDSCH transmission and the beginning of the associated HARQ-ACK feedback on PUCCH or PUSCH is fixed to 4 ms as shown in Figures 5.21 and 5.22. With a less rigid timing relationship, efficient scheduling of VoLTE transmissions is facilitated, especially in HD-FDD.

Therefore, Release 14 introduces a possibility for the base station to control this delay dynamically through a new 3-bit field in the DCI. This feature can be seen as a subset of the functionality introduced by the HARQ-ACK bundling feature described in Section 5.4.1.5. The interpretation of the 3-bit DCI field depends on the RRC configuration and it is either {4, 5, 6, 7, 8, 9, 10, 11} subframes or {4, 5, 7, 9, 11, 13, 15, 17} subframes, where the former range is intended for VoLTE scheduling with up to 4 MPDCCH repetitions (and for HARQ-ACK bundling) and the latter range is intended for VoLTE scheduling with more than 4 MPDCCH repetitions.

5.4.4.4 SRS Coverage Enhancement in TDD

All LTE-M physical channels support CE through repetition in Release 13 but the SRS does not. Release 14 introduces the possibility to use the UL part of the *special subframe* in TDD (described in Section 5.2.3.1) for transmission of SRS symbol repetitions. This can be used to improve the link adaptation in both UL and DL by exploiting the UL−DL channel reciprocity in TDD.

5.4.5 MOBILITY ENHANCEMENTS

LTE-M supports intrafrequency RSRP measurements in connected mode in Release 13. Release 14 introduces complete mobility support in connected mode in the form of intrafrequency RSRQ measurements and interfrequency RSRP/RSRQ measurements. These improvements are welcome for the more mobile and real-time use cases for LTE-M devices mentioned earlier, such as wearables and VoLTE.

REFERENCES

[1] Third Generation Partnership Project, Technical Report 36.888 v12.0.0, Study on Provision of Low-Cost Machine-Type Communications (MTC) User Equipments (UEs) Based on LTE, 2013.

[2] Vodafone Group, et al., RP-140522 Revised Work Item on Low Cost & Enhanced Coverage MTC UE for LTE, 3GPP RAN Meeting #63, 2014.

[3] Ericsson, et al., RP-141660 Further LTE Physical Layer Enhancements for MTC, 3GPP RAN Meeting #65, 2014.

[4] Ericsson, et al., RP-161321 New WI Proposal on Further Enhanced MTC, 3GPP RAN Meeting #72, 2016.

[5] Third Generation Partnership Project, Technical Specification 36.211 v14.0.0, Evolved Universal Terrestrial Radio Access (E-UTRA); Physical Channels and Modulation, 2016.

[6] Third Generation Partnership Project, Technical Specification 36.306 v14.0.0, Evolved Universal Terrestrial Radio Access (E-UTRA); User Equipment (UE) Radio Access Capabilities, 2016.

[7] Third Generation Partnership Project, Technical Specification 36.101 v14.0.0, Evolved Universal Terrestrial Radio Access (E-UTRA); User Equipment (UE) Radio Transmission and Reception, 2016.

[8] Third Generation Partnership Project, Technical Specification 36.300 v14.0.0, Evolved Universal Terrestrial Radio Access (E-UTRA) and Evolved Universal Terrestrial Radio Access Network (E-UTRAN); Overall Description; Stage 2, 2016.

[9] Third Generation Partnership Project, Technical Specification 36.321 v14.0.0, Evolved Universal Terrestrial Radio Access (E-UTRA); Medium Access Control (MAC) Protocol Specification, 2016.

[10] Third Generation Partnership Project, Technical Specification 36.213 v14.0.0, Evolved Universal Terrestrial Radio Access (E-UTRA); Physical Layer Procedures, 2016.

[11] Third Generation Partnership Project, Technical Specification 36.214 v14.0.0, Evolved Universal Terrestrial Radio Access (E-UTRA); Physical Layer; Measurements, 2016.

[12] Third Generation Partnership Project, Technical Specification 36.212 v14.0.0, Evolved Universal Terrestrial Radio Access (E-UTRA); Multiplexing and Channel Coding, 2016.

[13] Third Generation Partnership Project, Technical Specification 36.331 v14.0.0, Evolved Universal Terrestrial Radio Access (E-UTRA); Radio Resource Control (RRC); Protocol Specification, 2016.

[14] Third Generation Partnership Project, Technical Specification 36.304 v14.0.0, Evolved Universal Terrestrial Radio Access (E-UTRA); User Equipment (UE) Procedures in Idle Mode, 2016.

[15] Third Generation Partnership Project, Technical Specification 36.413 v14.0.0, Evolved Universal Terrestrial Radio Access Network (E-UTRAN); S1 Application Protocol (S1AP), 2016.

[16] Third Generation Partnership Project, Technical Specification 36.133 v14.0.0, Evolved Universal Terrestrial Radio Access (E-UTRA); Requirements for Support of Radio Resource Management, 2016.

[17] Intel Corporation, R1-1611936 Soft Buffer Requirement for feMTC UEs with Larger Max TBS, 3GPP RAN1 Meeting #87, 2016.

CHAPTER

LTE-M PERFORMANCE

6

CHAPTER OUTLINE

Abstract

This chapter presents Long-Term Evolution for Machine-Type Communications (LTE-M) performance in terms of coverage, data rate, latency, and system capacity based on the 3GPP Release 13 functionality described in Chapter 5. The methods and assumptions are to a large extent following those used for *Extended Coverage Global System for Mobile Communications Internet of Things* (EC-GSM-IoT) and *Narrowband-IoT* (NB-IoT) performance evaluations as described in Chapters 4 and 8. When there are differences, those are explicitly described in this chapter. The reduction in device complexity achieved by LTE-M compared to earlier 3GPP releases of LTE is also presented. While LTE-M has been specified for both half-duplex frequency-division duplexing (HD-FDD) and *full duplex FDD* (FD-FDD) operation as well as *time-division duplexing* (TDD) operation, this chapter focuses on the performance achievable for LTE-M in HD-FDD operation.

6.1 PERFORMANCE OBJECTIVES

The work on LTE-M started in the 3GPP *Technical Specification Group* (TSG) *Radio Access Network* (RAN) with the *Study on Provision of low-cost Machine-Type Communications (MTC) User Equipments (UEs) based on LTE* [1], referred to as the *LTE-M study item* in the following. The initial objective of the LTE-M study item was to find solutions providing reduced device complexity. Later a second objective was added to find solutions providing substantial coverage enhancement (CE). Some of the identified complexity reduction techniques were specified in Release 12 in the form of a new LTE device category (Cat-0). Release 13 enables further device cost reduction in the form of yet a new device category (Cat-M1) and specifies two CE modes targeting at least 15 dB CE compared to normal LTE coverage.

This work is briefly introduced in Section 2.2.4, where the LTE reference coverage is shown to correspond to 140.7 dB Maximum Coupling Loss (MCL) assuming base station and device noise figures of 5 and 9 dB, respectively.

In Release 13 3GPP TSG GERAN also carried out the study on *Cellular System Support for Ultra-low Complexity and Low Throughput Internet of Things* [2], in the following referred to as the *Cellular IoT study item* where the requirements went beyond reducing device cost and improving coverage. Additional objectives in terms of throughput, latency, battery life, and system capacity were also defined. These are the same requirements as targeted by EC-GSM-IoT and NB-IoT. Sections 1.2.3 and 2.2.5 in detail introduce these requirements. The following sections will go through the expected LTE-M HD-FDD performance for each of these performance objectives.

6.2 **COVERAGE**

LTE-M was initially intended to meet a coverage target of 155.7 dB MCL assuming noise figures of 5 dB in the base station and 9 dB in the device and a device power class of 20 dBm. EC-GSM-IoT and NB-IoT were on the other hand from the start designed to fulfill the Cellular IoT study item objective of 164 dB MCL assuming noise figures of 3 dB in the base station and 5 dB in the device. Assuming the same noise figures for the different technologies, the EC-GSM-IoT and NB-IoT design objective appears to be 6.3 dB more stringent in the uplink and 4.3 dB more stringent in the downlink compared to the LTE-M design objective. If a device power class of 23 dBm is assumed for LTE-M, the difference in the uplink shrinks to 3.3 dB.

LTE-M supports sufficiently many repetitions of the physical channels in downlink and uplink to ensure that the initial LTE-M coverage target is fulfilled with sufficient margin to further reduce the gap to the Cellular IoT coverage requirement. In Chapter 8, NB-IoT in-band performance is studied for a 10 MHz LTE carrier using 46 dBm output power combined with 6 dB power boosting, which results in a downlink power of 35 dBm per NB-IoT *Physical Resource Block* (PRB). If the same configuration is considered for LTE-M, then each LTE-M narrowband would be, without power boosting, transmitted using 36.8 dBm. To improve the LTE-M *Primary Synchronization Signal/Secondary Synchronization Signal* (PSS/SSS) detection and *Master Information Block* (MIB) acquisition, 3 dB power boosting may be assumed to be used during these transmissions. This gives a downlink power of 39.8 dBm which is close to 5 dB higher than what is assumed for NB-IoT in-band mode of operation. Finally, if it is assumed that, just as in the EC-GSM-IoT case, robust operation can be achieved for a targeted BLock Error Rate (BLER) of 10% on all physical channels, including the *Physical Random Access Channel* (PRACH), *Physical Uplink Control Channel* (PUCCH), and *MTC Physical Downlink Control Channel* (MPDCCH), then LTE-M completely closes the gap to the Cellular IoT coverage requirement and achieves a MCL of 164 dB. Table 6.1 presents the MCL calculations in detail.

Table 6.1 also summarizes the physical layer data rates achievable for LTE-M at the MCL of 164 dB. The downlink data rate of ~0.8 kbps is calculated over a scheduling cycle for which it is assumed that 64 repetitions is used on the MPDCCH scheduling the Physical Downlink Shared Channel (PDSCH), which is transmitted over 1024 repetitions. It is furthermore assumed that the device uses 32 repetitions when transmitting the PUCCH acknowledging the reception of the downlink transport block. The uplink data rate of 167 bps is calculated over a scheduling cycle for which it is assumed that 64 repetitions is used on the MPDCCH scheduling the PUSCH, which is transmitted over 2048 repetitions. The target set by the Cellular IoT study item is to provide a bit rate of 160 bps, which LTE-M thus (narrowly) achieves in the uplink and (with margin) in the downlink. For the MIB, SIB1-BR and

Table 6.1 LTE-M HD-FDD MCL [3,4]

#	Physical Channel Name	PUCCH	PRACH	PUSCH	PDSCH	MPDCCH	PBCH, MIB	PDSCH, SIB1-BR	PSS/SSS
1	BLER target [%]	10%	10%	10%	6%	10%	90th perc.	90th perc.	90th perc.
2	TBS [bits]	–	–	392	936	–	–	152	–
3	Repetitions	32	128	2048	1024	64	–	16	3
4	Data rate [bps], acquisition time [ms]	–	–	167 bps	0.8 kbps	–	640 ms	640 ms	460 ms
Transmitter									
5	Total Tx power [dBm]	23	23	23	46	46	46	46	46
6	Power boosting [dB]	–	–	–	–	–	3	–	3
7	Actual Tx power [dBm]	23	23	23	36.8	36.8	39.8	36.8	39.8
Receiver									
8	Thermal noise [dBm/Hz]	−174	−174	−174	−174	−174	−174	−174	−174
9	Receiver noise figure [dB]	3	3	3	5	5	5	5	5
10	Interference margin [dB]	0	0	0	0	0	0	0	0
11	Channel bandwidth [kHz]	180	1080	180	1080	1080	1080	1080	1080
12	Effective noise power [dBm] = (8) + (9) + (10) + $10 \log_{10}(11)$	−118.5	−110.7	−118.5	−108.7	−108.7	−108.7	−108.7	−108.7
13	Required SINR (dB)	−24	−31.2	−23.6	−18.5	−18.5	−15.5	−18.5	−16.2
14	Dual antenna receiver sensitivity [dBm] = (12) + (13)	−142.5	−141.9	−142.1	−127.2	−127.2	−124.2	−123.7	−124.9
15	MCL [dB] = (7) − (14)	165.5	164.9	165.1	164	164	164	164	164.7

PSS/SSS the acquisition times are presented in Table 6.1 to quantity the achieved performance. These results are used in Sections 6.4 and 6.5 to estimate LTE-M latency and battery life.

It should be noted that there are some differences in the evaluation assumptions used when deriving these results and the assumptions used in Chapters 4 and 8 for the EC-GSM-IoT and NB-IoT coverage evaluations, respectively. Table 6.2 summarizes some of the more important simulation assumptions. The initial PSS/SSS synchronization acquisition was evaluated assuming 1 kHz initial frequency offset. This is significantly lower than the assumption of 18 kHz initial frequency offset assumed for devices with 20 ppm frequency inaccuracy discussed in Chapters 4 and 8. Aligning this assumption to the frequency offset assumed for EC-GSM-IoT and NB-IoT may likely lead to an increased PSS/SSS detection time. In addition to the initial synchronization, a fairly modest frequency error of 25−30 Hz was assumed for all channels except for the Physical Broadcast Channel (PBCH) and PDSCH System Information Block 1 Bandwidth-Reduced (SIB1-BR) transmissions. For the PBCH and PDSCH SIB1-BR simulations a 50 Hz frequency error was modeled. Extended Pedestrian A (EPA) propagation conditions were used in the PDSCH and MPDCCH evaluations, whereas the Extended Typical Urban model (ETU) was used for all other channels. The difference between EPA and ETU is that EPA is supposed to model low delay spread environments, whereas ETU models high delay spread conditions [5]. The fading is in all case assumed to follow a Rayleigh distribution with a Doppler shift of 1 Hz.

Table 6.2 Assumptions made in the evaluations of LTE-M MCL

Parameter	Value
Frequency band	2 GHz
Propagation condition	PUCCH, PRACH, PUSCH, PBCH, PSS/SSS: ETU PDSCH, MPDCCH: EPA
Fading	Rayleigh, 1 Hz
Frequency error	PSS/SSS: 1 kHz PDSCH, MPDCCH, PUCCH, PRACH, PUSCH: 25 or 30 Hz PBCH, PDSCH SIB1-BR: 50 Hz
Device NF	5 dB
Device antenna configuration	One transmit antenna and one receive antenna
Device power class	23 dBm
Base station NF	3 dB
Base station antenna configuration	Two transmit antennas and two receive antennas
Base station power level	PSS, SSS, PBCH: 39.8 dBm per narrowband MPDCCH, PDSCH: 36.8 dBm per narrowband
Frequency hopping (FH)	PSS, SSS, PBCH: N/A MPDCCH, PDSCH, PUSCH, PUCCH, PRACH: FH enabled
Resource allocation	PSS, SSS, PBCH: N/A PDSCH, MPDCCH: 6 PRBs PUSCH, PUCCH: 1 PRB PRACH: 6 PRBs

Detailed link level performance for the different LTE-M physical channels and signals is provided in References [6,7].

6.3 DATA RATE

In this section the achievable data rates for the LTE-M device category Cat-M1 is examined in various scenarios. The maximum downlink physical layer data rate is achieved for LTE-M when three Hybrid Automatic Repeat Request (HARQ) processes are scheduled as shown in Figure 6.1. Although LTE-M supports up to eight HARQ processes the timing restrictions of the technique is such that, for HD-FDD three HARQ processes gives the maximum PDSCH data rate. In this example, the MPDCCH carrying the Downlink Control Information is mapped to 2 PRBs and schedules the PDSCH, containing the maximum 1000-bit transport block for Cat-M1, over 4 PRBs. Figure 6.1 illustrates an MPDCCH-to-PDSCH scheduling gap of 1 ms, a downlink-to-uplink switching gap of 1 ms, and a PDSCH-to-PUCCH gap of 3 ms. The PUCCH is transmitted on a single PRB location that is frequency hopping across the system bandwidth. This configuration gives a peak physical layer throughput of 300 kbps.

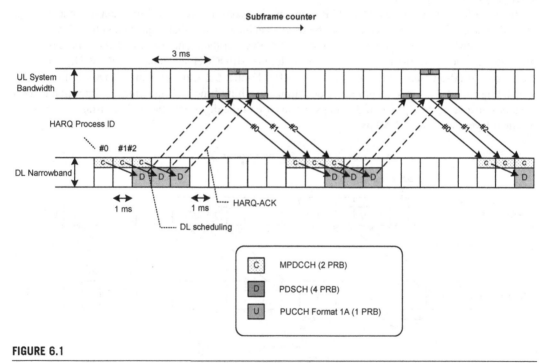

FIGURE 6.1

LTE-M PDSCH scheduling cycle.

Table 6.3 LTE-M HD-FDD PDSCH physical layer data rates for a 23-dBm device

Scenario	164 dB MCL	154 dB CL	144 dB CL	Peak	Instantaneous Peak
Data rate	0.8 kbps	9.9 kbps	76.6 kbps	300 kbps	1 Mbps

If only considering the instantaneous data rate, 1 Mbps can be achieved during the subframes where the transmission of the 1000-bit transport block takes place.

An achievable PDSCH data rate of 0.8 kbps at the MCL of 164 dB is presented in Section 6.2. At 154 dB, 64 PDSCH repetitions are sufficient to transmit a 936-bit transport block size (TBS) with 10% BLER, whereas 8 MPDCCH repetitions are sufficient to support the downlink scheduling. Using eight PUCCH repetitions a robust uplink Ack/Nack control signaling is supported. With these configurations a PDSCH physical layer data rate of 9.9 kbps is reached for a single HARQ process as shown in Table 6.3.

At 144 dB, using 4 PDSCH repetitions, one MPDCCH transmission and one PUCCH transmission to support data transmission with 10% BLER, a physical layer data rate of 76.6 kbps is reached for a single PDSCH HARQ process.

In all these cases, use of a single HARQ process is a conservative scheduling approach and use of more than a single HARQ processes can be used to achieve somewhat higher downlink data rates.

The maximum uplink data rate is achieved when three of the eight available HARQ processes are scheduled as shown in Figure 6.2. In this example, the MPDCCH schedules the PUSCH, containing the largest 1000-bit transport block, over 4 PRBs. Figure 6.2 illustrates an MPDCCH-to-PUSCH scheduling gap of 3 ms, and an uplink-to-downlink switching time of 1 ms. This configuration gives a peak physical layer throughput of 375 kbps. As in case of the downlink, the uplink instantaneous peak physical layer data rate is 1 Mbps.

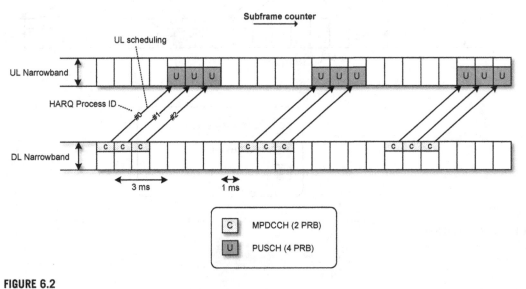

FIGURE 6.2

LTE-M PUSCH scheduling cycle.

Table 6.4 LTE-M HD-FDD physical layer data rates for a 23-dBm device

	164 dB MCL	154 dB CL	144 dB CL	Peak	Instantaneous Peak
Uplink	167 bps	3.1 kbps	40.1 kbps	375 kbps	1 Mbps

An achievable PUSCH data rate of 167 bps at the MCL of 164 dB is presented in Section 6.1. At 154 dB, if using 256 PUSCH repetitions to transmit a 936-bit TBS with 10% BLER, and using 8 MPDCCH repetitions to support a robust control signaling for the uplink scheduling, then, a physical layer data rate of 3.1 kbps is reached for a single HARQ process as shown in Table 6.4.

At 144 dB, using 16 PUSCH repetitions and 1 MPDCCH transmission, to support a data transmission with 10% BLER, a physical layer data rate of 40.1 kbps is reached for a single HARQ process.

An overview of the transmission configurations assumed in this section for deriving the data rates at the different coupling losses is provided in Reference [4].

Remember that the data rate estimates in this section concern Cat-M1 devices that only support HD-FDD operation. Higher data rates can be expected to be achieved by the following types of LTE-M devices:

- Cat-M1 devices supporting FD-FDD operation
- Higher LTE device categories (Cat-0, Cat-1, etc.) operating in CE mode A
- LTE-M devices supporting the Release 14 higher data rate features described in Section 5.4.1

6.4 LATENCY

In this section the latency for the transmission of a small uplink report is examined for the device category Cat-M1. Table 6.5 presents the content of the uplink report including the application data size and the overheads at application, transport and radio layers. The protocol overhead from higher layers is considered to be the same for LTE-M as for NB-IoT in Section 8.4.

Table 6.5 Latency evaluation packet definitions including application, security, transport, Internet Protocol (IP), and LTE-M protocol overheads

Type	Size [Byte]
Application data	20
CoAP	4
DTLS (Datagram Transport Layer Security)	13
UDP (User Datagram Protocol)	8
IP	40
PDCP (Packet Data Convergence Protocol)	5
RLC (Radio Link Control)	2
MAC	2
Total	94

The message flow during the transmission of the report is depicted in Figure 6.3. To optimize the latency it is assumed that the RRC Resume procedure is used to establish the connection to the cell. The latency is here defined from the time the cell synchronization is initiated until the PUSCH transmission delivers the uplink report. Also the overall message flow is similar for NB-IoT, shown in Figure 8.5, and LTE-M with one important difference: While an NB-IoT device can start the random access procedure after acquiring the MIB-NB, an LTE-M device needs to first read the MIB and then the SIB1-BR to acquire, for example, the barring status which is indicated by the presence of the SIB14 scheduling information in SIB1-BR. In addition, the actual message sizes differ somewhat between the two technologies due to minor differences in the NB-IoT and LTE-M RRC and MAC protocol stack designs.

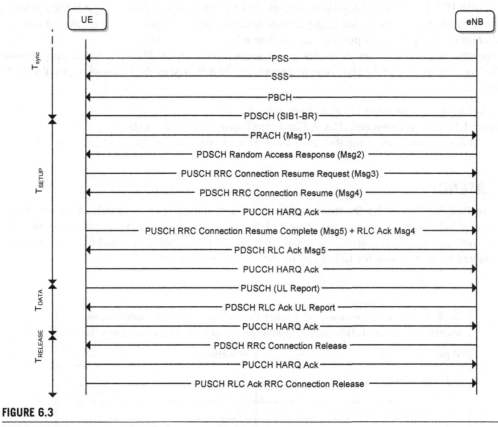

FIGURE 6.3

LTE-M RRC Resume procedure for latency evaluations.

The simulation assumptions presented in Table 6.2 apply to a large extent also to the simulations performed in the evaluation of the message latency. It is seen in Table 6.6 that the 94-byte uplink report for these simulation assumptions can be delivered with a latency between 0.2 and 8.5 seconds at the investigated coupling loss levels. At the MCL of 164 dB the latency is dominated by the PUSCH transmission of the report, which takes more than 4 seconds. The reason is the limited data rate of 167 bps offered at this deep coverage level. At the 144 dB coupling loss it is on the other hand the acquisition of the PSS, SSS, MIB, and SIB1-BR that limits the latency, as these initial steps are assumed to take 160 ms of the reported 200 ms latency.

Table 6.6 LTE-M exception report latency using the RRC resume procedure [4]	
Coupling Loss [dB]	**Latency [s]**
144	0.2
154	0.6
164	8.5

6.5 BATTERY LIFE

The battery life of LTE-M Cat-M1 is evaluated in this section. The evaluation follows the method used when evaluating EC-GSM-IoT and NB-IoT battery life in Chapters 4 and 8, respectively. The traffic model assumed for this investigation is based on the reporting intervals and packet sizes presented in Table 6.7, while the device power consumption levels used in the evaluations are presented in Table 6.8.

Table 6.7 Packet sizes on top of the PDCP layer for evaluation of battery life [4]			
Message Type	**UL Report**		**DL Application Acknowledgment**
Size	200 bytes	50 bytes	65 bytes
Arrival rate	Once every 2 h or once every day		

Table 6.8 NB-IoT power consumption [4]			
Tx, 23 dBm	**Rx**	**Light Sleep**	**Deep Sleep**
500 mW	80 mW	3 mW	0.015 mW

It should be noted that the power levels are reused from the NB-IoT evaluations and that reusing the 500 mW transmit power consumption level is optimistic because the NB-IoT uplink, in contrast to LTE-M, is designed with a close to constant envelope modulation to facilitate power amplifier efficiency as high as 45%–50% [2]. The RRC Resume procedure is, as in the latency evaluations, assumed for the connection setup procedure. As a result, the complete packet flow used in these evaluations and shown in Figure 6.4 closely follows the NB-IoT procedure depicted in Figure 8.6. Not depicted are MPDCCH transmissions, a 1 second period of light sleep in-between the end of the uplink report and the start of the downlink data transmission, and finally a 20 seconds period of light sleep after the connection release during which the device is available for mobile terminated traffic. As explained in Section 6.4, an LTE-M device needs to acquire the barring status on the SIB1-BR, which is not required for NB-IoT. Between the mobile originated events triggered every 2 or 24 hours, the device is assumed to use Power Saving Mode to optimize its power consumption.

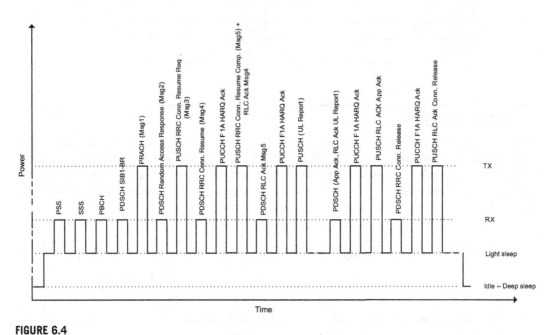

FIGURE 6.4

LTE-M RRC Resume procedure for battery life evaluations.

As for EC-GSM-IoT and NB-IoT, an ideal 5-Wh battery power source was assumed in the evaluations with no power leakage or other imperfections. Under these assumptions, the results presented in Table 6.9 are achievable. The general conclusion is that a 10 year battery life is close to feasible for a reporting interval of 24 hours. It is also clear that the 2 hour reporting interval is a too aggressive target when the device is at 164 dB MCL. Just as in the latency evaluations it is the limited PUSCH data rate that has the largest impact on the performance achievable at 164 dB MCL. At 144 dB coupling loss and 24-hours reporting interval, it is on the other hand the deep sleep power consumption of 0.015 mW that dominates the power consumption. The energy cost due to reception is in all cases small, relative to the cost of transmitting.

Table 6.9 LTE-M battery life [4]

Reporting Interval [Hours]	DL Packet Size [Bytes]	UL Packet Size [Bytes]	Battery Life [Years]		
			144 dB MCL	154 dB MCL	164 dB MCL
2	65	50	23.7	13.9	2
		200	22.3	8.7	0.9
24		50	36.5	33.4	15.5
		200	36.2	29.9	8.8

6.6 CAPACITY

3GPP has since its Release 13 set high requirements in terms of system capacity on the systems developed to support the Cellular Internet of things. Table 6.10 presents the 3GPP Release 13 and 14 scenarios for evaluating system capacity including the targeted load in terms of served number of devices per square kilometer. The Release 14 requirement corresponds to the requirement set on candidate technologies to qualify as a 5G system.

Table 6.10 3GPP assumption on required system capacity [8]

Assumed Scenario	Devices/km^2	Devices/cell
3GPP Release 13 assuming 40 devices per home in central London	60,680	52,547
3GPP Release 14 assuming large crowds in small areas such as a sport stadium	1,000,000	865,970

In this section it is shown that LTE-M meets these capacity requirements for a simulation setup closely following the assumptions described in Sections 4.6 and 8.6 for EC-GSM-IoT and NB-IoT. An important subset of the settings assumed in the LTE-M system level evaluations are captured in Table 6.11. Just as for NB-IoT, evaluated in Section 8.6, it is assumed that the base station is configured with 46 dBm power, which is equally divided across the 50 PRBs in the 10-MHz LTE system bandwidth. The studied LTE-M downlink narrowbands were assumed to be outside of the center 72 subcarriers, meaning that the narrowbands did not carry any load from mandatory PSS, SSS, and PBCH transmissions in the downlink. Similarly, the LTE-M uplink narrowbands in the simulation were assumed to not carry any load from the PRACH and this is a reasonable assumption when the LTE system bandwidth is 5 MHz or larger.

Table 6.11 System level simulation assumptions [8]

Parameter	Model
Cell structure	Hexagonal grid with 3 sectors per size
Cell intersite distance	1732 m
Frequency band	900 MHz
LTE system bandwidth	10 MHz
Frequency reuse	1
Base station transmit power	46 dBm
Power boosting	0 dB
Base station antenna gain	18 dBi
Device transmit power	23 dBm
Device antenna gain	−4 dBi
Device mobility	0 km/h
Path loss model	$120.9 + 37.6 \times \log_{10}(d)$, with d being the base station to device distance in km
Shadow fading standard deviation	8 dB
Shadow fading correlation distance	110 m

Figure 6.5 shows how the achieved average resource utilization increases more or less linearly on a narrowband as a function of the user arrival intensity. Although the traffic model as explained in Section 4.6.1 is uplink heavy the utilization is slightly higher on the downlink. This is explained by the downlink control signaling needed to assign uplink resources and confirm acknowledgment of received uplink data.

The x-axis in Figure 6.5 is defined in terms of device arrival rate in the system. As explained in Section 4.6.1.1, there is a linear relation between the arrival rate and the absolute system load. Three important reference points are as follows:

- 6.8 arrivals/s corresponds to a load of 60,680 users/km^2.
- 11.2 arrivals/s corresponds to a load of 100,000 users/km^2.
- 56.1 arrivals/s corresponds to a load of 500,000 users/km^2.

At a device arrival rate of 40.3 access attempts per second, the outage, that is the percentage of devices not served by the system, surpasses 1%. At this point, the LTE-M traffic capacity per narrowband equals 361,000 devices/km^2 (see Table 6.12) while the average utilization is seemingly low, as shown in Figure 6.5. However, the variation in load across the cells in the simulated system is high. It is the high resource utilization in a few cells that limits the overall capacity at the 1% outage criteria. If a higher outage than 1% is found acceptable a significantly higher system load is achievable.

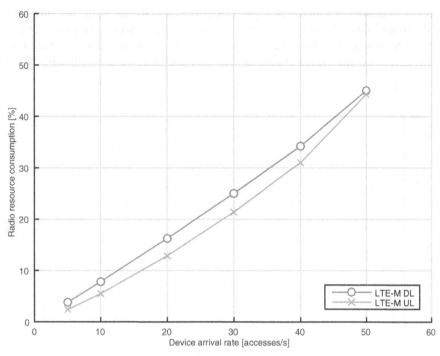

FIGURE 6.5

LTE-M average utilization of resources available for data transmission [8].

Table 6.12 LTE-M per narrowband capacity	
Connection Density at 1% Outage	**Arrival Rate at 1% Outage**
361,000 devices/km^2	40.3 access attempt/s

This high capacity achieved even for this moderate resource utilization indicates that LTE-M is capable of serving 1,000,000 devices/km^2 if three or more narrowbands are configured in a cell. This corresponds to the 3GPP Release 14 5G requirement in terms of supported connection density. It indicates that LTE-M is capable of meeting the 5G requirements to qualify as a 5G technology.

At the maximum system capacity also the service latency was recorded. Figure 6.6 shows the latency recorded in the system simulations counted from the time a device accesses the system on the PRACH to the point where the upper layer in the receiving side has correctly decoded the received transmission. It excludes the time needed to synchronize to the system and read the MIB and SIB1-BR.

Table 6.13 captures a few samples from the graphs in Figure 6.6 including the 99th percentile not visible in the figure due to the long tail of the Cumulative Distribution Function (CDF). Just as for NB-IoT, it is clear that at these extreme loads the last few percent of the users need to accept very high latency. This is a consequence of scheduling queues building up in certain cells where the load is high.

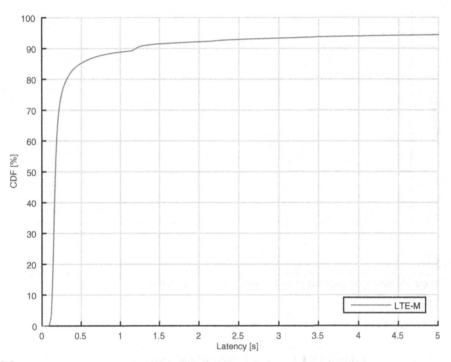

FIGURE 6.6

Service latency for LTE-M at the load presented in Table 6.12 [6].

Table 6.13 LTE-M service latency at the load presented in Table 6.12.

Percentile	Latency [s]
50th	0.2
90th	1.2
99th	45

6.7 DEVICE COMPLEXITY

NB-IoT and EC-GSM-IoT were both designed for ultra-low device complexity. This enables large scale deployments of IoT devices, and the systems to be competitive in the IoT landscape, competing, for example, with Low Power Wide Area Network alternatives in the unlicensed spectrum domain. The work on LTE-M was also triggered by a desired reduction in device cost, and the target was to go down significantly in complexity and cost relative earlier LTE device categories. LTE-M also intends to address a larger range of use cases than EC-GSM-IoT and NB-IoT, involving primarily higher throughput applications, which put higher requirements on computational complexity and memory requirements.

To get a better understanding of the LTE-M complexity, Table 6.14 summarizes some of the more important features of the LTE-M basic device Cat-M1 that was specified in Release 13.

Table 6.14 Overview of LTE-M device category M1	
Parameter	**Value**
Duplex modes	HD-FDD, FD-FDD, TDD
Half-duplex operation	Type B
Number of receive antennas	1
Transmit power class	20, 23 dBm
Maximum DL/UL bandwidth	6 PRB (1.080 MHz)
Highest DL/UL modulation order	16QAM
Maximum number of supported DL/UL spatial layers	1
Maximum DL/UL TBS	1000 bits
Peak DL/UL physical layer data rate	1 Mbps
DL/UL channel coding type	Turbo code
DL physical layer memory requirement	25,344 soft channel bits
Layer 2 memory requirement	20,000 bytes

To put the design parameters in Table 6.14 in a context, Table 6.15 estimates the modem cost reduction for the LTE-M device categories introduced in Release 12 (Cat-0) and Release 13 (Cat-M1) based on the cost reduction estimates in Table 7.1 in the LTE-M study item technical report [1]. The cost reductions are expressed in terms of modem cost reduction relative to the simplest LTE device available at the time of the LTE-M study item, which is a Cat-1 device supporting a single frequency band. The LTE-M study item concluded that the *Bill of Material* for a modem would need to be reduced to about 1/3 of that for a single-band LTE Cat-1 modem to be on par with that of an Enhanced General Packet Radio Service (EGPRS) modem, and as can be seen from the Table 6.15 Cat-M1 has the potential to reach even below this level.

Table 6.15 Overview of measures supporting an LTE-M modem cost reduction [1]

Combination of Modem Cost Reduction Techniques	Modem Cost Reduction
Single-band 23-dBm FD-FDD LTE Category 1 modem • Reference modem in the LTE-M study item	0%
Single-band 23-dBm FD-FDD LTE Category 1bis modem • Reduced number of receive antennas from 2 to 1	24%–29%
Single-band 23-dBm FD-FDD LTE Category 0 modem • Reduced peak rate from 10 to 1 Mbps • Reduced number of receive antennas from 2 to 1	42%
Single-band 23-dBm HD-FDD LTE Category 0 modem • Reduced peak rate from 10 to 1 Mbps • Reduced number of receive antennas from 2 to 1 • Half-duplex operation instead of full-duplex operation	49%–52%
Single-band 23-dBm FD-FDD LTE Category M1 modem • Reduced peak rate from 10 to 1 Mbps • Reduced number of receive antennas from 2 to 1 • Reduced bandwidth from 20 to 1.4 MHz	59%
Single-band 23-dBm HD-FDD LTE Category M1 modem • Reduced peak rate from 10 to 1 Mbps • Reduced number of receive antennas from 2 to 1 • Reduced bandwidth from 20 to 1.4 MHz • Half-duplex operation instead of full-duplex operation	66%–69%
Single-band 20-dBm FD-FDD LTE Category M1 modem • Reduced peak rate from 10 to 1 Mbps • Reduced number of receive antennas from 2 to 1 • Reduced bandwidth from 20 MHz to 1.4 MHz • Reduced transmit power from 23 to 20 dBm	69%–71%
Single-band 20-dBm HD-FDD LTE Category M1 modem • Reduced peak rate from 10 to 1 Mbps • Reduced number of receive antennas from 2 to 1 • Reduced bandwidth from 20 to 1.4 MHz • Half-duplex operation instead of full-duplex operation • Reduced transmit power from 23 to 20 dBm	76%–81%

As for EC-GSM-IoT it should be emphasized that the modem baseband and radio frequency cost is only one part of the total device cost. As pointed out in Section 4.7, also components to support peripherals, Real Time Clock, central processing unit, and power supply need to be taken into consideration to derive the total cost of a device. Also the potential of mass production is highly important. LTE-M, as all LTE-based technologies, has significant benefits in this area due to the widespread use of the technology.

REFERENCES

[1] Third Generation Partnership Project, Technical Report 36.888 v12.0.0, Study on Provision of Low-cost Machine-Type Communications (MTC) User Equipments (UEs) Based on LTE, 2013.
[2] Third Generation Partnership Project, Technical Report 45.820 v13.0.0, Cellular System Support for Ultra-Low Complexity and Low Throughput Internet of Things, 2016.
[3] Ericsson, et al., RP-170511 on eMTC 5G Requirement Fulfilment, 3GPP RAN Meeting #75, 2017.
[4] Ericsson, et al., R1-1706161 Early Data Transmission for MTC, 3GPP RAN1 Meeting #88bis, 2017.
[5] Third Generation Partnership Project, Technical Specification 36.101 v14.0.0, E-UTRA UE Radio Transmission and Reception, 2016.
[6] Saxena, et al., Reducing the Modem Complexity and Achieving Deep Coverage in LTE for Machine-Type Communications, Globecom 2016, 2016.
[7] Saxena, et al., On the Achievable Coverage and Uplink Capacity of Machine-Type Communications (MTC) in LTE Release 13, VTC 2016 Fall, 2016.
[8] Ericsson, et al., R1-1703865 on 5G mMTC Requirement Fulfilment, 3GPP RAN1 Meeting #88, 2017.

CHAPTER OUTLINE

Cellular Internet of Things. http://dx.doi.org/10.1016/B978-0-12-812458-1.00007-1

Abstract

This chapter presents the design of Narrowband Internet of Things (NB-IoT). The first part of this chapter describes the background behind the introduction of NB-IoT in the Third Generation Partnership Project (3GPP) specifications and the design principles of the technology. That is, what motivated 3GPP to embark on the standardization of yet a third cellular IoT technology, despite the ongoing activities of specifying both Extended Coverage Global System for Mobile Communications Internet of Things (EC-GSM-IoT) and Long-Term Evolution for Machine-Type Communications (LTE-M)? The second part of this chapter focuses on the physical channels with an emphasis on how these channels are designed to fulfill the objectives that NB-IoT is intended to achieve, namely, deployment flexibility, ubiquitous coverage, ultra-low device cost, long battery lifetime, and capacity sufficient for supporting a massive number of devices in a cell. Detailed descriptions are provided regarding both downlink and uplink transmission schemes and how each of the NB-IoT physical channels is mapped to radio resources in both the frequency and time dimensions. The third part of this chapter covers NB-IoT idle and connected mode procedures and the transition between these modes, including all activities from initial cell selection to completing a data transfer. The idle mode procedures include the initial cell selection, which is the procedure that a device has to go through when it is first switched on or is attempting to select a new cell to camp on. Idle mode activities also include paging and acquisition of Master and

System Information Blocks. After the idle mode procedures, the transition from idle to connected mode is described in a set of system access procedures. Then, descriptions of some fundamental connected mode procedures including scheduling, power control, and multicarrier operation are presented. Finally, a summary of the most recent improvements accomplished in Release 14 of the 3GPP specifications is presented.

7.1 BACKGROUND
7.1.1 3GPP STANDARDIZATION

In early 2015, the market for *Low-Power Wide Area Networks* (LPWAN) was rapidly developing. Sigfox was building out their *Ultra Narrowband Modulation* networks in France, Spain, the Netherlands, and the United Kingdom. The LoRa Alliance, founded with a clear ambition to provide IoT connectivity with wide-area coverage, released LoRaWAN R1.0 specification in June 2015 [1]. The Alliance at that point quickly gathered significant industry interest and strong membership growth. Up until then, Global System for Mobile Communications/General Packet Radio Service (GSM/ GPRS) had been the main cellular technology of choice for serving wide-area IoT use cases, thanks to it being a mature technology with low modem cost (at least when compared with 3G and 4G). This position was challenged by the emerging LPWAN technologies that presented an alternative technology choice to many of the IoT verticals served by GSM/GPRS.

Anticipating the new competition, 3GPP (see Section 2.1) started a feasibility study on *Cellular System Support for Ultra-low Complexity and Low Throughput Internet of Things* [2], referred to as the *Cellular IoT study* for short in the following sections. As explained in Sections 1.2.3 and 2.2.5, challenging objectives on coverage, capacity, and battery lifetime were set up, together with a more relaxed objective of a maximum system latency. All these performance objectives offer major improvements over GSM and GPRS, as specified at that time, toward better serving the IoT verticals. One additional objective was that it should be possible to introduce the IoT features to the existing GSM networks through software upgrade. Building out a national network takes many years and requires substantial investment up front. With software upgrade, however, the well-established cellular network can be upgraded overnight to meet all the key performance requirements of the IoT market.

Among the solutions proposed to the *Cellular IoT study*, some were backward compatible with GSM/GPRS and were developed based on the evolution of the existing GSM/GPRS specifications. EC-GSM-IoT described in Chapters 3 and 4 is the solution eventually standardized in 3GPP Release 13.

Historically, the group carrying out the study, 3GPP TSG GERAN (Technical Specifications Group GSM/EDGE Radio Access Network), had focused on the evolution of GSM/GPRS technologies, developing features for meeting the need of GSM operators. Certain GSM operators, however, at that point considered refarming their GSM spectrum to Long-Term Evolution (LTE) as well as to LPWAN dedicated for IoT services. This consideration triggered the study on non−GSM backward compatible technologies, referred to as *clean-slate* solutions. Although none of the clean-slate solutions in the study were specified, it provided a firm ground for the NB-IoT technology that emerged after study completion and was standardized in 3GPP Release 13. As described later in this chapter, the entire NB-IoT system is supported in a bandwidth of 180 kHz, for each of the downlink and uplink. This allows for deployment in refarmed GSM spectrum as well as within an LTE carrier. NB-IoT is part of the 3GPP LTE specifications and employs many technical components already defined for LTE. This approach reduced the standardization process and leveraged the LTE eco-system to ensure a fast time-to-market. It also possibly allows NB-IoT to be introduced through a software upgrade of the existing LTE networks. The normative work of developing the core specifications of NB-IoT took only a few months and was completed in June 2016 [3]. Then, within a year from the completion of the core specifications, mobile network operators and technology vendors have been launching NB-IoT commercial networks and devices.

7.1.2 RADIO ACCESS DESIGN PRINCIPLES

NB-IoT is designed for ultra-low-cost massive Machine-Type Communications, aiming to support a massive number of devices in a cell. Low device complexity is one of the main design objectives, enabling low module cost. Furthermore, it is designed to offer substantial coverage improvements over GPRS as well as for enabling long battery lifetime. Finally, NB-IoT has been designed to give maximal deployment flexibility. In this section, we will highlight the design principles adopted in NB-IoT to achieve these objectives.

7.1.2.1 Low Device Complexity and Cost

Device modem complexity and cost are primarily related to the complexity of baseband processing, memory consumption, and radio-frequency (RF) requirements. Regarding baseband processing, NB-IoT is designed for allowing low-complexity receiver processing during initial cell selection and during connection. For initial cell selection, a device needs to search for only one synchronization sequence for establishing basic time and frequency synchronization to the network. The device can use a low sampling rate (e.g., 240 kHz) and take advantage of the synchronization sequence properties to minimize memory and complexity.

During connected mode, low device complexity is facilitated by restricting the DL transport block size (TBS) to be no larger than 680 bits in Release 13 and relaxing the processing time requirements compared with LTE. For channel coding, instead of using the LTE turbo code [4], which requires iterative receiver processing, NB-IoT adopts a simple convolutional code, i.e., the LTE *tail-biting convolution code* (TBCC) [4], in the DL channels. In addition, NB-IoT does not use higher-order modulations or multilayer *multiple-input multiple-output* transmissions. Furthermore, a device needs to support only half-duplex operation and is not required to listen to the DL, while transmitting in the UL, and vice versa.

Regarding RF, all the performance objectives of NB-IoT can be achieved with one transmit-and-receive antenna in the device. That is, neither DL receiver diversity nor UL transmit diversity is required in the device. NB-IoT is designed for allowing relaxed oscillator accuracy in the device. For example, a device can achieve initial acquisition when its oscillator inaccuracy is as large as 20 *parts per million* (ppm). During a data session, the transmission scheme is designed for the device to easily track its frequency drift. Because a device is not required to simultaneously transmit and receive, a duplexer is not needed in the RF front end of the device. The maximum transmit power level of an NB-IoT device is either 20 or 23 dBm in Release 13. This allows on-chip integration of the *power amplifier* (PA), which can contribute to the device cost reduction.

Economy of scale is yet another contributor to cost reduction. Thanks to its deployment flexibility and low minimum system bandwidth requirement, it is expected that NB-IoT will be made globally available in many networks. This will help to increase the economy of scale of NB-IoT.

7.1.2.2 Coverage Enhancement (CE)

Coverage Enhancement (CE) is mainly achieved by trading off data rate for coverage. Like EC-GSM-IoT and LTE-M, repetitions are used to ensure that devices in coverage challenging locations can still have reliable communications with the network, although at a reduced data rate. Furthermore, NB-IoT has been designed to use a close to constant envelope waveform in the UL. This is an important factor for devices in extreme coverage- and power-limited situations because it minimizes the need to back off the output power from the maximum configurable level. Minimizing the power backoff helps preserve the best coverage possible for a given PA capability.

7.1.2.3 Long Device Battery Lifetime

Minimizing power backoff also gives rise to higher PA power efficiency, which helps extend device battery lifetime. Device battery lifetime, however, depends heavily on how the device behaves when it does not have an active data session. In most use cases, the device actually spends the vast majority of its lifetime in idle mode as most of the IoT applications only require infrequent transmission of short packets. Traditionally, an idle device needs to monitor paging and perform mobility measurements. Although the energy consumption during idle mode is much lower compared with during connected mode, further energy saving can be achieved by simply increasing the periodicity between paging occasions or not requiring the device to monitor paging at all. As elaborated on in Section 2.2, 3GPP Releases 12 and 13 introduced both *extended Discontinuous Reception* (eDRX) and *Power-Saving Mode* (PSM) to support this type of operation and optimize device power consumption. In essence, a device can shut down its transceiver and only keep a basic oscillator running for the sake of keeping a time reference to know when it should come out of the PSM or eDRX. The reachability during PSM is set by the *Tracking Area Update* (TAU) timer with the maximum settable value exceeding 1 year [5]. eDRX can be configured with a DRX cycle just below 3 h [6].

During these power-saving states, both device and network maintain device context, saving the need for unnecessary signaling when the device comes back to connected mode. This optimizes the signaling and power consumption when making the transition from idle to connected mode.

In addition to PSM and eDRX, NB-IoT also adopts *connected mode DRX* (cDRX) as a major tool for achieving energy efficiency. In Release 13, the cDRX cycles were extended from 2.56 to 10.24 s for NB-IoT [7].

7.1.2.4 Support of Massive Number of Devices

NB-IoT achieves high capacity in terms of number of devices that can be supported on one single NB-IoT carrier. This is made possible by introducing spectrally efficient transmission scheme in the UL for devices in extreme coverage-limited situation.

Shannon's well-known channel capacity theorem [8] establishes a relationship between bandwidth, power, and capacity in an *additive white Gaussian noise* channel as

$$C = W \log_2 \left(1 + \frac{S}{N} \right) = W \log_2 \left(1 + \frac{S}{N_0 W} \right), \qquad (7.1)$$

where C is the channel capacity (bits/s), S is the received desired signal power, N is the noise power, which is determined by the product of noise bandwidth (W) and one-sided noise power spectral density (N_0). The noise bandwidth is identical to the signal bandwidth if Nyquist pulse shaping function is used. At extreme coverage-limited situation, $\frac{S}{N} \ll 1$. Using the approximation $\ln(1 + x) \approx x$, for $x \ll 1$, it can be shown that the channel capacity in very low Signal-to-Noise-power Ratio (SNR) regime is

$$C = \frac{S}{N_0} \log_2(e). \qquad (7.2)$$

In this regime, the bandwidth dependency vanishes, and therefore channel capacity, in terms of bits per second, is only determined by the ratio of S and N_0. Thus, in theory the coverage for a target data rate $R = C$ depends only on the received signal power level and not on the signal bandwidth. This implies that because the data rate at extreme coverage-limited situation does not scale according to the device bandwidth allocation, for the sake of spectral efficiency, it is advantageous to allocate small bandwidth for devices in bad coverage. NB-IoT UL waveforms include various bandwidth options. While a waveform of wide bandwidth (e.g., 180 kHz) is beneficial for devices in good coverage, waveforms of small bandwidths are more spectrally efficient from the system point of view for serving devices in bad coverage. This will be illustrated by the coverage results presented in Chapter 8.

7.1.2.5 Deployment Flexibility

To support maximum flexibility of deployment and prepare for refarming scenarios, NB-IoT supports three modes of operation, stand-alone, in-band, and guard-band.

7.1.2.5.1 Stand-alone Mode of Operation

NB-IoT may be deployed as a stand-alone carrier using any available spectrum with bandwidth larger than 180 kHz. This is referred to as the stand-alone deployment. A use case of the stand-alone deployment is for a GSM operator to deploy NB-IoT in its GSM band by refarming part of its GSM spectrum. In this case, however, additional guard-band is needed between a GSM carrier and the NB-IoT carrier. Based on the coexistence requirements in [9], 200 kHz guard-band is recommended, which means that a GSM carrier should be left empty on one side of an NB-IoT carrier between two operators. In case of the same operator deploying both GSM and NB-IoT, a guard-band of 100 kHz is recommended based on the studies in [2], and hence an operator needs to refarm at least two consecutive GSM carriers for NB-IoT deployment. An example is illustrated in Figure 7.1. Here, NB-IoT bandwidth is shown as 200 kHz. This is because that NB-IoT needs to meet the GSM spectral mask when deployed using refarmed GSM spectrum. The GSM spectral mask is specified according to 200 kHz channelization.

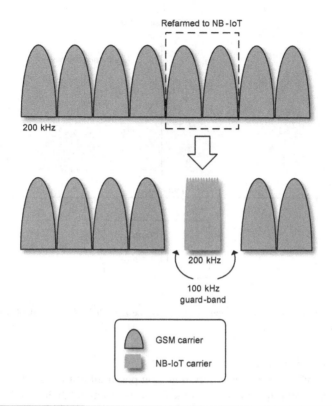

FIGURE 7.1

NB-IoT stand-alone deployment using refarmed GSM spectrum.

7.1.2.5.2 In-Band and Guard-Band Modes of Operation

NB-IoT is also designed to be possible for deployment in the existing LTE networks, either using one of the LTE *Physical Resource Blocks* (PRBs) or using the LTE guard-band. These two deployment scenarios are referred to as in-band and guard-band deployments, respectively. An example is illustrated in Figure 7.2. An LTE carrier with a number of PRBs is shown. NB-IoT can be deployed using one LTE PRB or using the unused bandwidth in the guard-band. The guard-band deployment makes use of the fact that the occupied bandwidth of the LTE signal is roughly 90% of channel bandwidth when the LTE carrier bandwidth is 3, 5, 10, 15, or 20 MHz [10]. Hence, there is roughly 5% of the LTE channel bandwidth on each side available as guard-band.

Yet another possible deployment scenario is to have NB-IoT in-band deployment on an LTE carrier that supports LTE-M features. The concept of *narrowband* used in LTE-M is explained in Section 5.2.3.2. Some of these LTE-M narrowbands (NB) are not used for transmitting LTE-M *System Information Block Type 1* (SIB1) and thus can be used for deploying NB-IoT. More details about this deployment scenario are given in Section 7.2.2.

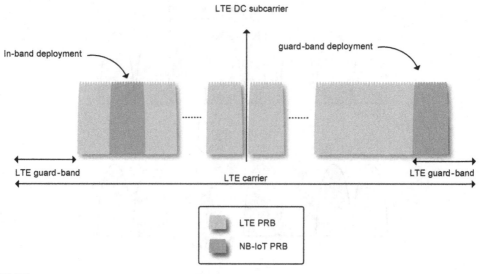

FIGURE 7.2

NB-IoT deployment inside an LTE carrier, either using one of the LTE PRBs (in-band deployment) or using the LTE guard-band (guard-band deployment).

7.1.2.5.3 Spectrum Refarming

NB-IoT is intended to offer flexible spectrum migration possibilities to a GSM operator. An operator can take an initial step to refarm a small part of the GSM spectrum to NB-IoT as the example shown in the top part of Figure 7.3. Thanks to the support of LTE in-band and guard-band deployment, such an initial migration step will not result in spectrum fragmentation to make the eventual migration of the entire GSM spectrum to LTE more difficult. As illustrated in Figure 7.3, the NB-IoT carrier already deployed in the GSM network as a stand-alone deployment may become an LTE in-band or guard-band deployment when the entire GSM spectrum is migrated to LTE. This high flexibility is also expected to secure NB-IoT deployments when LTE is later on refarmed to 5G.

7.2 PHYSICAL LAYER

In this section, we describe NB-IoT physical layer design with an emphasis on how these channels are designed to fulfill the objectives that NB-IoT targets, namely, deployment flexibility, ubiquitous coverage, ultra-low device cost, long battery lifetime, and capacity sufficient for supporting a massive number of devices in a cell.

7.2.1 GUIDING PRINCIPLES

When designing a new radio access technology, the degree of freedom is higher compared with when basing the design on an existing technology. Therefore NB-IoT could with limited restrictions be designed from ground-up with the intention to follow the radio access design principles outlined in Section 7.1.2.

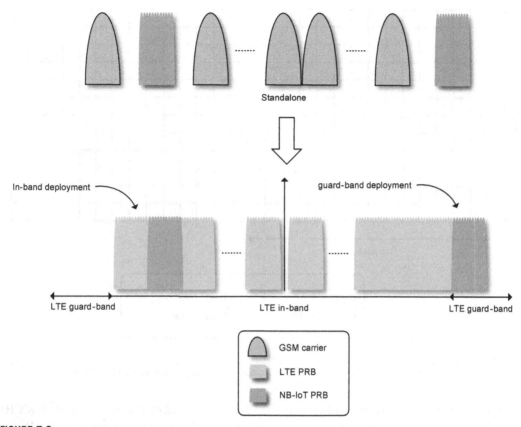

Standalone

In-band deployment

guard-band deployment

LTE guard-band   LTE in-band   LTE guard-band

GSM carrier

LTE PRB

NB-IoT PRB

FIGURE 7.3

Partial GSM spectrum to NB-IoT introduction as an initial spectrum refarming step, followed by eventual total migration to LTE.

The ambition to provide a radio access technology with high deployment flexibility and the capability to operate both in a refarmed GSM spectrum and inside an LTE carrier did, however, impose certain guiding principles, onto the design of the technology.

While the stand-alone deployment in the GSM spectrum, discussed in Section 7.1.2, is facilitated by the introduction of a guard-band between the NB-IoT and GSM carriers, the expectations on close coexistence with LTE were higher. NB-IoT deployment inside an LTE carrier was hence required to be supported without any guard-band between NB-IoT and LTE PRBs. To minimize the impact on the existing LTE deployments and devices, this imposed requirements on the NB-IoT physical layer waveforms to preserve orthogonality with the LTE signal in adjacent PRBs. It also implies that NB-IoT should be able to share the same time-frequency resource grids as LTE the same way as different LTE physical channels share time-frequency resources. Last but not least, because legacy LTE devices will not be aware of the NB-IoT operation, NB-IoT transmissions should not collide with essential LTE transmissions.

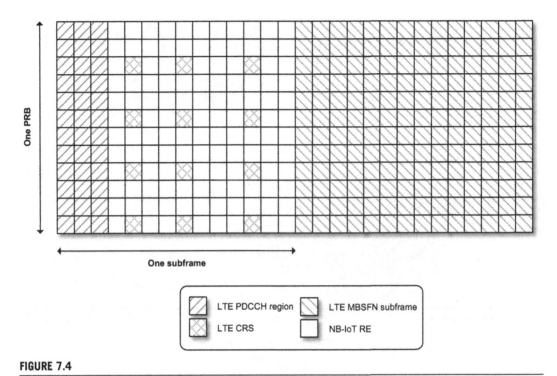

One PRB

One subframe

| LTE PDCCH region | LTE MBSFN subframe |
| LTE CRS | NB-IoT RE |

FIGURE 7.4

An illustration of how NB-IoT physical channels of the in-band mode share REs with LTE in the DL.

Among the essential LTE transmissions are the *Physical Downlink Control Channel* (PDCCH) transmissions for scheduling information, paging indicators, and random access response (RAR), etc. The PDCCH spans over the entire range of LTE PRBs in frequency and may span up to the first three Orthogonal Frequency-Division Multiplexing (OFDM) symbols in every subframe in time. As illustrated in Figure 7.4, the resource elements (REs) of the first three OFDM symbols in every subframe may therefore not be used by NB-IoT DL channels.

Additional essential LTE physical channels and signals are the *Cell-specific Reference Signal* (CRS), *Physical Control Format Indicator Channel*, *Physical Hybrid-ARQ Indicator Channel*, *Channel State Information Reference Signal*, and *Multicast-Broadcast Single-Frequency Network* (MBSFN) signal. The REs used by these channels and signals are also preserved and not mapped to any NB-IoT physical channels. MBSFN signal, for example, spans one entire subframe in the time dimension and all PRBs in the frequency dimension, and thus if one subframe is configured as an LTE MBSFN subframe, that subframe is not used by NB-IoT. Figure 7.4 illustrates such an example.

Furthermore, resources used by LTE *Primary Synchronization Signal* (PSS), *Secondary Synchronization Signal* (SSS), and *Physical Broadcast Channel* (PBCH) are protected by NB-IoT, avoiding to use any of the middle six PRBs in an LTE carrier in case of in-band deployment.

Following these guiding principles will naturally mean that the physical layer to a large extent is inspired by LTE. At the same time, changes are required to meet the design objectives listed in Section 7.1.2.

7.2.2 PHYSICAL LAYER NUMEROLOGY

7.2.2.1 Channel Raster

An NB-IoT carrier carrying essential physical signals that allow a device to perform cell selection is referred to as an *anchor* carrier. The carrier frequency of an NB-IoT anchor carrier is determined by the *E-UTRA Absolute Radio Frequency Channel Number* (EARFCN) [11]. The placement of an NB-IoT anchor carrier in frequency corresponding to a certain EARFCN is just as for LTE, based on a 100 kHz channel raster [11]. Contrary to LTE, an NB-IoT anchor carrier can, however, be located slightly off the 100 kHz channel raster. The need for this offset is justified by the required NB-IoT deployment flexibility as explained next.

The deployment scenario chosen by the operator, stand-alone, in-band, or guard-band mode of operation should be transparent to a device when it is first turned on and searches for an NB-IoT carrier on its supported frequency bands. In case of stand-alone deployment, because GSM uses 200 kHz channel raster, the NB-IoT anchor carrier can always be placed on a refarmed GSM 200 kHz channel. However, for LTE in-band and guard-band deployments, it is not possible to place the NB-IoT anchor carrier with a center frequency exactly on the 100 kHz grid. An example is given in Table 7.1 based on a 3 MHz LTE carrier, which has 15 PRBs. The PRB frequency offsets, i.e., the PRB center frequencies relative to the LTE DC subcarrier, are listed. Because the LTE DC subcarrier is placed on the 100 kHz raster, the relative frequencies, listed in Table 7.1, give indications of the PRB centers relative to the 100 kHz raster. We will refer to this as raster offset in the discussion below.

As can be seen, no PRB has a center frequency that exactly falls on the raster, i.e., has a PRB frequency offset that equals a multiple of 100 kHz. However, PRB#2 and PRB#12 in this example have the smallest raster offset of 7.5 kHz. To facilitate an efficient initial cell selection for NB-IoT, in case of LTE in-band deployment, an anchor carrier is required to be configured on a PRB that has the smallest raster offset. It turns out that for LTE carrier bandwidth of 3, 5, and 15 MHz, the smallest raster offset is 7.5 kHz, whereas for LTE carrier bandwidth of 10 and 20 MHz, the smallest raster offset is 2.5 kHz. A full list of PRB indexes for in-band deployment is given in Table 7.2. The underlined PRB indexes correspond to PRBs located above the LTE DC subcarrier.

As discussed in Section 7.2.1, NB-IoT avoids the middle six LTE PRBs because these PRBs overlap with LTE PSS, SSS, and PBCH. Because a 1.4 MHz LTE carrier has only six PRBs and very small guard-bands, in-band and guard-band deployments of NB-IoT are not supported on a 1.4 MHz LTE carrier.

Table 7.1 LTE PRB center frequencies relative to the DC subcarrier for 3 MHz LTE bandwidth

PRB Index	0	1	2	3	4	5	6	7
PRB frequency offset [kHz]	−1267.5	−1087.5	−907.5	−727.5	−547.5	−367.5	−187.5	0
PRB Index	8	9	10	11	12	13	14	
PRB frequency offset [kHz]	187.5	367.5	547.5	727.5	907.5	1087.5	1267.5	

The PRB indexing starts from index 0 for the PRB occupying the lowest frequency within the LTE carrier.

Table 7.2 Suitable PRB indexes for NB-IoT anchor carrier in the in-band deployment.

LTE Bandwidth [MHz]	Allowed PRB Indexes for NB-IoT Anchor Carrier	Raster Offset
1.4	Not supported	Not applicable
3	2, 12	7.5 kHz
5	2, 7, 17, 22	7.5 kHz
10	4, 9, 14, 19, 30, 35, 40, 45	2.5 kHz
15	2, 7, 12, 17, 22, 27, 32, 42, 47, 52, 57, 62, 67, 72	7.5 kHz
20	4, 9, 14, 19, 24, 29, 34, 39, 44, 55, 60, 65, 70, 75, 80, 85, 90, 95	2.5 kHz

Underlined PRB indexes correspond to PRBs located above the LTE DC subcarrier.

Similar to the in-band deployment, an NB-IoT anchor carrier in the guard-band deployment needs to have raster offset as small as possible. For LTE carrier bandwidth of 10 and 20 MHz, the smallest raster offset for the guard-band deployment is 2.5 kHz. An example is shown in Figure 7.5, in which NB-IoT is deployed immediately adjacent to the edge LTE PRB, i.e., PRB#49 on a 10 MHz LTE carrier. The center of the NB-IoT guard-band PRB is 4597.5 kHz from the DC subcarrier, giving rise to a 2.5 kHz raster offset. For LTE carrier bandwidth of 3, 5, and 15 MHz, an anchor carrier cannot be placed immediately adjacent to the edge LTE PRB as the raster offset would become too large. However, shifting the NB-IoT anchor by additional three subcarriers away from the edge LTE PRB gives rise to a raster offset of 7.5 kHz.

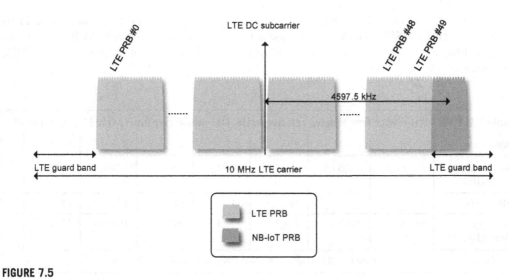

FIGURE 7.5

Deployment of NB-IoT in the guard-band of a 10 MHz LTE carrier.

Table 7.3 Suitable PRB indexes for NB-IoT in-band deployment on an LTE carrier that supports LTE-M

LTE Bandwidth [MHz]	Allowed PRB Indexes for NB-IoT Carrier
1.4	Not supported
3	Anchor carrier: none; nonanchor: 0 or 14
5	Anchor carrier: 7, 17; nonanchor carrier: 6, 7, 8, 16, 17, 18
10	Anchor carrier: 19, 30; nonanchor carrier: 0, 19, 20, 21, 28, 29, 30, 49
15	Anchor carrier: 32, 42; nonanchor carrier: 0, 31, 32, 33, 41, 42, 43, 74
20	Anchor carrier: 44, 55; nonanchor carrier: 0, 1, 44, 45, 53, 54, 55, 98, 99

Multicarrier operation of NB-IoT is supported. Because NB-IoT suffices to have one anchor carrier for facilitating device initial cell selection, the additional carriers may be located with an offset of up to 47.5 kHz outside the 100 kHz raster grid (a total of 21 offset values are defined) [9]. These additional carriers are referred to as nonanchor or secondary carriers. A nonanchor carrier does not carry the physical channels that are required for device initial cell selection.

As mentioned in Section 7.1.2.5.2, NB-IoT can be deployed as an LTE in-band deployment together with LTE-M on the same LTE carrier. The PRB indexes that can be used for deploying an NB-IoT anchor carrier for such scenarios without collisions with LTE-M NB are shown in Table 7.3. Furthermore, the PRB indexes that can be used for deploying NB-IoT nonanchor carriers are also shown in Table 7.3. As indicated in Table 7.3, deploying both NB-IoT anchor carrier in-band and LTE-M on a 3-MHz LTE carrier is not possible without collision with the LTE-M NB. This is because that LTE-M defines two NB on a 3-MHz LTE carrier, as illustrated in Figure 7.6, leaving one PRB on the center and both edge PRBs unused. However, none of these three unused PRBs can be used as an NB-IoT anchor because of raster offset restriction. The edge PRBs can nevertheless be used as a nonanchor carrier. But, the center PRB cannot be used as a nonanchor carrier either, as it is one of the center 6 PRBs within an LTE carrier.

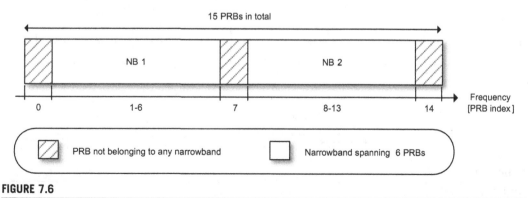

FIGURE 7.6

Definition of LTE-M NB on a 3 MHz LTE carrier.

7.2.2.2 Frame Structure

The overall time frame structure on the NB-IoT access stratum is illustrated in Figure 7.7. On the highest level, one hyperframe cycle has 1024 hyperframes each consisting of 1024 frames. One frame consists of 10 subframes, each dividable into two slots of 0.5 ms. Hyperframe and frame are each labeled with a hyper system frame number (H-SFN) and a system frame number (SFN), respectively. Each subframe can therefore be uniquely identified by an H-SFN, an SFN, and a subframe number (SN). The ranges of H-SFN, SFN, and SN are 0–1023, 0–1023, and 0–9, respectively.

In the DL and UL, the NB-IoT design supports a subcarrier spacing of 15 kHz, for which each frame contains 20 slots as illustrated in Figure 7.7. In the UL, the technology supports an additional subcarrier spacing of 3.75 kHz. For this alternative subcarrier spacing, each frame is directly divided into five slots, each of 2 ms as shown in Figure 7.8.

7.2.2.3 Resource Grid

In the DL, the concept of PRBs is used for specifying the mapping of physical channels and signals onto REs. The RE is the smallest physical channel unit, each uniquely identifiable by its subcarrier index k and symbol index l within the PRB. A PRB spans 12 subcarriers over 7 OFDM symbols and in total $12 \times 7 = 84$ REs. One PRB pair is the smallest schedulable unit in most DL cases, as of Release 14, and fits into two consecutive slots as shown in Figure 7.9.

For the UL, the *resource unit* (RU) is used to specify the mapping of the UL physical channels onto REs. The definition of the RU depends on the configured subcarrier spacing and the number of sub-carriers allocated to the UL transmission. In the basic case where 12 subcarriers using a spacing of

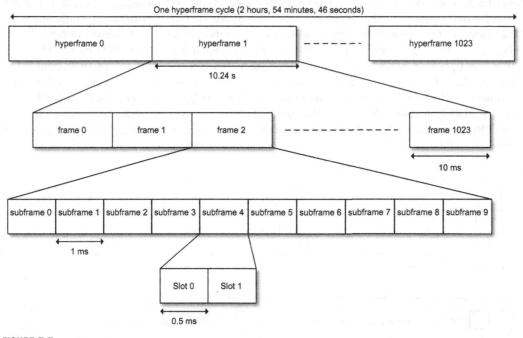

FIGURE 7.7

NB-IoT frame structure for 15 kHz subcarrier spacing.

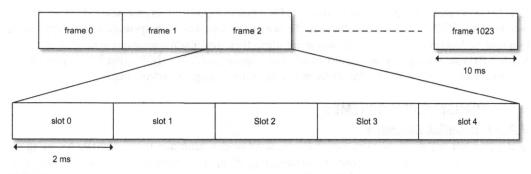

FIGURE 7.8

NB-IoT frame structure for 3.75 kHz subcarrier spacing. (UL only)

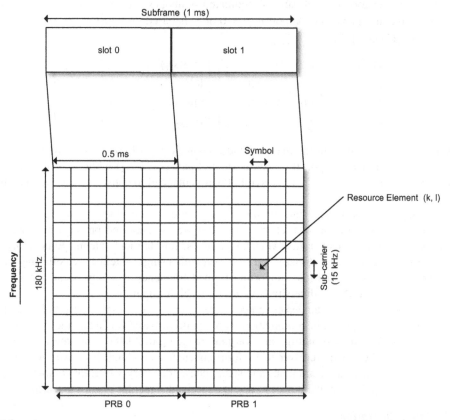

FIGURE 7.9

PRB pair in NB-IoT DL.

15 kHz are allocated, the RU corresponds to the PRB pair in Figure 7.9. In case of *sub-PRB* scheduling assignments of 6, 3, or 1 subcarrier, then the RU is expanded in time to compensate for the diminishing frequency allocation. For the single subcarrier allocation, also known as *single-tone* allocation, the NB-IoT RU concept supports an additional subcarrier spacing of 3.75 kHz. Section 7.2.5.2 presents the available formats of the RU for the different UL transmit configurations in detail.

7.2.3 TRANSMISSION SCHEMES

7.2.3.1 Downlink Operation

In the DL, NB-IoT employs *Orthogonal Frequency-Division Multiple-Access* (OFDMA), using the same numerologies as LTE in terms of subcarrier spacing, slot, subframe, and frame durations as presented in Sections 7.2.2.2 and 7.2.2.3. The slot format and OFDM symbol and *cyclic prefix* (CP) durations are also identical to those defined for LTE normal CP, i.e., 4.7 µs. The DL waveform is defined for 12 subcarriers and is identical for stand-alone, guard-band, and in-band modes of operation. Section 7.2.6 describes the baseband signal generation in detail.

In the DL, support for one or two *logical antenna ports* has been specified. In case of two logical antenna ports, the transformation of pairs of modulated symbols s_{2i}, s_{2i+1} to sets of precoded symbols y_{2i}^p, y_{2i+1}^p presented at logical antenna port p, $p = 0$ or 1, is based on transmit diversity using *Space-Frequency Block Coding* [12]:

$$\begin{bmatrix} y_{2i}^0 \\ y_{2i}^1 \\ y_{2i+1}^0 \\ y_{2i+1}^1 \end{bmatrix} = \frac{1}{\sqrt{2}} \begin{bmatrix} 1 & 0 & j & 0 \\ 0 & -1 & 0 & j \\ 0 & 1 & 0 & j \\ 1 & 0 & -j & 0 \end{bmatrix} \begin{bmatrix} \text{Re}(s_{2i}) \\ \text{Re}(s_{2i+1}) \\ \text{Im}(s_{2i}) \\ \text{Im}(s_{2i+1}) \end{bmatrix} = \frac{1}{\sqrt{2}} \begin{bmatrix} s_{2i} \\ -s_{2i+1}^* \\ s_{2i+1} \\ s_{2i}^* \end{bmatrix} \tag{7.3}$$

Here Re (x) and Im (x) represent the real part and imaginary part of complex symbol x.

In essence, logical antenna port 0 transmits symbol pair (s_{2i}, s_{2i+1}) while logical antenna port 1 transmits symbol pair $\left(-s_{2i+1}^*, s_{2i}^* \right)$. Each symbol pair is mapped to consecutively available two REs within an OFDM symbol.

The mapping from the logical antenna ports to the base station physical antenna ports is up to implementation, and there is no restriction in the number of supported physical antennas. This is important for NB-IoT in-band operation when LTE may use four or more antenna ports. Many NB-IoT devices are also expected to be stationary and equipped with a single antenna, i.e., offering low spatial diversity. The system is, in addition, supporting limited frequency diversity because of the narrow system bandwidth. The transmit diversity is therefore of high importance.

To support operation in an extended coverage range, mapping of a single transport block (TB) over up to 10 subframes is supported in combination with repetition-based link adaptation. A consequence of this is that a single transmission time interval may range up to 20,480 subframes.

Finally, to allow the device receiver to perform coherent combining, it is expected that the DL waveform is transmitted with a continuous and stable phase trajectory. Coherent combining optimizes receive performance and allows the device to detect a received signal far below the thermal noise floor.

7.2.3.2 Uplink Operation

NB-IoT UL uses *Single-Carrier Frequency-Division Multiple-Access* (SC-FDMA) with 15 kHz subcarrier spacing for multitone transmissions. In this case, the same numerologies as NB-IoT

DL are used. Multitone transmissions may use 12, 6, or 3 subcarriers. In addition, single-tone transmissions are supported, and in that case, the time–frequency resource grids can be based on 15 or 3.75 kHz subcarrier spacing.

The single-tone waveform has been designed with a close to constant envelope modulation to allow transmit operation without any power backoff to optimize coverage as well as PA efficiency. The 3.75 kHz subcarrier spacing allows for increased system capacity in the extended coverage domain where the data rate is power limited and not bandwidth limited.

Besides the definition of the RU, single-tone transmissions with 15 kHz subcarrier spacing use the same numerologies as multitone transmissions and thus achieve the best coexistence performance with multitone transmissions as well as with LTE. As for the DL, the UL supports mapping of a TB over multiple consecutive RUs in combination with an extensive set of repetitions to achieve operation in extended coverage.

SC-FDMA with single-tone is mathematically identical to OFDM, as the *Discrete Fourier Transform* (DFT) precoding step can be omitted.

Just as in case of the DL, also the UL waveform is expected to be transmitted with a continuous phase trajectory to allow the base station receiver to perform coherent combining of the received waveform at low signal-to-noise ratios.

7.2.4 DOWNLINK PHYSICAL CHANNELS AND SIGNALS

NB-IoT supports the set of DL physical channels and signals depicted in Figure 7.10.

At a high level, the DL physical channels and signals are time-multiplexed, except for *Narrowband Reference Signal* (NRS), which are present in every subframe carrying *Narrowband Physical Broadcast Channel* (NPBCH), *Narrowband Physical Downlink Control Channel* (NPDCCH), or *Narrowband Physical Downlink Shared Channel* (NPDSCH). Figure 7.11 shows time-multiplexing of different mandatory DL physical channels in a 20-ms period. The same pattern is repeated in subsequent periods. As shown, NPBCH and *Narrowband Primary Synchronization Signal* (NPSS) are transmitted in subframes 0 and 5 in every frame, respectively, and *Narrowband Secondary Synchronization Signal* (NSSS) is transmitted in subframe 9 in every other frame. The remaining subframes may be used to transmit NPDCCH or NPDSCH.

7.2.4.1 NB-IoT Subframes

Some of the subframes not carrying NPBCH, NPSS, or NSSS may be declared as *invalid subframes* according to the *Narrowband System Information Block Type 1* (NB-SIB1) described in Section 7.3.1.2. Those are not considered to belong to the set of NB-IoT subframes and are skipped when the NB-IoT NPDSCH and NPDCCH transmissions are mapped onto subframes. A device monitoring the DL will skip these invalid subframes.

The notion of invalid subframe is especially useful when NB-IoT is deployed inside an LTE carrier having MBSFN subframes configured [10]. An MBSFN subframe uses all PRBs within the LTE carrier, making the resource elements in the subframe not available for NB-IoT. The notion of invalid

FIGURE 7.10

DL physical channels and signals used in NB-IoT.

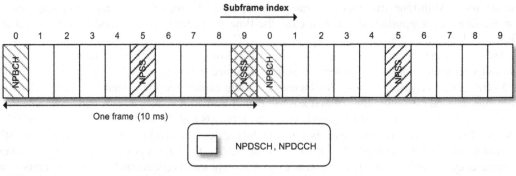

FIGURE 7.11

Time-multiplexing of downlink physical channels on an NB-IoT anchor carrier.

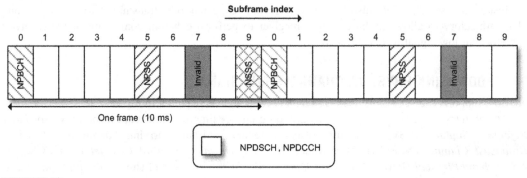

FIGURE 7.12

Time-multiplexing of downlink physical channels, with invalid subframes declared through system information (SI).

subframe is also useful when NB-IoT is deployed in the guard band of an LTE carrier configured with MBSFN subframes. An MBSFN subframe has a different subframe structure than a regular subframe, making it hard to ensure coexistence performance between NB-IoT and LTE MBSFN subframes on adjacent PRBs. This is due to that the CP used in an MBSFN subframe is longer than the normal CP, resulting in a different OFDM symbol duration compared with that used for NB-IoT. In this case, it is convenient to simply declare the subframes used as LTE MBSFN subframes as invalid subframes for NB-IoT and avoid transmitting NB-IoT DL signals in these subframes. An example is given in Figure 7.12, in which subframes #7 are declared as invalid subframes and not available for NPDCCH or NPDSCH.

7.2.4.2 Synchronization Signals

The NPSS and NSSS allow a device to synchronize to an NB-IoT cell. They are transmitted in certain subframes based on an 80-ms repetition interval as illustrated in Figure 7.13. By synchronizing to NPSS and NSSS, the device will be able to detect the cell identity number and identify the framing information within an 80-ms NPSS and NSSS repetition interval.

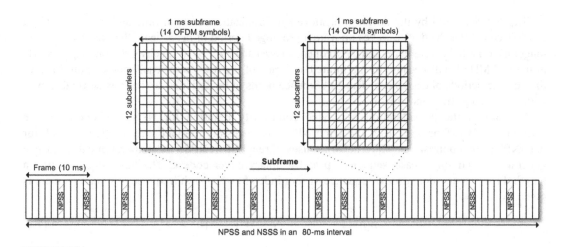

FIGURE 7.13

NPSS and NSSS transmissions within an 80-ms repetition interval.

NPSS and NSSS are designed to allow a device to use a unified synchronization algorithm during initial acquisition without knowing the NB-IoT deployment mode. This is achieved by avoiding collision with REs used by LTE as much as possible. For example, numbering subframes within a frame from 0 to 9, LTE may use any of subframes 1, 2, 3, 6, 7, or 8 as an MBSFN subframe. However, during initial cell acquisition the device does not know the deployment mode and whether any of these subframes is used as an MBSFN subframe. Collision avoidance with any possible LTE MBSFN subframe is achieved by using subframe 5 for NPSS and subframe 9 for NSSS as illustrated in Figures. 7.12 and 7.13. Furthermore, LTE may use up to the first three OFDM symbols in every subframe for PDCCH. To avoid potential collision with LTE PDCCH, the first three OFDM symbols are not used in the subframes that carry either NPSS or NSSS. This leaves only 11 OFDM symbols per subframe available for NPSS and NSSS (see Figure 7.13). These symbols are generated from frequency-domain Zadoff-Chu (ZC) sequences, as described in next two sections, and finally modulated into OFDM waveforms according to Section 7.2.6.2.

7.2.4.2.1 NPSS

Subframe	5
Subframe periodicity	10 ms
Sequence pattern periodicity	10 ms
Basic TTI	1 ms
Subcarrier spacing	15 kHz
Bandwidth	180 kHz
Carrier	Anchor

The NPSS is used by the device to achieve synchronization, in both time and frequency, to an NB-IoT cell. After the device wakes up from a prolonged period of deep sleep, the time base will no longer have a reliable reference, and the frequency base can be off by as much as 20 ppm (e.g., 18 kHz in the 900 MHz band) because of the limited accuracy of the low-power oscillator keeping track of time during periods of deep sleep. The NPSS hence needs to be designed so that it is detectable even with a very large frequency offset.

Because of the consideration of device complexity required for NPSS detection, all the cells in an NB-IoT network uses the same NPSS. As a result, a device only needs to search for one NPSS. In contrast, an LTE network uses three PSSs. NPSS is a hierarchical sequence generated based on a base sequence $\mathbf{p}$ and a binary cover code $\mathbf{c}$. The base sequence $\mathbf{p}$ is a length-11 frequency-domain ZC sequence of root index 5, whose nth frequency-domain element is given by

$$p(n) = e^{\frac{-j5\pi n(n+1)}{11}}, \quad n = 0, 1, ..., 10. \tag{7.4}$$

The binary cover code is $\mathbf{c} = (1,1,1,1,-1,-1,1,1,1,-1,1)$.

Each of the 11 OFDM symbols in an NPSS subframe carries a copy of the base sequence, either $\mathbf{p}$ or $-\mathbf{p}$, based on the binary cover code. The same NPSS sequence is repeated in all subframes that are designated to transmit NPSS. The hierarchical sequence design reduces the device complexity in searching for an NPSS subframe.

Resource mapping within an NPSS subframe is shown in Figure 7.14. For the in-band mode, a number of NPSS REs overlap with LTE CRS. The NPSS frequency-domain symbols on those REs will be punctured by the LTE CRS. However, the device performing NPSS detection does not need to be aware of such puncturing. For example, the device can simply correlate the receive signal with an unpunctured NPSS. Although there is a mismatch between the NPSS transmitted from the base station and the device locally generated NPSS in the in-band mode, the impact on NPSS detection performance is small because there is only a small percentage of NPSS symbols punctured.

7.2.4.2.2 NSSS

Subframe	9
Subframe periodicity	20 ms
Sequence pattern periodicity	80 ms
Subcarrier spacing	15 kHz
Bandwidth	180 kHz
Carrier	Anchor

After the device has performed coarse synchronization in time and frequency when acquiring the NPSS, it turns to NSSS to detect the cell identity and acquire more information about the frame structure.

NB-IoT supports 504 unique *physical cell identities* (PCIDs) indicated by NSSS. NSSS has an 80-ms repetition interval, within which four NSSS sequences are transmitted as shown in Figure 7.13.

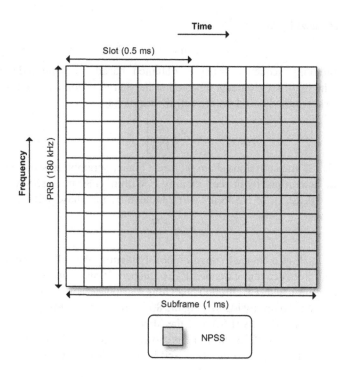

FIGURE 7.14

Resource mapping in a NPSS subframe.

The four NSSS sequences transmitted in an 80-ms repetition interval are all different; however, the same set of four sequences are repeated in every 80-ms repetition interval. As explained earlier, only the last 11 OFDM symbols in an NSSS subframe are used to carry NSSS. However, compared with NPSS, NSSS is mapped to all 12 subcarriers of the PRB resulting in 132 REs in an NSSS subframe for NSSS.

The frequency-domain symbols for these 132 NSSS REs are determined according to the sequence described below.

$$s(n) = b_q(n)e^{-j2\pi\theta_l n}\widetilde{z}_u(n), \quad n = 0, 1, ..., 131. \tag{7.5}$$

In essence, the NSSS for a cell with PCID k is determined by an extended ZC sequence $\widetilde{z}_u(n)$, a binary scrambling sequence $b_q(n)$, and a phase shift θ_l. The extended ZC sequence $\widetilde{z}_u(n)$ is obtained by first generating a length 131 ZC sequence of root u:

$$z_u(n) = e^{\frac{-ju\pi n(n+1)}{131}}, \quad n = 0, 1, ..., 130. \tag{7.6}$$

That in a second step is extended to length 132 by repetition of the first element:

$$\widetilde{z}_u(n) = z_u(n \bmod 131), \quad n = 0, 1, ..., 131. \tag{7.7}$$

The root is determined by the cell identity k as:

$$u = (k \bmod 126) + 3. \tag{7.8}$$

The binary scrambling sequence $b_q(n)$ is obtained based on length-128 Walsh–Hadamard sequence [12] with the first four elements repeated at the end to become a length-132 sequence. The sequence index q is determined based on the cell identity k as:

$$q = \left\lfloor \frac{k}{126} \right\rfloor. \tag{7.9}$$

In a cell, all the NSSS transmissions share the same binary scrambling sequence and extended ZC sequence as these are determined by the cell identity k. Within an 80-ms NSSS repetition interval, the four occurrences, $l \in \{0,1,2,3\}$, of NSSS are differentiated by the phase shift θ_l defined as:

$$\theta_l = \frac{33l}{132}. \tag{7.10}$$

NSSS is designed to allow a device to unambiguously identify the cell identity k through matching the binary scrambling sequence and extended ZC sequence. It also supports frame synchronization within the 80-ms repetition interval through matching the phase shift term. Note that because the duration of a radio frame is 10 ms, by identifying the 80 ms framing information, the device essentially knows the three least significant bits (LSBs) of the SFN.

Resource mapping within an NSSS subframe is shown in Figure 7.15. For the in-band mode, NSSS frequency-domain symbols mapped to the REs used by LTE CRS are punctured by CRS.

7.2.4.3 NRS

Subframe	Any
Basic TTI	1 ms
Sequence pattern periodicity	10 ms
Subcarrier spacing	15 kHz
Bandwidth	180 kHz
Carrier	Any

TTI, *transmission time interval.*

The NRS is used to allow the device to estimate the DL propagation channel coefficients and perform DL signal strength and quality measurements both in idle and connected mode procedures. It is mapped to certain subcarriers in the last two OFDM symbols in every slot within a subframe that carries NPBCH, NPDCCH, or NPDSCH. NRS may be transmitted also in subframes that do not have any NPDCCH or NPDSCH scheduled. For both channel estimation and DL measurement, it is important that there is no ambiguity regarding which subframes that a device can assume the presence of NRS. In Reference [12], NRS presence in different operation scenarios is elaborated. We highlight the important rules below.

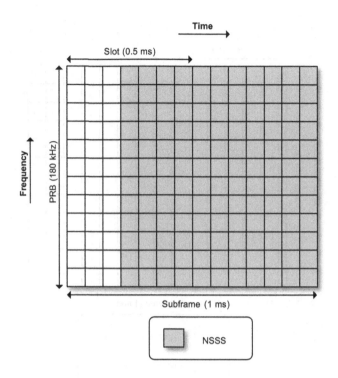

FIGURE 7.15

Resource mapping in a NSSS subframe.

- In all operation modes, NRS is present in subframes 0 and 4 as well as in subframe 9 not containing NSSS.
- In stand-alone and guard-band modes, NRS is also present in subframes 1 and 3.
- In all operation modes, NRS is present in all valid NB-IoT DL subframes (see Section 7.2.4.1).

Based on these rules, a device without knowing the operation mode can only assume the presence of NRS in subframes 0 and 4 as well as in subframe 9 not containing NSSS.

The exact subcarriers that NRS is mapped onto depend on the cell identity and logical antenna port number. NB-IoT supports one or two logical antenna ports using space-frequency transmit diversity (see Section 7.2.3.1). An example is shown in Figure 7.16, where NRS for the first antenna port and the second antenna port is illustrated. The NRS resource mapping pattern may shift up or down in the frequency domain depending on the cell identity. This is to allow adjacent cells to use orthogonal NRS resources to avoid mutual interference between NRS.

The NRS symbol sequence is generated based on cell identity and port number. In essence, a pseudorandom QPSK sequence is used for randomizing interference between cells. The NRS sequence repeats itself in every 10 ms radio frame.

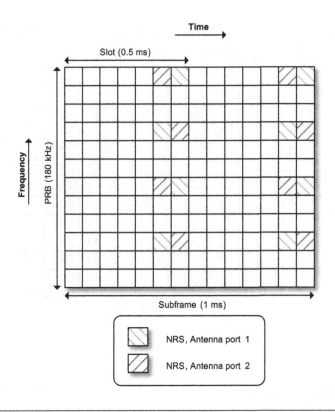

FIGURE 7.16

A resource mapping example for NRS in a subframe carrying NPBCH, NPDCCH, or NPDSCH, or valid NB-IoT DL subframe.

7.2.4.4 NPBCH

Subframe	0
Subframe periodicity	10 ms
Basic TTI	640 ms
Subcarrier spacing	15 kHz
Bandwidth	180 kHz
Carrier	Anchor

The NPBCH is used to deliver the NB-IoT *Master Information Block* (MIB), which provides essential information for the device to operate in the NB-IoT network. NPBCH uses a 640 ms TTI, but within the TTI only subframe 0 is used in every radio frame. A resource mapping example is shown in Figure 7.17. As mentioned earlier, the subcarriers used for NRS depends on NB-IoT cell identity. Thus, the NRS REs may shift up or down in the frequency domain. Some REs are reserved because they may be used by LTE, in case of in-band deployment. The LTE CRS may use REs

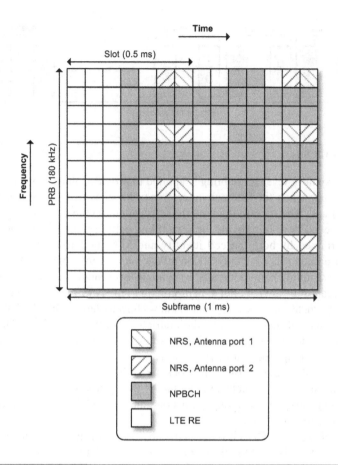

FIGURE 7.17

A resource mapping example for NPBCH.

outside of the potential LTE control region, i.e., the first three OFDM symbols. These REs may also shift up and down depending on the LTE cell identity. NB-IoT requires that the in-band deployment uses a cell identity that results in an NRS subcarrier set, e.g., subcarriers 2, 5, 8, and 11 in Figure 7.17, which is identical to the subcarrier set used by CRS of the hosting LTE cell. Thus when the device knows the NB-IoT cell identity, it also knows which REs are used for LTE CRS in case of in-band deployment.

As shown in Figure 7.17, there are 100 REs available for NPBCH in a subframe. NPBCH uses QPSK modulation, and thus 200 encoded bits can be carried in an NPBCH subframe. As stated above, NPBCH will only be transmitted in subframe 0 of the radio frame.

The TBS of NPBCH is 34 bits. A 16-bit CRC is attached to the TB. Together, these 50 bits are encoded using the LTE TBCC [4] and rate-matched to generate 1600 encoded bits. The encoded bits are segmented into eight *code subblocks* (CSBs), each of which is 200-bit long and mapped to 100 QPSK symbols. The OFDM modulation of the QPSK symbols is presented in Section 7.2.6.2 in detail.

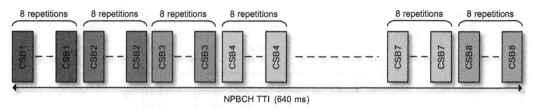

FIGURE 7.18

Transmission of NPBCH CSBs.

On each CSB, a symbol-level scrambling is applied to provide robust protection against intercell interference, especially when the intercell interference happens to be dominated by the NPBCH signal from another cell. The scrambling pattern is dependent on the cell identity and system frame number. This is achieved by re-initializing the scrambling pattern at the start of each radio frame with a seed that is determined by both the cell identity and SFN modulo 8. As a result, each CSB is scrambled to eight unique sets of 100 QPSK symbols mapped to subframe 0 in eight consecutive radio frames. The scrambling is implemented as symbol-level rotations that can easily be undone at the receiver side because the device after synchronizing to the NSSS knows the cell identity and the frame structure within an 80-ms interval. Thus, it knows how to descramble the NPBCH symbols and obtain the subframe symbol sequences over the multiple repetitions of a CSB. This facilitates, e.g., coherent combining of the repeated CSBs. Correlating repeated CSBs, after descrambling, is also a powerful tool for performing frequency offset estimation.

The transmission of the eight NPBCH subblocks in an NPBCH TTI is illustrated in Figure 7.18. Each NPBCH subframe is self-decodable, but all the NPBCH subframes can also be jointly decoded. For certain devices in good coverage, a single transmission of one CSB may be sufficient to decode the NPBCH information correctly.

The CRC attached to an NPBCH TB is masked with a sequence that is dependent on the number of NRS antenna ports (1 or 2). This allows the device to detect the number of NRS antenna ports through blind decoding. In the case of two NRS ports, Space-Frequency Block Code (SFBC) is used [12,13] (see Section 7.2.3.1). Every two consecutive NPBCH REs along the frequency dimension form an SFBC pair. While antenna port 0 transmits QPSK symbol pair (s_1, s_2) in these two REs, antenna port 1 transmits QPSK symbol pair $\left(-s_2^*, s_1^* \right)$.

7.2.4.5 NPDCCH

Subframe	Any
Basic TTI	1 ms
Repetitions	1, 2, 4, 8, 16, 32, 64, 128, 256, 512, 1024, 2048
Subcarrier spacing	15 kHz
Bandwidth	90 or 180 kHz
Carrier	Any

TTI, *transmission time interval.*

The NPDCCH is used to carry *Downlink Control Information* (DCI). A device needs to monitor NPDCCH for three types of information mentioned below.

- UL grant information (DCI format N0, 23 bits)
- DL scheduling information (DCI format N1, 23 bits)
- Indicator of paging or SI update (DCI format N2, 15 bits)

An NPDCCH subframe is divided into two narrowband control channel elements (NCCEs). NCCE0 takes the lowest six subcarriers and NCCE1 take the highest six subcarriers. The number of REs available for one NCCE depends on NB-IoT deployment modes and the number of logical antenna ports, and for the in-band deployment it further depends on the configuration of the LTE cell. Two examples are shown in Figure 7.19.

After cell selection and SI acquisition, the device will understand the exact mapping of NPDCCH REs in an NPDCCH subframe (see Section 7.3.1.2.2). For example, for the in-band deployment, NPDCCH is not mapped to the first few OFDM symbols in a subframe. This is to avoid the LTE DL

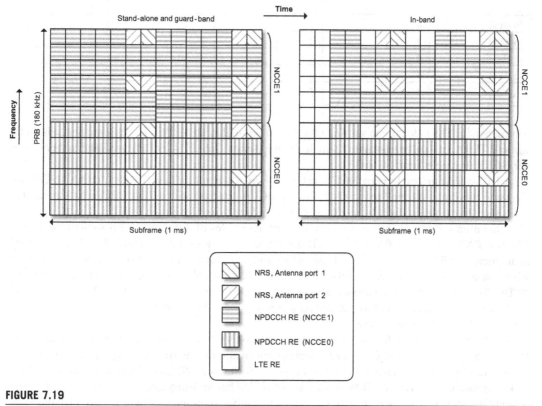

FIGURE 7.19

Two NPDCCH resource mapping examples. On the left: Stand-alone and guard-band deployment with two NRS ports. On the right: In-band deployment with two NRS ports, four CRS ports, and two OFDM symbols for LTE downlink control region.

Table 7.4 Number of REs available per NCCE

Operation Mode	Number of LTE CRS Antenna Ports	Number of OFDM Symbols for LTE Control Region	Number of NRS Antenna Ports	Number of REs per NCCE
Stand-alone, guard-band	N/A	0	1	80
Stand-alone, guard-band	N/A	0	2	76
In-band	2	1	1	68
In-band	2	1	2	64
In-band	2	2	1	62
In-band	2	2	2	58
In-band	2	3	1	56
In-band	2	3	2	52
In-band	4	1	1	66
In-band	4	1	2	62
In-band	4	2	1	60
In-band	4	2	2	56
In-band	4	3	1	54
In-band	4	3	2	50

control region. The index of the starting OFDM symbol within an NPDCCH subframe depends on the size of LTE DL control region and is signaled to the device. Table 7.4 lists all possible number of REs per NCCE. It ranges from 50 to 80.

A DCI can be mapped to one NCCE, referred to as *Aggregation Level* (AL) 1, or both NCCEs in the same subframe, referred to as AL 2. First, a DCI is attached with 16-bit CRC, which is masked with a sequence determined by a *Radio Network Temporary Identifier* (RNTI). The RNTI is an identifier used for addressing one or more devices. Examples of different RNTIs such as the *Paging RNTI* (P-RNTI) and *Cell RNTI* (C-RNTI) are made in Sections 7.3.1.4 and 7.3.2. After the CRC attachment and RNTI masking, TBCC encoding and rate-matching is used to generate a code word with a length matched to the number of encoded bits available. QPSK modulation is used for NPDCCH, and thus the code word length ranges from 100 to 160 for AL 1, or from 200 to 320 for AL 2. The baseband signal generation described in Section 7.2.6.2 takes the QPSK symbols as input and generates the baseband waveform.

NPDCCH AL 2 is used for increasing the coverage of NPDCCH. More REs used for transmitting a DCI message give rise to a higher energy level per information bit. Further coverage enhancements can be provided by subframe-level repetitions. Figure 7.20 shows an NPDCCH transmission using AL 2 and 8 repetitions. The NPDCCH bit stream is scrambled before being mapped to symbols. The same scrambling sequence is used in sets of four consecutive subframes, which means that the same NPDCCH symbols are transmitted in each of the four subframes. This allows the devices to optimize their performance through the use of coherent combining for received power estimation and for

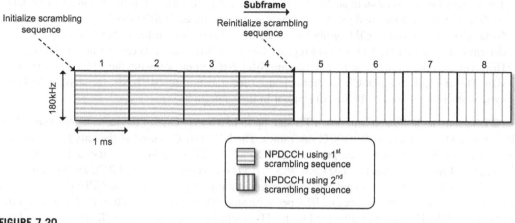

FIGURE 7.20

An example of NPDCCH transmission configured with eight repetitions.

frequency offset estimation. The scrambling sequence is reinitialized after transmitting every four repetitions as indicated by the figure to provide randomization of the transmitted waveform.

NB-IoT allows the same NPDCCH to be repeated up to 2048 times. Thus, in the worst case, a DCI is transmitted in 2048 subframes. To avoid a DCI blocking DL NPDCCH/NPDSCH resources for an extended period of time, a transmission gap can be configured through Radio Resource Configuration (RRC) signaling. More details about DL transmission gap will be discussed in Section 7.2.7.1.

7.2.4.6 NPDSCH

Subframe	Any
Basic TTI	1, 2, 3, 4, 5, 6, 8, 10 SF
Repetitions	1, 2, 4, 8, 16, 32, 64, 128, 192, 256, 384, 512, 768, 1024, 1536, 2048
Subcarrier spacing	15 kHz
Bandwidth	180 kHz
Carrier	Any

The NPDSCH is used to transmit unicast data. The data packet from high layers is segmented into one or more TBs, and NPDSCH transmits one TB at a time. NPDSCH is also used to transmit broadcast information such as system information (SI) messages.

NPDSCH has a similar subframe-level resource mapping as NPDCCH. There are, however, two differences:

- NPDCCH may multiplex resources in a subframe to transmit two DCI messages. One NPDSCH subframe, however, can at most carry one TB. That is, the basic RU for NPDSCH is one PRB pair.

- The starting OFDM symbol in an NPDSCH subframe may be different from that in an NPDCCH subframe in the in-band mode if the subframe is used to transmit SIB1-NB. Like in the case of NPDCCH, the starting OFDM symbol in an NPDSCH subframe in the in-band mode is determined based on the LTE control region size. Such information is carried in SIB1-NB. However, a device needs to be able to acquire SIB1-NB without knowing the LTE control region size. Therefore, if an NPDSCH subframe is used for transmitting SIB1-NB, the starting OFDM symbol position is always the fourth symbol in the subframe.

NPDSCH uses QPSK and supports a TB size up to 680 bits in case of device category *Cat-N1*. A TB is mapped to a number of NPDSCH subframes. The LTE TBCC is used as the only *forward-error correcting* code for NPDSCH. The processing of NPDSCH TB is as follows. First, a 24-bit CRC is calculated and attached to the TB. The CRC-attached TB is encoded using the TBCC encoder and rate-matched according to the code-word length determined jointly by the number of NPDSCH subframes allocated to the TB and the number of REs per subframe. Thus, the combination of TB size and the number of NPDSCH subframes allocated to the TB determines the coding rate. Table 7.5 lists all the combinations of TB size and resource allocation. The last two rows in Table 7.5 are not used for in-band deployments. This is because that there are fewer REs available to NPDSCH in a subframe in the in-band deployment compared to stand-alone and guard-band deployments. Using the last two rows in Table 7.5 may result in too high code rates in certain in-band configurations.

To limit the requirement on the receiver, only a single redundancy version is specified for the NPDSCH encoding.

For each of the combinations listed in Table 7.5, the code rate for a stand-alone deployment with two NRS ports is shown in Table 7.6. For a TB size, better coverage is achieved by allocating more NPDSCH subframes, giving a higher energy level per information bit and, in most cases, a higher coding gain as well.

Table 7.5 Combinations of NPDSCH TBS and number of NPDSCH subframes

NPDSCH TB Size Index	Number of Subframes (N_{SF})							
	1	2	3	4	5	6	8	10
0	16	32	56	88	120	152	208	256
1	24	56	88	144	176	208	256	344
2	32	72	144	176	208	256	328	424
3	40	104	176	208	256	328	440	568
4	56	120	208	256	328	408	552	680
5	72	144	224	328	424	504	680	Not used
6	88	176	256	392	504	600	Not used	Not used
7	104	224	328	472	584	680	Not used	Not used
8	120	256	392	536	680	Not used	Not used	Not used
9	136	296	456	616	Not used	Not used	Not used	Not used
10	144	328	504	680	Not used	Not used	Not used	Not used
11	176	376	584	Not used	Not used	Not used	Not used	Not used
12	208	440	680	Not used	Not used	Not used	Not used	Not used

Table 7.6 NPDSCH code rates for stand-alone deployment with two NRS ports

| NPDSCH TB Size Index | Number of NPDSCH Subframes (N_{SF}) | | | | | | | |
	1	2	3	4	5	6	8	10
0	0.13	0.09	0.09	0.09	0.09	0.10	0.10	0.09
1	0.16	0.13	0.12	0.14	0.13	0.13	0.12	0.12
2	0.18	0.16	0.18	0.16	0.15	0.15	0.14	0.15
3	0.21	0.21	0.22	0.19	0.18	0.19	0.19	0.19
4	0.26	0.24	0.25	0.23	0.23	0.24	0.24	0.23
5	0.32	0.28	0.27	0.29	0.29	0.29	0.29	Not used
6	0.37	0.33	0.31	0.34	0.35	0.34	Not used	Not used
7	0.42	0.41	0.39	0.41	0.40	0.39	Not used	Not used
8	0.47	0.46	0.46	0.46	0.46	Not used	Not used	Not used
9	0.53	0.53	0.53	0.53	Not used	Not used	Not used	Not used
10	0.55	0.58	0.58	0.58	Not used	Not used	Not used	Not used
11	0.66	0.66	0.67	Not used	Not used	Not used	Not used	Not used
12	0.76	0.76	0.77	Not used	Not used	Not used	Not used	Not used

As seen in Table 7.5, up to $N_{SF} = 10$ NPDSCH subframes can be used to carry one NPDSCH TB. Before mapping the bits of the encoded TB to QPSK symbols, the bits are scrambled. The scrambling is reinitialized every $\min(N_{REP}, 4)$ repetition of the code word, where N_{REP} is the number of configured repetitions. At most 2048 repetitions can be transmitted. After mapping the NPDSCH code word on a subframe, the subframe is repeated $\min(N_{REP}, 4)$ times before the mapping of the code word continues. Figure 7.21 shows the transmission of a two-subframe TB configured for eight repetitions. The first subframe is repeated four times before the mapping continues to the second subframe. After the second subframe has been repeated four times, the scrambling is reinitialized and the procedure is repeated once to complete eight repetitions of the code word in total.

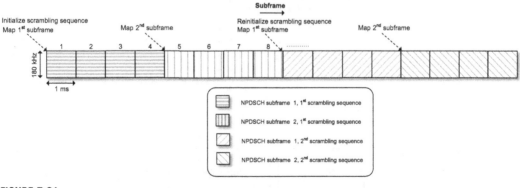

FIGURE 7.21

Transmission of a TB mapped to two subframes and configured with eight repetitions.

The repeated subframes do, just as in case of the NPDCCH, allow coherent combining for received power estimation and for frequency offset estimation. It also allows a device to attempt decoding of the code word before the transmission has completed. The example shown in Figure 7.21 does, for example, support decoding of the full code word already after eight subframes.

Because of the high number of supported repetitions, an NPDSCH TB may be mapped over 20,480 NPDSCH subframes. To avoid one long NPDSCH transmission blocking other NPDCCH or NPDSCH transmissions, NPDSCH transmission gaps can be configured. The concept is similar to the NPDCCH transmission gap and will be described in more detail in Section 7.2.7.1.

3GPP Release 14 introduces a new device category Cat-N2, which supports TBSs up to 2536 bits. The merits of Cat-N2 are presented in Section 7.4 in more detail.

As already indicated, the NPDSCH is also used for transmission of the SIBs. These transmissions are, however, following slightly different principles compared with those just described. For the SIB1-NB the TBS is always configured for $N_{SF} = 8$ subframes, and the available set of TBSs are limited to the set of $\{208, 328, 440, 680\}$ bits. For the other SIBs the TBS is configured for $N_{SF} = 2$ or 8 subframes, and the available set of TBSs are limited to the set of $\{56, 120, 208, 256, 328, 440, 552, 680\}$ bits. The two lowest TBSs use $N_{SF} = 2$ subframes while the six larger options are mapped to $N_{SF} = 8$ subframes.

The scrambling of the NPDSCH carrying a SIB is reinitialized for every repetition. The full code word is mapped to the full set of configured subframes N_{SF} before the repetition starts. This allows all devices in good coverage to decode a SIB already after the first transmission.

The mapping and scheduling of the SIB repetitions on the DL frame structure follows specific rules described in detail in Section 7.3.1.2.3.

The baseband generation of the NPDSCH signal uses the QPSK-modulated symbols as input and generates the transmitted waveform as described in Section 7.2.6.2.

7.2.5 UPLINK PHYSICAL CHANNELS AND SIGNALS

NB-IoT supports the set of UL physical channels and signals depicted in Figure 7.22.

7.2.5.1 Narrowband Physical Random Access Channel (NPRACH)

Subframe	Any
Basic TTI	5.6 or 6.4 ms
Repetitions	1, 2, 4, 8, 16, 32, 64, 128
Subcarrier spacing	3.75 kHz
Bandwidth	3.75 kHz
Carrier	Anchor

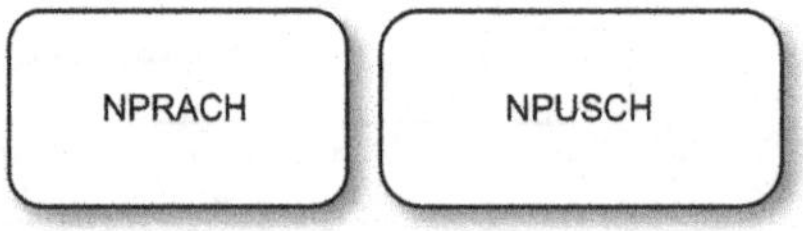

FIGURE 7.22

UL physical channels and signals used in NB-IoT.

Like the *Physical Random Access Channel* (PRACH) in LTE, the NPRACH in NB-IoT is used by the device to initialize connection and allows the serving base station to estimate the time of arrival (ToA) of the received NPRACH signal. The ToA of the received NPRACH signal reflects the round-trip propagation delay between the base station and device. Because NB-IoT UL employs OFDM-like transmission scheme (e.g., SC-FDMA or single-tone transmission with CP), it is important to align the received signals from multiple devices so that the orthogonality between different frequency-division multiplexed devices can be preserved. The ToA estimate helps the base station determine the *timing advance* (TA) for aligning the received signal for each device.

The LTE PRACH preambles are based on ZC sequences, spanning ~ 1 MHz bandwidth in frequency, much larger than the carrier bandwidth of NB-IoT. This motivates using a new NPRACH preamble waveform compared with LTE. Another important consideration is PA backoff and efficiency, which have profound impact on coverage and battery efficiency, respectively. A ZC sequence, although having a constant envelope as a time-domain discrete sequence, after transmit chain processing such as upsampling, *discrete-to-analog conversion*, and filtering, often ends up having a time-continuous waveform that has a *peak-to-average power ratio* (PAPR) greater than 3 dB [14]. There are a number of techniques developed to reduce the PAPR for a ZC sequence in the upsampling process [14,15]; however, the PAPR with such techniques is still not close to 0 dB. This is undesirable as it is advantageous to keep power backoff as low as possible. Power backoff results in a compromise on UL coverage as the PA cannot be used at its maximum configurable output power level. Power backoff also reduces PA efficiency, resulting in a compromise of battery lifetime. NPRACH preamble design targets a time-continuous waveform that has a PAPR close to 0 dB. As described in this section, NPRACH preambles are based on single-tone frequency-hopping waveforms, which indeed achieve a PAPR very close to 0 dB.

In NB-IoT, up to three NPRACH configurations can be used in a cell to support devices in different *coverage classes*. Different configurations are separated by using different time–frequency resources. Before we describe how NPRACH is configured, we first describe the waveform adopted for the NPRACH physical layer random access preambles.

NPRACH preambles use single-tone transmission with frequency hopping. The very basic unit in an NPRACH preamble is a *symbol group*, illustrated in Figure 7.23, which consists of a CP plus five single-tone symbols of tone frequency $n\Delta f_{NPRACH}$, where n is an integer number and Δf_{NPRACH} is 3.75 kHz, which is the NPRACH tone spacing. The value n is fixed within a symbol group. The

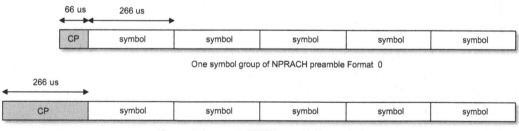

One symbol group of NPRACH preamble Format 0

One symbol group of NPRACH preamble Format 1

FIGURE 7.23

NPRACH symbol groups for Format 0 and Format 1, respectively.

resulting waveform of an NPRACH symbol group is thus a continuous phase sinusoidal of baseband frequency $n\Delta f_{\text{NPRACH}}$.

NB-IoT supports two NPRACH *formats*: NPRACH Format 0 using a CP duration 66.7 μs and NPRACH Format 1 using a CP duration 266.67 μs for supporting cell radii up to at least 10 and 40 km, respectively.

The basic NPRACH repetition unit consists of four symbol groups, with special relationship between tone frequencies within a repetition unit as the example illustrated in Figure 7.24. The example shown in Figure 7.24 uses the deterministic tone hopping pattern between the four symbol groups in a repetition unit; that is, the second symbol group uses a tone that is right above the tone used by the first symbol group, the third symbol group uses a tone that is six tones above the tone used by the second symbol group, and the fourth symbol group uses a tone that is right below the tone used by the third symbol group.

The deterministic tone hopping pattern within a repetition unit is designed for the base station to estimate the TA in the presence of an unknown device residual frequency offset. Note that as mentioned earlier, within a symbol group, however, the same frequency tone is used. In essence, the time–frequency relationship between the four symbol groups allows the base station to solve for the two unknowns of TA and device residual frequency offset. The hopping pattern uses a band of 12 tones, and the tones within the band can be indexed by 0,1,…,11. There are four possible deterministic frequency-hopping patterns within an NPRACH repetition unit. These are shown in Table 7.7.

These deterministic hopping patterns create a set of 12 orthogonal NPRACH preambles within a repetition unit as summarized in Table 7.8. The baseband signal generation of the NPRACH preambles is explained in Section 7.2.6.1 in detail, which also illustrates the use of the parameters presented in Table 7.8.

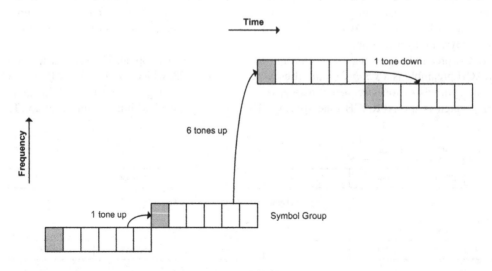

FIGURE 7.24

One NPRACH preamble repetition unit and an example of the tone relationship between the four symbol groups.

Table 7.7 Deterministic hopping patterns within an NPRACH repetition unit

Index of the Tone Used by the First Symbol Group	Deterministic Hopping Patterns Within a Repetition Unit
0, 2, 4	{+1, +6, −1}
1, 3, 5	{−1, +6, +1}
6, 8, 10	{+1, −6, −1}
7, 9, 11	{−1, −6, +1}

Table 7.8 An example of NPRACH preambles defined for a repetition unit

NPRACH Preamble	Tone Index $k(l)$ for Symbol Group l			
0	0	1	7	6
1	1	0	6	7
2	2	3	9	8
3	3	2	8	9
4	4	5	11	10
5	5	4	10	11
6	6	7	1	0
7	7	6	0	1
8	8	9	3	2
9	9	8	2	3
10	10	11	5	4
11	11	10	4	5

Using one NPRACH repetition unit alone, however, is not sufficient for meeting the aggressive NB-IoT coverage target. To ensure that NPRACH signals from devices in coverage challenging locations can be detected reliably at the base station, NPRACH preambles may be configured with 1, 2, 4, 8, 16, 32, 64, or 128 repetition units. When the number of NPRACH repetitions increases, it is desirable to avoid persistent interference between NPRACH preambles in different cells. To achieve this objective, pseudorandom hopping is introduced between different repetition units. This is done by applying a pseudorandom integer tone offset χ to each and every one of the tone indexes in Table 7.8. Because the hopping range needs to be within 12 tones, the offset is applied in the modulo-12 sense, keeping the tone indexes in the set of {0,1,2,…,11}. The pseudorandom tone offset χ is determined by both the cell identity and the repetition index. Thus, the NPRACH preambles in different cells will not end up having the same hopping pattern over the entirety of its transmission interval, thereby avoiding persistent interference.

An NB-IoT cell can configure up to three NPRACH configurations, each supporting a set of preambles and repetition units. Each configuration targets a specific coupling loss. The number of preambles corresponds to the number of supported access attempts. Information about NPRACH configurations is provided in the SI. Parameters in an NPRACH configuration include the number of repetitions, number of NPRACH preambles, the time periodicity by which they are reoccurring, a signal-level threshold, etc. A few configurations are illustrated in Section 7.2.5.4, and the random access procedure including the meaning of these parameters is described in Section 7.3.1.6.

7.2.5.2 Narrowband Physical Uplink Shared Channel (NPUSCH)

Subframe	Any
RU	1, 2, 4, 8, 32 ms
Repetitions	1, 2, 4, 8, 16, 32, 64, 128
Subcarrier spacing	3.75, 15 kHz
Bandwidth	3.75, 15, 45, 90, 180 kHz
Carrier	Any

NPUSCH is used to carry UL user data and control information from higher layers. Additionally, NPUSCH also carries Hybrid Automatic Repeat Request (HARQ) acknowledgment for NPDSCH. The waveform adopted by NPUSCH is in principle the same as the LTE SC-FDMA waveform. However, in LTE, SC-FDMA supports device bandwidth with a granularity of one PRB, i.e., 12 subcarriers. A device may be scheduled for $12K$ subcarriers, where K is a positive integer. Thus, the minimum device-scheduled bandwidth allocation in LTE is one PRB or 12 subcarriers. However, because NB-IoT uses only one PRB, the maximum device-scheduled bandwidth can be only one PRB. The considerations below motivate NB-IoT to include lower device-scheduled bandwidth options:

- NB-IoT targets ultra-low-end IoT use cases, and it is envisioned that such use cases often have small data packets. Thus, in many cases, a device may not need to use the entire radio resources of one PRB (180 kHz).
- NB-IoT targets devices in coverage-limited scenarios. These devices operate in power limited regime, rather than in bandwidth limited regime, and thus do not benefit from having higher device bandwidth (see the discussion on this aspect in Sections 7.1.2.4 and 8.2.3).
- As mentioned in the NPRACH discussion above, having a low PAPR waveform is important for coverage and battery lifetime, mainly for devices at the edge of the network coverage. NPUSCH needs to include a waveform that has a PAPR close to 0 dB to best serve devices at the edge of coverage.

NPUSCH thus adds sub-PRB scheduled bandwidth options, including two special cases of single-tone transmissions, which have the advantages of having a close to 0 dB PAPR.

NPUSCH employs two transmission formats depending on the data it carries. Format 1 is used for UL data transfer and uses the same turbo code as used in LTE [4] for error correction. The maximum TBS of NPUSCH. Format 1 is 1000 bits in case of device category Cat-N1. Format 2 is used for signaling HARQ feedback for NPDSCH and uses a repetition code for error correction. Both Format 1 and Format 2 use SC-FDMA waveform involving DFT-precoding and CP insertion in waveform generation process. As mentioned earlier, NPUSCH supports both multitone and single-tone device

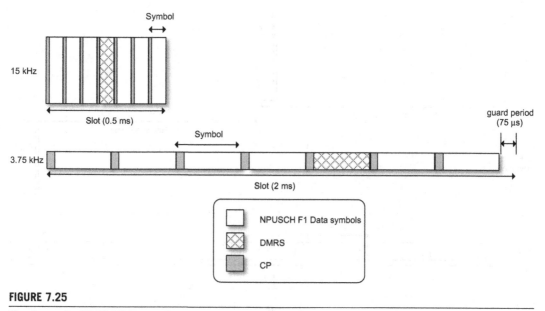

FIGURE 7.25

Slot format for NPUSCH Format 1.

scheduling bandwidth options. Waveform generation in the single-tone case may omit the DFT-precoding. All the multitone NPUSCH transmissions are based on 15 kHz subcarrier spacing; however, the single-tone transmissions use either 15 or 3.75 kHz subcarrier spacing.

The slot formats applicable to NPUSCH Format 1 are illustrated in Figure 7.25. The definition of CP and OFDM symbol durations for 15 kHz subcarrier spacing are identical to those in LTE, as described in Section 7.2.3.2. For single-tone transmissions with 3.75 kHz subcarrier spacing, NB-IoT introduces new numerologies as shown in Figure 7.25. The slot duration in this case is 2 ms, consisting of seven SC-FDMA symbols and a guard period at the end. Each SC-FDMA symbol is 275 μs, including a CP of 8.33 μs. The guard period is created to avoid collision with LTE *Sounding Reference Signal* (SRS), which is a signal transmitted by an LTE device to facilitate the base station to estimate the channel quality. An LTE device may be configured to transmit SRS on a PRB that is used as an NB-IoT carrier. SRS can only use the last OFDM symbol in an LTE subframe and can be configured with a periodicity as small as 2 ms and as large as 320 ms [12]. An SRS collision example is illustrated in Figure 7.26. LTE devices 1 and 2 are configured with SRS transmission. The collision with the SRS transmitted by LTE device 2 is avoided by having the guard period in the slot format of NPUSCH with 3.75 kHz subcarrier spacing. However, SRS collision may still occur on the fourth OFDM symbol, as shown in Figure 7.26. Collisions between SRS and NPUSCH symbols are avoided by puncturing the NPUSCH symbols.

For 15 kHz subcarrier spacing, the middle OFDM symbol in each slot is used as demodulation reference symbol (DMRS), which allows the base station to estimate the UL propagation conditions. The DMRS design is described in Section 7.2.5.3. For 3.75 kHz subcarrier spacing, the placement of

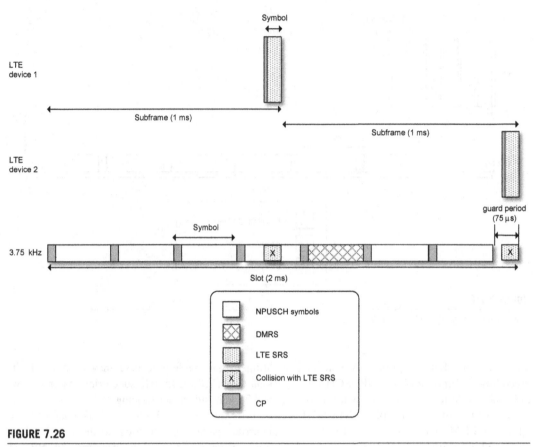

FIGURE 7.26

An example of NPUSCH collision with LTE SRS.

the DMRS is shifted to the fifth OFDM symbol in the slot. This is also for the consideration of SRS avoidance. It can be seen in Figure 7.26 that the fourth symbol in the slot may have collision with SRS.

NPUSCH Format 2 uses the same slot formats; however, three OFDM symbols are used as DMRS, leaving only four information-bearing symbols per slot, as illustrated in Figure 7.27. Like in the case of NPUSCH Format 1, the placements of DMRS are designed to avoid collision with LTE SRS. For 15 kHz subcarrier spacing, using the middle 3 OFDM symbols avoids collision with LTE SRS, and for 3.75 kHz, using the first 3 OFDM symbols achieves the same.

The basic NPUSCH time scheduling unit is referred to as RU. It is specified in terms of number of slots and is dependent on user bandwidth allocation and NPUSCH format. Table 7.9 summarizes the definition of RU for various NPUSCH configurations. A device may be scheduled with 1, 2, 3, 4, 5, 6, 8, or 10 RUs per repetition, resulting in a transmission interval per repetition as short as 1 ms, using one RU for 180 kHz scheduled bandwidth, and as high as 320 ms, using 10 RUs for 3.75 kHz scheduled bandwidth. The numbers of data symbols per RU for the various NPUSCH configurations are shown in Table 7.9.

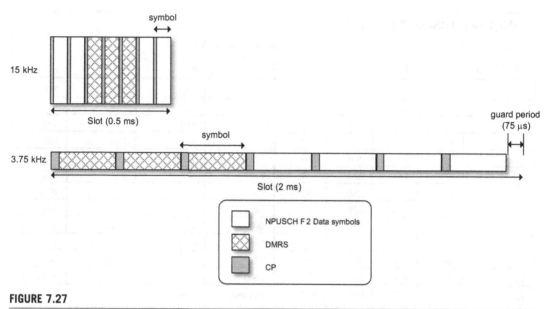

FIGURE 7.27

Slot format for NPUSCH Format 2.

Table 7.9 Number of slots and number of data symbols per NPUSCH RU

NPUSCH Format	Device Scheduled Bandwidth (kHz)	Number of Slots per RU N_{SLOTS}	Number of OFDM Symbols per Slot N_{SYMB}	Length of RU [ms]	Number of REs for Data per RU
Format 1	180	2	7	1	144
	90	4	7	2	144
	45	8	7	4	144
	15	16	7	8	96
	3.75	16	7	32	96
Format 2	15	4	7	2	16
	3.75	4	7	8	16

The TBSs supported for NPUSCH Format 1 are shown in Table 7.10A. The smallest TBS is 16-bit long and the largest is 1000-bit long. The TBSs given in the last two rows in Table 7.10A are used only for multitone transmissions because the block sizes are too large for single-tone transmission. All the other rows are used for both single-tone and multitone transmissions. A 24-bit CRC is calculated and attached to a TB. Afterward, the LTE turbo code is used for encoding, based on the same LTE mother code and rate-matching scheme [4]. NPUSCH Format 1 supports incremental redundancy; however, only redundancy versions 0 and 2 as defined in LTE [4] are used. Combination of redundancy

Table 7.10A TBSs for NPUSCH Format 1

TBS (I_{TBS})	Number of RUs (N_{RU})							
	1	2	3	4	5	6	8	10
0	16	32	56	88	120	152	208	256
1	24	56	88	144	176	208	256	344
2	32	72	144	176	208	256	328	424
3	40	104	176	208	256	328	440	568
4	56	120	208	256	328	408	552	696
5	72	144	224	328	424	504	680	872
6	88	176	256	392	504	600	808	1000
7	104	224	328	472	584	712	1000	Not used
8	120	256	392	536	680	808	Not used	Not used
9	136	296	456	616	776	936	Not used	Not used
10	144	328	504	680	872	1000	Not used	Not used
11	176	376	584	776	1000	Not used	Not used	Not used
12	208	440	680	1000	Not used	Not used	Not used	Not used

Table 7.10B Code rates for multitone NPUSCH Format 1 transmissions

NPUSCH TB size index	Number of RUs							
	1	2	3	4	5	6	8	10
0	0.14	0.10	0.09	0.10	0.10	0.10	0.10	0.10
1	0.17	0.14	0.13	0.15	0.14	0.13	0.12	0.13
2	0.19	0.17	0.19	0.17	0.16	0.16	0.15	0.16
3	0.22	0.22	0.23	0.20	0.19	0.20	0.20	0.21
4	0.28	0.25	0.27	0.24	0.24	0.25	0.25	0.25
5	0.33	0.29	0.29	0.31	0.31	0.31	0.31	0.31
6	0.39	0.35	0.32	0.36	0.37	0.36	0.36	0.36
7	0.44	0.43	0.41	0.43	0.42	0.43	0.44	Not used
8	0.50	0.49	0.48	0.49	0.49	0.48	Not used	Not used
9	0.56	0.56	0.56	0.56	0.56	0.56	Not used	Not used
10	0.58	0.61	0.61	0.61	0.62	0.59	Not used	Not used
11	0.69	0.69	0.70	0.69	0.71	Not used	Not used	Not used
12	0.81	0.81	0.81	0.89	Not used	Not used	Not used	Not used

versions 0 and 2 gives a higher coding gain than versions 0 and 1. Rate-matching is done based on the modulation scheme and number of data symbols available in a transmission interval per repetition. For all the multitone transmissions, QPSK is used. For single-tone transmissions, both 15 and 3.75 kHz numerologies, either BPSK or QPSK, are used, depending on the TBS index. BPSK is used for $I_{TBS} = 0$, or 2, and all the other I_{TBS} values use QPSK. The reason to extend BPSK to $I_{TBS} = 2$ is to allow BPSK to be used for higher TBS, up to 424 bits. Similarly, using QPSK for $I_{TBS} = 1$ allows QPSK to be extended for the use of smaller TBSs. The BPSK and QPSK modulations are later converted to $\pi/2$-BPSK and $\pi/4$-QPSK, respectively, in the baseband signal generation process (see Section 7.2.6.1). The $\pi/2$ and $\pi/4$ rotations for single-tone transmissions are intended for reducing the PAPR, aiming to improve the PA efficiency. For each of the combinations listed in Table 7.10A, the code rate for multitone NPUSCH Format 1 transmissions is shown in Table 7.10B. For a TB size, better coverage is achieved by allocating more RUs, giving a higher energy level per information bit, and in most cases, a higher coding gain as well.

The processing of NPUSCH Format 2 differs from NPUSCH Format 1 in that there is no CRC attachment and it is only based on a simple repetition code, which is essentially repeating the HARQ feedback bit by 16 times as shown in Table 7.11. NPUSCH Format 2 uses only BPSK and like the case of Format 1 converted to $\pi/2$-BPSK in the baseband signal generation process to help reduce the PAPR.

Before modulating the bits of the encoded TB to BPSK or QPSK symbols, the bits are scrambled. If a single-tone transmission is used, then the scrambling is reinitialized for every repetition. For multitone transmission, the scrambling is reinitialized every $\min(N_{REP}/2, 4)$ repetition of the code word, where N_{REP} is the number of configured repetitions. Also, the redundancy version is changed at the same time. At most 128 repetitions can be transmitted.

For the 15 kHz subcarrier spacing, after mapping the code word on a pair of slots, the pair of slots are repeated $\min(N_{REP}/2, 4)$ times before the mapping of the code word continues. In case of 3.75 kHz transmission the mapping is done on a single slot before repeating.

Figure 7.28 shows the transmission of a TB configured on 12 subcarriers, 2 RUs, and 8 repetitions. The first pair of timeslots 1, 2 are repeated four times before the mapping continues to the second pair of timeslots 3, 4. After 16 timeslots the full code word has been repeated four times and the scrambling is reinitialized and the redundancy version is updated. The procedure is then repeated once to complete eight repetitions of the TB in total.

The repeated slots do, just as in case of the NPDCCH and NPDSCH, allow for coherent combining for received power estimation and for frequency offset estimation. It also allows a base station to attempt decoding of the code word before the transmission has completed. The example shown in Figure 7.28 supports decoding of the full code word already after 16 timeslots.

Table 7.11 Repetition code used for NPUSCH Format 2	
HARQ ACK Bit	**NPUSCH Format 2 Code Word**
0	0,0,0,0,0,0,0,0,0,0,0,0,0,0,0,0
1	1,1,1,1,1,1,1,1,1,1,1,1,1,1,1,1

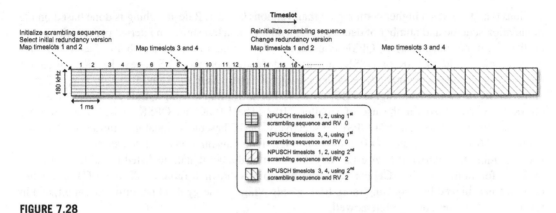

FIGURE 7.28

NPUSCH F1 transmission configured with 12 subcarriers, two RUs and eight repetitions.

7.2.5.3 Demodulation Reference Signal (DMRS)

Subframe	Any
TTI	Same as associated NPUSCH
Repetitions	Same as associated NPUSCH
Subcarrier spacing	3.75, 15 kHz
Bandwidth	Same as associated NPUSCH
Carrier	Any

DMRS is always associated with NPUSCH, either Format 1 or 2, and is transmitted in every NPUSCH slot, as shown in Figures 7.25 and 7.27. The bandwidth of DMRS is identical to the associated NPUSCH.

NB-IoT DMRS of 180 kHz bandwidth reuses the LTE DMRS sequences defined for one PRB [12]. New DMRS sequences are introduced to support NB-IoT DMRS with bandwidth smaller than 180 kHz. For all the multitone formats, the DMRS sequences are QPSK sequences, and for the single-tone formats, either 15 or 3.75 kHz bandwidth, BPSK sequences are used. We will use the NPUSCH Format 1 three-tone DMRS as an example to describe the key features of DMRS. Interested readers can refer to Reference[12] for more details.

The base DMRS sequences for three-tone transmissions are associated with a length-3 frequency-domain sequence of the form $(e^{j\phi(0)\pi/4}, e^{j\phi(1)\pi/4}, e^{j\phi(2)\pi/4})$ with $\phi(n)$ determined by one of the 12 base sequences shown in Table 7.12. Each element of the sequence is mapped to an RE in the frequency domain, i.e., to one of the three transmitted tones. Because $\phi(n) \in \{\pm 1, \pm 3\}$, each element in the basic sequence is essentially a QPSK symbol. A device is given a base sequence index through a higher layer parameter, or, if such a higher layer parameter is not given, the device determines the base sequence index by PCID, i.e., $u = $ PCID mod 12.

A base sequence can be applied with cyclic shift α. The cyclic shift α therefore gives rise to a DMRS sequence of the form $(e^{j\phi(0)\pi/4}, e^{j\alpha}e^{j\phi(1)\pi/4}, e^{j2\alpha}e^{j\phi(2)\pi/4})$. The value of α is given to the device through a higher layer parameter.

Table 7.12 DMRS basic sequences for three-tone transmissions

Base Sequence Index, u	$\phi(0)$	$\phi(1)$	$\phi(2)$
0	1	−3	−3
1	1	−3	−1
2	1	−3	3
3	1	−1	−1
4	1	−1	1
5	1	−1	3
6	1	1	−3
7	1	1	−1
8	1	1	3
9	1	3	−1
10	1	3	1
11	1	3	3

To randomize interference, DMRS sequence hopping can be optionally used when it is associated with NPUSCH Format 1. In that case, the base sequence index varies from slot to slot following a pseudorandom pattern. The pseudorandom pattern is cell specific and depends on PCID. If DMRS sequence hopping is not activated, the DMRS symbols are repeated across the full length of the NPUSCH Format 1 transmission.

7.2.5.4 NPRACH and NPUSCH Multiplexing

The resources for NPRACH and NPUSCH are time and frequency multiplexed. An example is given in Figure 7.29 where NPRACH resources reserved for coverage classes 0, 1, and 2 are identified. Coverage class 0 can be said to correspond to normal coverage, while coverage classes 1 and 2 are facilitating system access for users in extended and extreme coverage, respectively. Coverage class 0 may not need to provide NPRACH repetitions and is therefore of short time duration. Because most users are located in normal coverage, it spans many tones, or preambles, and appears with a short periodicity to facilitate large capacity. Coverage classes 1 and 2 are configured for support of NPRACH repetitions to facilitate system access, for example, from deep indoors. They are expected to require less tones than coverage class 0 because of relatively fewer devices in the extended and extreme coverage domains.

Except for the resources carved out for NPRACH, most of the UL resources are available for NPUSCH transmissions. The NPRACH resources are signaled in one of the SIBs, and thus the device knows exactly which resources are set aside for NPRACH. During an NPUSCH transmission, if an overlap with NPRACH resources occurs, NPUSCH transmission needs to be postponed until the first available UL slot without overlap.

The NPRACH is further described in the context of the random access procedure in Section 7.3.1.6.

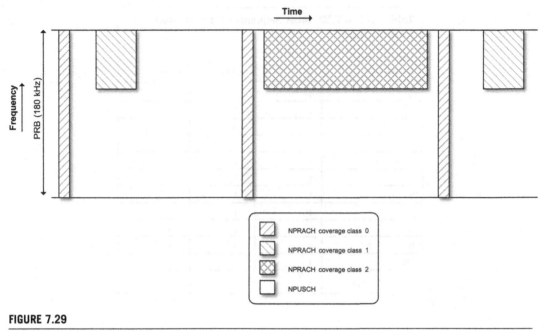

FIGURE 7.29

Multiplexing of NPRACH and NPUSCH resources.

7.2.6 BASEBAND SIGNAL GENERATION

7.2.6.1 Uplink

7.2.6.1.1 Multitone NPUSCH

The multitone NPUSCH baseband signal generation is based on the same principle that is used for LTE PUSCH. The data symbols are DFT precoded into a set of frequency-domain complex symbols $a(k, l)$ [12]. Elements of the DMRS sequence, however, are directly used as frequency-domain complex symbols $a(k, l)$. The symbol $a(k, l)$ defines the transmitted waveform as it modulates the k-th tone of the l-th OFDM symbol through the *inverse DFT* (IDFT):

$$s_l(t) = \sum_{k=0}^{11} a(k,l)e^{j2\pi(k-5.5)\Delta f(t-N_{cp}(l)T_s)}, \quad 0 \le t \le (N + N_{cp}(l))T_s \quad (7.11)$$

where Δf is the subcarrier spacing of 15 kHz, $N_{cp}(l)$ is the number of samples for the CP of the l-th OFDM symbol, T_s is the *basic time unit*, and N is 2048. The frequency grid illustrated in Figure 7.30 defines the tone-index k and the absolute tone frequency $(k - 5.5)\Delta f$. Note that in Eq. (7.11) if a tone with index k is not allocated to the device, $a(k, l)$ is 0.

The basic time unit in Eq. (7.11) is specified as in LTE assuming a sampling rate of 30.72 MHz, i.e., $T_s = 1/30.72$ µs. In practice though, NB-IoT UL baseband signal generation can be based on a much lower sampling rate as the signal bandwidth is not higher than 180 kHz. One straightforward approach is to use a sampling rate of 1.92 MHz, as at this sampling rate the CP durations in different OFDM

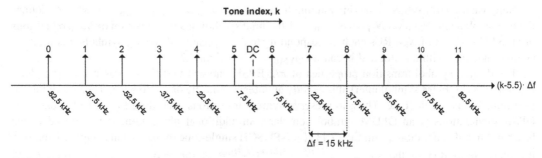

FIGURE 7.30

Frequency grids used for uplink baseband signal generation (15 kHz subcarrier spacing).

Table 7.13 CP duration in terms of number of samples at different sampling rates for the 15 kHz numerology		
Parameter	**First OFDM Symbol in a Slot**	**Other OFDM Symbols in a Slot**
Time	5.21 μs	4.69 μs
Samples at 30.72 MHz	160 samples	144 samples
Samples at 1.92 MHz	10 samples	9 samples

symbol periods amount to integer numbers. CP duration in terms of number of samples at different sampling rates is summarized in Table 7.13.

7.2.6.1.2 Single-Tone NPUSCH

For single-tone NPUSCH with $\Delta f = 15$ kHz, one obvious approach is to take only one k term in Eq. (7.11) as the baseband signal. However, one desired characteristic for single-tone baseband waveform is low PAPR. To achieve this, $\pi/2$ or $\pi/4$ rotation is introduced in the baseband signal generation process, which essentially converts BPSK and QPSK to $\pi/2$-rotated BPSK and $\pi/4$-rotated QPSK modulation constellations, respectively. This allows NPUSCH to avoid having symbol transitions going through the origin of the in-phase and quadrature-phase (IQ) plane, as such transitions increase the PAPR. The rotation can be expressed as a modification of the QPSK or BPSK symbol $a(k, l)$ modulating the k-th tone of the l-th OFDM symbol:

$$\tilde{a}(k, l) = a(k, l)e^{j\varphi_{l'}}$$

$$\varphi_{l'} = \rho \times (l' \bmod 2)$$

$$\rho = \begin{cases} \dfrac{\pi}{2} & \text{for BPSK} \\[2mm] \dfrac{\pi}{4} & \text{for QPSK} \end{cases} \tag{7.12}$$

Here, we use l' for symbol indexing ranging from 0 to $N_{REP}N_{RU}N_{SLOTS}N_{SYMB} - 1$, where 0 defines the first symbol and $N_{REP}N_{RU}N_{SLOTS}N_{SYMB} - 1$ the final symbol in a transmission of N_{REP} repetitions of an NPUSCH TB of N_{RU} RUs each corresponding to N_{SLOTS} slots of N_{SYMB} symbols. l is used for symbol indexing within a slot and because $N_{SYMB} = 7$, $l = l'$ mod 7.

The desired symbol transition properties of $\pi/2$-BPSK and $\pi/4$-QPSK are preserved if the phase rotation in one OFDM symbol interval is exactly an integer multiple of π. However, because of the CP insertion, this is not the case. Thus, to preserve symbol transitions according to $\pi/2$-BPSK and $\pi/4$-QPSK modulations at an OFDM symbol boundary, an additional phase term is introduced in the baseband signal generation. From Eq. (7.11) for NPUSCH single-tone transmission using tone-index k, the phase term due to the sinusoid $e^{j2\pi(k-5.5)\Delta f(t-N_{cp}(l)T_s)}$ at the end of symbol 0 (i.e., $l = 0$, $t = (N + N_{cp}(0))T_s$) is $\phi_e(0) = 2\pi(k - 5.5)\Delta f N T_s$ and the phase term at the beginning of symbol 1 (i.e., $l = 1, t = 0$) is $\phi_b(1) = -2\pi(k - 5.5)\Delta f N_{cp}(1)T_s$. Therefore, the additional phase term that needs to be introduced to compensate for the phase discontinuity at the boundary between symbols 0 and 1 is

$$\phi(1) = \phi_e(0) - \phi_b(1) = 2\pi(k - 5.5)\Delta f(N + N_{cp}(1))T_s. \tag{7.13}$$

To compensate for the phase discontinuities at subsequent symbol boundaries, the phase term needs to be accumulated over subsequent symbol periods. Thus, $\phi(0) = 0$, and

$$\varphi(l') = \varphi(l' - 1) + 2\pi(k - 5.5)\Delta f\big(N + N_{cp}(l' \bmod 7)\big)T_s. \tag{7.14}$$

The baseband waveform for single-tone NPUSCH using tone-index k can be described mathematically as

$$s_{k,l'}(t) = \tilde{a}(k, l')e^{\varphi(l')}e^{j2\pi(k-K)\Delta f\big(t-N_{cp}(l' \bmod 7)T_s\big)}, \quad 0 \le t \le \big(N + N_{cp}(l' \bmod 7)\big)T_s, \tag{7.15}$$

where $a(k, l')$ is the modulation value of the l'-th symbol according to either $\pi/2$-BPSK or $\pi/4$-QPSK modulation. In case of $\Delta f = 15$ kHz, $K = 5.5$ and the tone-index k is defined according to Figure 7.30. Eq. (7.15) can be reused for single-tone NPUSCH with $\Delta f = 3.75$ kHz, by using $K = 23.5$ and the tone-index k is defined according to Figure 7.31.

7.2.6.1.3 NPRACH
The NPRACH symbol group baseband signal of unit power is given by:

$$s_l(t) = e^{j2\pi(k(l)-23.5)\Delta f(t-T_{cp})}, \quad 0 \le t \le T \tag{7.16}$$

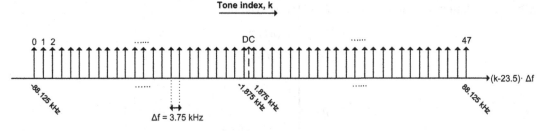

FIGURE 7.31

Frequency grids used for uplink baseband signal generation. (3.75 kHz subcarrier spacing)

Table 7.14 NPRACH Format 1 and 2

	NPRACH Format 0	NPRACH Format 1
Supported cell size	10 km	40 km
Symbol group length, T	1396 μs	1596 μs
Cyclic prefix, T_{cp}	66 μs	266 μs

where $k(l)$ is the tone from the frequency grid depicted in Figure 7.31 selected for transmission of the NPRACH symbol group l in an NPRACH repetition unit. The relation between k and l is exemplified in Table 7.8. Δf equals the NPRACH subcarrier spacing of 3.75 kHz, while the length T and CP T_{cp} are presented in Table 7.14.

When comparing Eq. (7.16) with the NPUSCH definition in Eq. (7.15) it is seen that the additional phase term $e^{\phi(l')}$ is not implemented for NPRACH. For NPRACH preamble Format 1, the phase rotation over a symbol group is equal to an integer multiple of 2π. For NPRACH Format 0, it can be shown that the transition between the symbol groups will not go through the origin.

7.2.6.2 Downlink

In the DL case a unified baseband definition applies for the NPSS, NSSS, NPBCH, NPDSCH, and NPDCCH. The baseband signal generation is based on the frequency grids with 15 kHz subcarrier spacing illustrated in Figure 7.30 with the DC carrier located in between the two center subcarriers. The baseband waveform is also the same across all three modes of operation to allow for a single device receiver implementation regardless of the network mode of operation.

The actual implementation of the baseband signal generation may, however, take different forms depending on the deployment scenario. For the stand-alone deployment scenario where the base station transmitter is likely only generating the NB-IoT signal, the baseband implementation may follow Eq. (7.11). Eq. (7.11) may also be used for guard-band and in-band deployments, when the NB-IoT cell and LTE cell are configured with different cell identities implying that there is no specified relation between the NRS and LTE CRS.

In the in-band scenario it is, however, convenient to generate the LTE and NB-IoT signals jointly. This can be achieved by using the LTE baseband definition to basically generate both LTE and NB-IoT signal using the same IDFT process. Consider, e.g., a 20 MHz LTE carrier containing 100 PRBs plus one extra subcarrier located in the center of the carrier to eliminate the impact from the DC component. An NB-IoT nonanchor carrier located at PRB 0 and spanning the first 12 of the 1201 subcarriers would then be generated as:

$$s_l(t) = \sum_{k=0}^{11} a(k,l)e^{j2\pi(k-600)\Delta f(t-N_{cp}(l)T_s)}, \quad 0 \le t \le (N+N_{cp}(l))T_s \qquad (7.17)$$

When comparing Eqs. (7.17) and (7.11), it becomes evident that the joint LTE and NB-IoT generation in Eq. (7.17) introduces a frequency offset $\Delta f_{\text{NB-IoT}} = (5.5 - 600)\Delta f$ for each tone k. This is automatically compensated for when the NB-IoT carrier is upconverted to its RF carrier frequency. But the frequency offset also introduces a phase offset at start of each symbol l:

$$\theta_{k,l} = 2\pi\Delta f_{\text{NB-IoT}}\left(lN - \sum_{i=0}^{l}N_{cp}(l \bmod 7)\right)T_s \qquad (7.18)$$

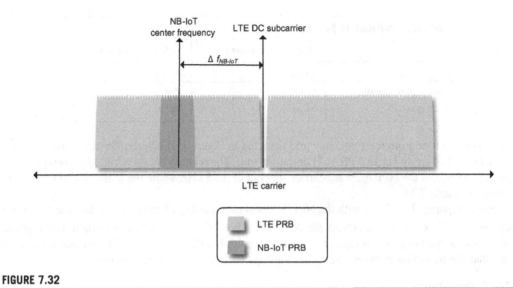

FIGURE 7.32

Frequency offset between the centers of NB-IoT and LTE.

To allow joint IDFT generating the LTE and NB-IoT signal, this phase shift must be compensated for as well. A detailed description of this compensation is found in [12] where the herein made example is generalized by defining $\Delta f_{NB\text{-}IoT}$ as the center frequency of the NB-IoT PRB minus the LTE center frequency as illustrated in Figure 7.32.

The phase compensation in Eq. (7.18) determines a fix relation between NB-IoT NRS and LTE CRS symbols that is utilized by the system in case of in-band operation when the NB-IoT cell and LTE cell are configured with the same cell identity and the same number of logical antenna ports to allow devices to use the LTE CRS, in addition to the NRS, for estimating the NB-IoT DL radio channel.

7.2.7 TRANSMISSION GAP

7.2.7.1 Downlink Transmission Gap

Transmission gaps are introduced in the DL and UL for different reasons. In the DL case, NPDCCH and NPDSCH transmissions serving a device in extreme coverage can take a long time. For example, if the maximum repetitions of 2048 are used for NPDCCH, the total transmission time is more than 2 s. For NPDSCH, it can be more than 20 s if basic transmission requires 10 subframes. Thus, for NPDCCH and NPDSCH transmissions with large repetitions, it helps to define transmission gaps so that the network can serve other devices during a transmission gap. The higher layer can signal a threshold applied to the number of repetitions. If the number of repetitions is less than such a threshold, no transmission gaps are configured. An example is illustrated in Figure 7.33. It is shown that transmissions gaps are configured for devices in extreme coverage as these devices require a large number of repetitions for NPDCCH and NPDSCH transmissions. In the example of Figure 7.33, during a transmission gap, the NPDCCH or NPSDCH transmission to device 1 is postponed and the base station can serve device 2 and device 3 in normal coverage during the transmission gap of device 1.

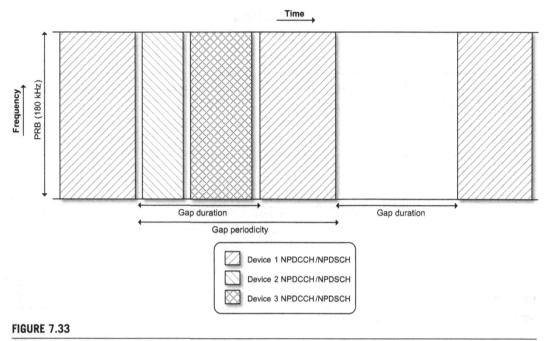

FIGURE 7.33

NPDCCH and NPDSCH transmission gaps.

The transmission gap parameters such as gap periodicity and gap duration are signaled by the higher layers. The transmission time between gaps can be calculated as the gap periodicity minus the gap duration. The transmission time, gap periodicity, and gap duration are all given in terms of NPDCCH or NPDSCH subframes. Therefore, subframes used by other DL physical channels (e.g., NPSS, NSSS, NPBCH) or invalid subframes cannot be counted when mapping the NPDCCH or NPDSCH on the subframe structure.

7.2.7.2 Uplink Transmission Gap

In the UL case, devices in extreme coverage are most likely scheduled with single-tone transmissions. Thus, they will not block the radio resources of the entire carrier because other subcarriers can be used to serve other devices. Low-cost oscillators that are not temperature-compensated can have frequency drifts resulting from self-heating. Thus, after certain continuous transmission time, self-heating will cause a frequency drift. UL transmission gaps therefore are introduced, not for the consideration of avoiding blocking other devices. Rather, they are introduced to allow the device to have an opportunity to recalibrate its frequency and time references by resynchronizing to the DL reference signals, e.g., NPSS, NSSS, NRS. For NPUSCH transmissions, for every 256 ms continuous transmission, a 40 ms gap is introduced. For NPRACH, every 64 NPRACH preamble repetitions, corresponding to 410 ms in case of NPRACH Format 1, a 40 ms gap is introduced. Illustration of NPUSCH and NPRACH transmission gaps are shown in Figure 7.34.

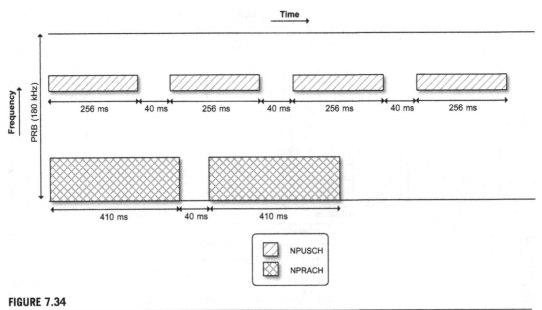

FIGURE 7.34

NPUSCH and NPRACH transmission gaps.

7.3 IDLE AND CONNECTED MODE PROCEDURES

In this section, we describe NB-IoT idle and connected mode procedures, including all activities from initial cell selection to completing a data transfer. Before the device has established a connection with the network, it is considered to be in idle mode. The idle mode procedures include the initial cell selection, which is the procedure that a device must go through when it is first switched on or is attempting to select a new cell to camp on. Idle mode activities also include procedures for acquisition of MIB and SIB, paging, and mobility. At the end of the idle mode procedures section, the transition from idle to connected mode is described in terms of the random access procedure and access control. For NB-IoT the random access procedure can also be triggered in connected mode, but here we focus on the use case in idle mode. Finally, descriptions of some fundamental connected mode procedures including scheduling, HARQ, power control, and multicarrier operation are elaborated on.

7.3.1 IDLE MODE PROCEDURES

7.3.1.1 Cell Selection

The main purpose of cell selection is to identify, synchronize to, and determine the suitability of an NB-IoT cell. Besides the *initial cell selection*, there are also the procedures of *noninitial cell selection* and *cell reselection*. For NB-IoT, cell reselection is used to support idle mode mobility and is described in Section 7.3.1.3. From a physical layer perspective, one of the main differences between initial and noninitial cell searches is the magnitude of carrier frequency offset (CFO) that the device has to deal with when synchronizing to a cell.

The initial cell selection is carried out by the device before it possesses any knowledge of the network and before any prior synchronization to the network has taken place. This corresponds to, e.g., the *public land mobile network* (PLMN) and cell search performed by the device on being switched on the first time. In this case, cell selection needs to be achieved in the presence of a large CFO because of the possible initial oscillator inaccuracy of the device. The initial oscillator inaccuracy for a low-cost device module may be as high as 20 ppm. Thus, for a 900 MHz band, the CFO may be as high as 18 kHz ($900e6 \times 20e^{-6}$). Furthermore, as explained in Section 7.2.1, as the device searches for a cell on the 100 kHz raster grid, there is a raster offset of ±2.5 or ±7.5 kHz for the in-band and guard-band deployments. Thus, the magnitude of total initial frequency offset for in-band and guard-band deployments can be as high as 25.5 kHz.

The noninitial cell selection, also known as *stored information cell selection*, is carried out by the device after previous synchronization to the network has taken place, and the device possesses stored knowledge of the network. After the device has synchronized to the network, it has, e.g., resolved the raster offset and has corrected its initial oscillator inaccuracy. In this case, the CFO may be smaller compared with that during initial cell selection. One example for the device to perform the noninitial cell selection procedure is when the device's connection to the current cell has failed and it needs to select a new cell or when the device wakes up from sleep.

The general steps in the NB-IoT cell selection procedure are as follows:

- Search for the NPSS to identify the presence of an NB-IoT cell.
- Synchronize in time and frequency to the NPSS to identify the carrier frequency and the subframe structure within a frame.
- Identify the PCID and the three LSBs of the SFN by using the NSSS.
- Acquire the MIB to identify the complete SFN as well as the two LSBs of H-SFN, and resolve the frequency raster offset.
- Acquire the SIB1-NB to identify the complete H-SFN, the PLMN, tracking area, and cell identity and to prepare for verification of the cell suitability.

These procedures are described in the next few sections in detail.

7.3.1.1.1 Time and Frequency Synchronization

The first two steps in the initial cell selection procedure aim to time-synchronize to NPSS and obtain a CFO estimation. In principle, they can be combined into one step of joint time and frequency synchronization. However, joint time and frequency synchronization is more costly in terms of receiver complexity. For low-end IoT devices, it is easier to achieve NPSS time synchronization first, in the presence of CFO, and once the NPSS time synchronization is achieved, the device can use additional occurrences of NPSS for CFO estimation. As shown in Figure 7.11, NPSS is transmitted in subframe 5 in every frame, and by time synchronizing to NPSS the device detects subframe 5 and consequently all the subframe numbering within a frame. NPSS synchronization can be achieved by correlating the received signal with the known NPSS sequence or by exploiting the autocorrelation properties of NPSS. As described in Section 7.2.4.2.1, NPSS uses a hierarchical sequence structure with a base sequence repeated based on a cover code. The NPSS detection algorithm can be designed to exploit such structure to achieve time synchronization to NPSS. Interested readers can refer to [16] for more details. Once the device achieves time synchronization to NPSS, it can find NPSS in the next frame and

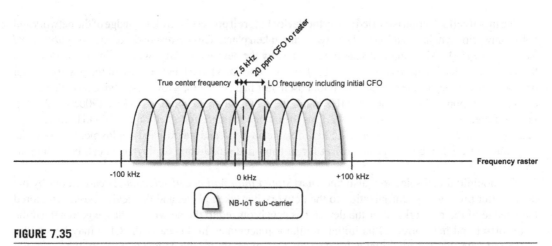

FIGURE 7.35

Illustration of over-CFO estimation due to the raster offset in the in-band and guard-band deployments.

use it for CFO estimation. In coverage-limited condition, these two steps may rely on accumulating detection metrics over many NPSS subframes.

As described in Section 7.2.2.1, for in-band and guard-band deployments, there is a frequency raster offset referred to as the frequency separation between the 100 kHz raster grid, which is the basis of device searching for an NB-IoT carrier, and the actual center frequency of an NB-IoT anchor carrier. This frequency raster offset is, however, unknown to the device before the initial cell selection. The CFO contributed by the oscillator inaccuracy is, at this stage, relative to the raster grid. As illustrated in Figure 7.35, this may result in a needed correction of the local oscillator exceeding the oscillator-induced CFO. The additional correction is equivalent to the raster offset and thus has a magnitude 2.5 or 7.5 kHz for in-band and guard-band deployments. This means that the initial cell selection algorithm needs to be robust in the presence of a raster offset up to 7.5 kHz.

7.3.1.1.2 Physical Cell Identification and Initial Frame Synchronization
The NSSS transmissions are mapped to subframe 9 in even-numbered radio frames. After NPSS synchronization, the device knows where subframe 9 is, but does not know whether a frame is even- or odd-numbered. As described in Section 7.2.4.2.2, the NSSS waveform also depends on the SFN, i.e., SFN mod 8 = 0, 2, 4, or 6. Furthermore, NSSS waveforms depend on the PCID. A straightforward NSSS detection algorithm is therefore to form $504 \times 8 = 4032$ hypotheses, where 504 equals the number of PCIDs used in a NB-IoT network. Each hypothesis corresponds to a hypothesized NSSS waveform during the NSSS detection period. Correlating the received signal with the NSSS waveforms based on each of these hypotheses would allow the device to detect the PCID and the three LSBs of SFN, essentially the 80 ms framing structure. In coverage-limited condition, NSSS detection may rely on accumulating detection metrics over multiple NSSS repetition intervals.

7.3.1.1.3 MIB Acquisition
After acquiring the PCID, the device knows the NRS placement within a resource block as the sub-carriers that NRS REs are mapped to are determined by PCID. It can thus demodulate and decode

NPBCH, which carries the NB-IoT MIB, often denoted as MIB-MB. One of the information elements carried in the MIB-NB is the four most significant bits (MSBs) of the SFN. Because the SFN is 10-bit long, the four MSBs of SFN change every 64 frames, i.e., 640 ms. As a result, the TTI of NPBCH is 640 ms. A MIB-NB is encoded to an NPBCH code block, consisting of eight CSBs. NPBCH is transmitted in subframe 0, and each NPBCH subframe carries a CSB. A CSB is repeated in eight consecutive NPBCH subframes as explained in Section 7.2.4.4. Thus, by knowing the 80 ms frame block structure, the device knows which NPBCH subframes carry identical CSB. It can combine these identical CSBs to improve detection performance in coverage extension scenarios. However, the device does not know which subblock is transmitted in a specific 80 ms interval. Therefore, the device needs to form eight hypotheses to decode a MIB-NB during the cell selection process. This is referred to as NPBCH blind decoding. In addition, to correctly decode the MIB-NB CRC, the device needs to hypothesize whether one or two antenna ports are used for transmitting NPBCH. Therefore, in total 16 blind decoding trials are needed. A successful MIB-NB decoding is indicated by having a correct CRC. When the device can successfully decode NPBCH, it acquires the 640 ms NPBCH TTI boundaries. Afterward, from MIB-NB the device also acquires the information listed below:

- Operation mode (stand-alone, in-band, guard-band).
- In case of in-band and guard-band, the frequency raster offset (± 2.5, ± 7.5 kHz).
- Four MSBs of the SFN.
- Two LSBs of the H-SFN.
- Information about SIB1-NB scheduling.
- SI value tag, which is essentially a version number of the SI. It is common for all SIBs except for *System Information Block Type 14* (SIB14-NB) and *System Information Block Type 16* (SIB16-NB).
- Access barring (AB) information which indicates whether AB is enabled, and in that case, the device shall acquire a specific SI (i.e., SIB14-NB, see Section 7.3.1.2.3) before initiating RRC connection establishment or resume.

The operation mode information further indicates, in the in-band case, how the NB-IoT cell is configured compared with the LTE cell. Such an indication is referred to as *same PCI indicator*. If the *same PCI indicator* is set true, the NB-IoT and LTE cells share the same PCID, and NRS and CRS have the same number of antenna ports. One implication is that in this case, the device can also use the LTE CRS for channel estimation. As a result, when the *same PCI indicator* is set true, MIB-NB further provides information about the LTE CRS sequence. On the other hand, if the *same PCI indicator* is set false, the device still needs to know where the CRSs are in a PRB. For the in-band mode, there is a required relationship between the PCIDs of LTE and NB-IoT in that they need to point to the same subcarrier indexes for CRS and NRS. Thus, the device already knows the subcarrier indexes for CRS by knowing the PCID of the NB-IoT cell. However, it does not know whether the LTE CRS has the same number of antenna ports as NRS or four antenna ports. Knowing this is important as NPDCCH and NPDSCH are rate-matched according to the number of CRS REs in a PRB. The information concerning whether the LTE cell uses four antenna ports or not is carried in the MIB-NB.

7.3.1.1.4 Cell Identity and H-SFN Acquisition

After acquiring the MIB-NB, including the scheduling information about SIB1-NB, a device is able to locate and decode SIB1-NB. We will describe more about how the device acquires SIB1-NB in

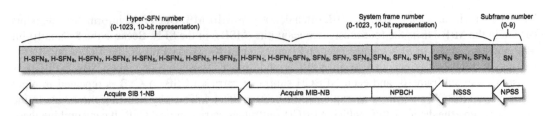

FIGURE 7.36

Illustration of how the device acquires complete timing information during the initial cell selection.

Section 7.3.1.2.1. From a cell selection perspective, it is important to know that the SIB1-NB carries the eight MSBs of the H-SFN, the PLMN, tracking area, and a 28-bit long cell identity, which is used to unambiguously identify a cell within a PLMN. Thus, after acquiring SIB1-NB, the device has achieved complete synchronization to the frame structure shown in Figure 7.7. Based on the cell identifiers, the device is able to determine if it is allowed to attach to the cell. It will finally be able to evaluate the suitability of the cell against a pair of minimum required signal strength and signal quality threshold parameters broadcasted in SIB1-NB. A suitable cell is a cell that is sufficiently good to camp on, but it is not necessarily the best cell. The cell reselection procedure described in Section 7.3.1.3 supports the selection of the best available cell for optimizing link and system capacity.

Figure 7.36 illustrates and summarizes how the device acquires complete framing information during the initial cell selection procedure.

After completing the initial cell selection, the device is expected to have a time-accuracy within a few microseconds and a residual frequency offset within 50 Hz. In essence, the device has achieved time and frequency synchronization with residual errors that will not result in significant performance degradation in subsequent transmission and reception during connected and idle mode operations.

7.3.1.2 System Information Acquisition

After selecting a suitable cell to camp on, a device needs to acquire the full set of SI messages. This procedure and the information associated with the seven NB-IoT SI messages are presented in the next few sections.

7.3.1.2.1 SIB1

The content and importance of SIB1-NB has already been indicated in Section 7.3.1.1.4. Table 7.15 presents in more detail the information acquired by the device on reading SIB1-NB.

Table 7.15 System Information Block	
System Information Block	**Content**
SIB1-NB	Hyperframe information, network information such as PLMN, tracking area and cell identities, access barring status, thresholds for evaluating cell suitability, valid subframe bitmap and scheduling information regarding other System Information Blocks.

The scheduling information of SIB1-NB carried in MIB-NB describes the TBS (208, 328, 440, or 680 bits) and number of repetitions (4, 8, or 16) used for SIB1-NB transmissions. With such information, the device knows how to receive SIB1-NB.

The transmission of SIB1-NB is illustrated in Figure 7.37. A SIB1-NB TB is carried in eight SIB1-NB subframes, mapped to subframe 4 in every other frame during 16 frames. These 16 frames are repeated 4, 8, or 16 times. The repetitions are evenly spread over the SIB1-NB transmission interval, which is defined as 256 frames, i.e., 2.56 s.

There is a notion of SIB1-NB modification period, which equals to 40.96 s. This means that SIB1-NB content is not supposed to change within a SIB1-NB modification period. Thus, the same SIB1-NB TB is repeated in all SIB1-NB transmission periods within a modification period. In the next SIB1-NB modification period, the content is allowed to change. However, in practice, excluding the changes of H-SFN bits, such changes occur rarely.

The starting frame for SIB1-NB in a transmission period depends on the PCID as well as the aforementioned SIB1-NB repetition factor signaled in MIB-NB. The motivation of having different starting frames in different cells is to randomize interference and avoid persistent intercell interference between the transmissions of SIB1-NB in different cells.

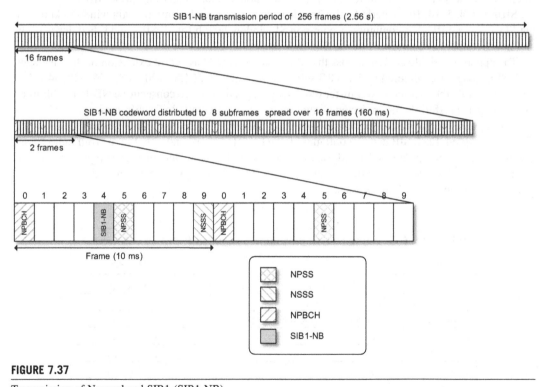

FIGURE 7.37

Transmission of Narrowband SIB1 (SIB1-NB).

7.3.1.2.2 Information Specific to In-Band Mode of Operation

For the in-band mode, an NB-IoT device needs to acquire certain LTE carrier parameters to be able to know which resources are already taken by LTE, and in cases when CRS may be used for assisting measurements and channel estimation, the device needs to know the relative power between CRS and NRS as well as the sequence information of CRS. As explained in Section 7.3.1.1.3, some of this information is provided by MIB-NB. Additional information is carried in SIB1-NB. By acquiring SIB1-NB, the device is able to have the complete information it needs to figure out the resource mapping within a PRB for the case of in-band operation. Table 7.16 provides a list of such information.

The valid subframe bitmap as explained in Section 7.2.4.1 is used to indicate which DL subframes within a 10 or 40 subframes interval can be used for NB-IoT. This can be used, for example, to avoid collision with LTE MBSFN.

An example is given in Figure 7.38. The NB-IoT devices use the provided information about LTE configuration in terms of control region, antenna ports, and CRS to black out resources taken by LTE. The remaining resources can be used for NB-IoT DL.

7.3.1.2.3 System Information Blocks 2, 3, 4, 5, 14, 16

After reading the SI scheduling information in SIB1-NB, the device is ready to acquire the full set of SI messages. In addition to SIB1-NB, NB-IoT defines six additional types of SIBs as listed in Table 7.17. Interested readers can refer to Reference [7] for additional details regarding these SIBs.

SIBs 2, 3, 4, 5, 14, 16 are periodically broadcasted during specific time-domain windows known as the *SI windows*. SIB1-NB configures a common SI window length for all SIBs and schedules periodic and nonoverlapping occurrences of the SI windows.

To support a variable content across the SI messages as well as future extension of the messages, each SI message is configured with a TBS selected from the set of {56, 120, 208, 256, 328, 440, 552, 680} bits. While the two smallest transfer blocks are mapped over two consecutive NB-IoT subframes, the six larger transfer blocks are mapped over eight consecutive NB-IoT subframes.

Furthermore, to support operation in extended coverage, a configurable repetition level of the SIBs is supported. Each SIB can be configured to be repeated every 2nd, 4th, 8th, or 16th radio frame. The total number of repetitions depends on the configured SI window length and the repetition pattern. For the largest SI window of 160 frames, up to 80 repetitions are supported. Figure 7.39 illustrates the transmission of an NB-IoT SI message.

Table 7.16 LTE configuration parameters signaled to NB-IoT Devices	
Information	**Message**
Whether the NB-IoT cell is configured as in-band mode	MIB-NB
If same PCID is used for NB-IoT and LTE cells	MIB-NB
Number of LTE antenna ports (in case different PCIDs)	MIB-NB
CRS sequence information	MIB-NB
Valid subframe bitmap	SIB1-NB
LTE control region size	SIB1-NB
NRS to CRS power offset	SIB1-NB

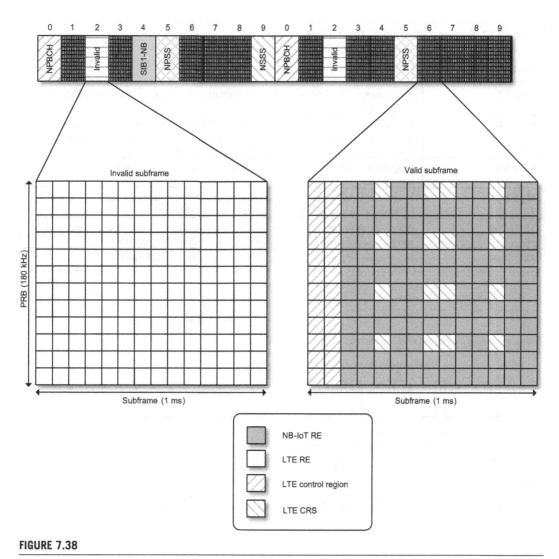

FIGURE 7.38

An example of LTE configuration information provided to NB-IoT devices for determining the resource mapping at the physical layer.

SIB1-NB indicates the latest status of each of the scheduled SIBs. A change in the content of a SIB is indicated in the MIB, with a few exceptions, as described in Section 7.3.1.2.4, and may optionally also be indicated in SIB1-NB. This allows a device to determine if a specific block needs to be reacquired.

Table 7.17 System Information Blocks 2, 3, 4, 5, 14, 16

System Information Block	Content
SIB2-NB	Radio resource configuration (RRC) information for all physical channels that is common for all devices.
SIB3-NB	Cell reselection information that is common for intrafrequency and interfrequency cell reselection. It further provides additional information specific to intrafrequency cell reselection such as cell suitability related information.
SIB4-NB	Neighboring cell-related information, e.g., cell identities, relevant only for intrafrequency cell reselection.
SIB5-NB	Neighboring cell-related information, e.g., cell identities and cell suitability-related information, relevant only for interfrequency cell reselection.
SIB14-NB	Access class barring information per PLMN. Contains a specific flag for barring of a specific access class, It also indicates barring of exception reporting.
SIB16-NB	Information related to GPS time and Coordinated Universal Time (UTC).

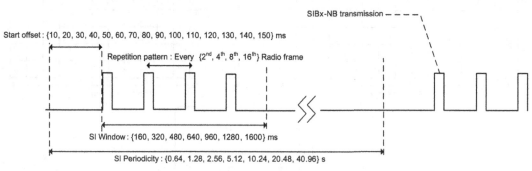

FIGURE 7.39

Illustration of Narrowband System Information Block Type-x (SIBx-NB) transmission.

7.3.1.2.4 System Information Update

The value tag (5 bits) provided in MIB-NB serves as content version number and is to be considered valid for 24 h by the devices. Different version numbers correspond to different system information contents. When the SI has changed, the network can explicitly notify the devices. A device in the idle mode with a DRX cycle shorter than the modification period has the option to monitor the paging messages for SI change notifications. Paging will be discussed in Section 7.3.1.4. There might be cases where the device has not been notified about the SI changes before it attempts to access the network, e.g., in case the device has been configured with a DRX cycle longer than the modification period, in case the device uses PSM during idle mode, or in case the base station does not transmit any paging at SI change. Such a potential problem is addressed by requiring the device to always read the MIB-NB on an access attempt. By reading the MIB-NB, the device will realize that there is an SI change through the SI value tag. This also allows the device to acquire information regarding the barring status as explained in Section 7.3.1.7. Note, however, that changes in SIB14-NB (AB parameters) and SIB16-NB (information related to GPS time and UTC) will not result in an SI value tag change.

7.3.1.3 Cell Reselection

After selecting a cell, a device is mandated to monitor up to 16 intrafrequency neighbor cells and up to 16 neighbor cells on an interfrequency. In simple words, in case the device detects that a neighbor cell has become stronger in terms of the reference signal received power (RSRP) than the currently serving cell, then the cell reselection procedure is triggered. Devices that are in good coverage, i.e., experience a sufficiently high RSRP level in the serving cell, can be excluded from measuring for cell reselection. This helps improve the battery life of these devices.

Besides securing that a device camps on the best cell, the cell reselection procedure is the main mechanism for supporting idle mode mobility. During the connected mode, the device does not need to perform mobility measurements on the serving or on the neighboring cells. In case the signal quality of the serving cell becomes very poor, resulting in persistent link-level failures, the device will invoke the link-layer failure procedure, which in essence moves it from the connected mode back to the idle mode. In the idle mode, the device can use the cell reselection mechanism to find a new serving cell. After establishing a new serving cell, the device can start a random access procedure (see Section 7.3.1.6) to get back to the connected mode to complete its data reception and transmission.

7.3.1.4 Paging and eDRX

The monitoring of paging during the idle mode has implications on device battery lifetime and the latency of DL data delivery to the device. A key to determining the impact is how often a device monitors paging. NB-IoT does just as LTE uses *search spaces* for defining paging transmission opportunities. The search space concept, including the *Type-1 Common Search Space* (CSS) implementation used for paging indication, is covered in detail in Section 7.3.2.1.

For now, it is sufficient to note that a device monitors a set of subframes defined by the Type-1 CSS to detect an NPDCCH containing a DCI of format N2 that schedules a subsequent NPDSCH containing a paging message addressed to the device. The P-RNTI is the identifier used to address a device for the purpose of paging and is, as described in Section 7.2.4.5, used to mask the NPDCCH CRC.

The starting subframe for the Type-1 CSS candidates is determined from the location of NB-IoT paging opportunity (PO) subframe, which is determined based on configured DRX cycle [17]. If the starting subframe is not a valid NB-IoT DL subframe, then the first valid NB-IoT DL subframe after the PO is the starting subframe of the NPDCCH repetitions. The Type-1 CSS candidates are based on only NPDCCH AL 2 described in Section 7.2.4.5. A search space contains NPDCCH candidates defined for repetition levels R up to a configured maximum NPDCCH repetition level R_{max}. R_{max} is typically configured to secure that all devices in a cell can be reached by the paging mechanism, and the relation between the possible repetition levels R for a certain R_{max} is given by Table 7.18.

Figure 7.40 illustrates possible paging configurations in NB-IoT. Either DRX or eDRX can be used. In case of DRX the paging occasions occur with a periodicity of at most 10.24 s. For eDRX, the longest eDRX period is 2 h, 54 min, and 46 s, which corresponds to one hyperframe cycle. After each eDRX cycle a *paging transmission window* starts during which DL reachability is achieved through the configured DRX cycle.

Table 7.18 NPDCCH Type-1 common search space candidates

R_{max}	R
1	1
2	1, 2
4	1, 2, 4
8	1, 2, 4, 8
16	1, 2, 4, 8, 16
32	1, 2, 4, 8, 16, 32
64	1, 2, 4, 8, 16, 32, 64
128	1, 2, 4, 8, 16, 32, 64, 128
256	1, 4, 8, 16, 32, 64, 128, 256
512	1, 4, 16, 32, 64, 128, 256, 512
1024	1, 8, 32, 64, 128, 256, 512, 1024
2048	1, 8, 64, 128, 256, 512, 1024, 2048

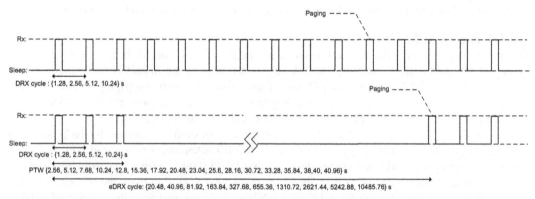

FIGURE 7.40

Illustration of possible Discontinuous Reception (DRX) and extended DRX (eDRX) paging configuration in NB-IoT.

7.3.1.5 PSM

Some applications have very relaxed requirements on mobile terminated reachability (e.g. longer than a day) and then power consumption can be further reduced compared with using eDRX, which has a longer cycle close to 3 h. For this type of devices, the most energy-efficient *sleep mode* is the *PSM*. PSM, introduced in 3GPP Release 12 as general improvement for devices with high requirements on long battery life, is described in Section 2.2.3.

After monitoring paging for a short period of time, the device can enter the PSM, during which it uses the smallest possible amount of energy, essentially only needing to leave its Real Time Clock running for keeping track of time and scheduled events. The device aborts all idle mode operations in PSM and will not transmit or monitor paging during PSM and need not keep an up-to-date synchronization to the network. The duration of PSM is configurable and may be more than a year in the most extreme configuration. The device exits the PSM once it has UL data to transmit or is mandated to send TAU, which informs the network of the whereabouts of the device. After its UL transmission, the device may enter DRX mode for a short period of time, configured by the active timer, for monitoring paging to enable mobile terminated reachability. After such a short duration in which the device monitors paging, the device enters the next PSM period. This procedure is illustrated in Figure 2.3.

7.3.1.6 Random Access Procedure

NB-IoT random access procedure is generally the same as LTE [18]. We will mainly highlight aspects that are unique to NB-IoT.

The random access procedure is illustrated in Figure 7.41. After synchronizing to the network and confirming that access is not barred, the device sends a random access preamble using NPRACH.

The device needs to determine an appropriate NPRACH configuration according to its coverage class estimation. Recall that SIB2-NB carries RRC information as explained in Section 7.3.1.2.3. One of the RRC information elements is the NPRACH configuration. The cell can configure up to two RSRP thresholds that are used by the device to select the NPRACH configuration appropriate for its *coverage class*. An example is given in Figure 7.42, in which two RSRP thresholds are configured and therefore there are three NPRACH configurations for three CE levels, respectively. Essentially, the network uses these RSRP thresholds to configure the MCLs of the different CE levels. If the network does not configure any RSRP threshold, the cell supports only a single NPRACH configuration used by all devices regardless of their actual path loss to the serving base station.

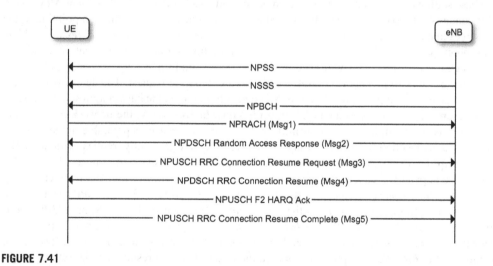

FIGURE 7.41

NB-IoT random access procedure, using the Radio Resource Control (RRC) resume procedure.

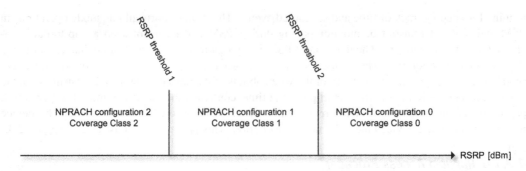

FIGURE 7.42

NPRACH configurations and RSRP thresholds.

The SIB2-NB NPRACH configuration information for each CE level includes the time–frequency resource allocation. The resource allocation in the frequency domain is a set of starting preambles. Each starting preamble is equivalent to the first NPRACH symbol group and associated with a specific 3.75 kHz tone. The set of starting preambles is determined by a subcarrier offset and a number of spanned subcarriers. The time-domain allocation is defined by a periodicity, a starting time with the period, and the number of repetitions associated with the NPRACH resource. This is exemplified in Figure 7.43 where an NPRACH configuration intended to support a high access load is illustrated.

This set of starting preambles may further be partitioned into two subsets. The first subset is used by devices that do not support multitone NPUSCH transmissions, whereas the second subset is used by devices with multitone capability. In essence, the device signals its NPUSCH multitone support by selecting the NPRACH starting preamble according to its capability.

If the base station detects an NPRACH preamble, it sends back a RAR, also known as *Message 2*. The RAR contains a TA parameter. The RAR further contains scheduling information pointing to the radio resources that the device can use to transmit a request to connect, also known as *Message 3*. Note that at this point the base station already knows the device multitone transmission capability, and thus resource allocation for Message 3 will account for the device multitone transmission capability. In Message 3, the device will include its identity as well as a scheduling request. The device will always include its buffer status and power headroom in Message 3 to facilitate the base station scheduling and power allocation decision for subsequent UL transmissions. In *Message 4*, the network resolves any contention because of multiple devices transmitting the same random access preamble in the first step and transmits a connection setup or resume message. The device finally replies with an RRC connection setup, or resume, complete message to complete the transition to connected state. The device may also append UL data to this message to optimize the latency of the data transfer.

To adapt the connection setup to the short data transfers expected for NB-IoT, 3GPP specified two alternative sets of optimizations known as the *User Plane CIoT* Evolved Packet System (*EPS*) optimization procedure and the *Control Plane CIoT EPS optimization* procedure. Both aims to optimize or reduce the message exchanges that are needed before a device transmits its data. Part of the user plane optimizations is the *RRC Resume* procedure where the basic idea is to resume configurations established in a previous connection. This is achieved by using a *Resume Identity* that is used by the network

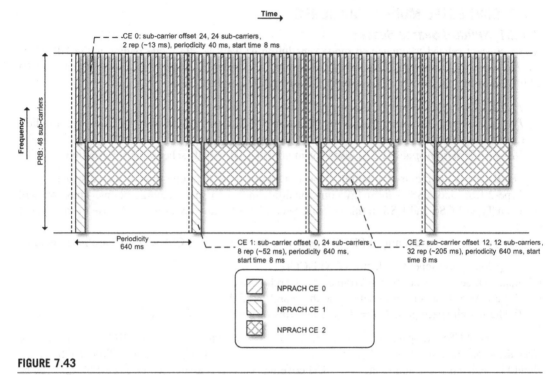

FIGURE 7.43

NPRACH configurations. (an example)

to retrieve the device context stored in the network. The RRC Resume procedure minimizes the signaling needed to setup the *data radio bearer*. The *Control Plane CIoT EPS optimization* procedure allows data to be transmitted over a *signaling radio bearer* in a Non-Access Stratum container. This allows data transmission before setting up a data radio bearer.

7.3.1.7 Access Control

Access Barring (AB) is an access control mechanism adopted in NB-IoT; it closely follows the Access Class Barring functionality described in Section 2.2 and allows PLMN-specific barring across 10 normal and 5 special access classes. It also supports special treatment of devices intending to transmit an *Exception report*. The Exception report concept was introduced in 3GPP Release 13 to allow a network to prioritize reports of high urgency transmitted by a device.

An AB flag is provided in MIB-NB. If it is set false, then all devices are allowed to access the network. If the AB flag is set true, then the device must read SIB14-NB before it attempts to access the network, which provides the just introduced access class—specific barring information. The device needs to check whether its access class is allowed to access the network. In case the device is barred, it should back off and then reattempt access at a later point in time.

It should be noted that when the MIB-NB AB flag toggles, there is no impact on the SI value tag.

7.3.2 CONNECTED MODE PROCEDURES

7.3.2.1 NPDCCH Search Spaces

A key concept related to connected mode scheduling as well as idle mode paging is the NPDCCH *search spaces*. A search space consists of one or more subframes in which a device may search for DCI addressed to the device. There are three types of search spaces defined:

- *Type-1 CSS*, used for monitoring paging
- *Type-2 CSS*, used for monitoring RAR, Message 3 HARQ retransmissions and Message 4 radio resource assignments
- *UE-specific search space* (USS), used for monitoring DL or UL scheduling information

The device is not required to simultaneously monitor more than one type of search space.

Type-2 CSS and USS share many commonalities in search space configurations. Thus, we will focus on Type-2 CSS and USS in this section. Section 7.3.1.4 already presents the use of the Type-1 CSS.

Key parameters for defining NPDCCH search spaces for Type-2 CSS and USS are listed below:

- R_{max}: Maximum repetition factor of NPDCCH
- α_{offset}: Offset of the starting subframe in a search period
- G: Parameter that is used to determine the search period
- T: The search space period, $T = R_{max}G$ in terms of number of subframes

For Type-2 CSS, the parameters R_{max}, α_{offset}, and G are signaled in the SI SIB2-NB, whereas for USS these parameters are signaled through device-specific RRC signaling. For Type-2 CSS, R_{max} should be adapted according to the NPRACH coverage class it is associated to. For USS, R_{max} can be optimized to serve the coverage of the connected device. There is a restriction that a search period must be more than four subframes, i.e., $T > 4$.

Within a search period, the number of subframes that the device needs to monitor is R_{max} and the number of search space candidates defined is also based on R_{max}. Note that the R_{max} subframes that the device needs to monitor within a search period have to exclude the subframes used for transmitting NPBCH, NPSS, NSSS, and SI. Furthermore, these subframes need to be NB-IoT subframes according to the valid subframe bitmap described in Sections 7.3.1.2.1 and 7.3.1.2.2. Table 7.19 shows the USS candidates for different R_{max} values and NCCE ALs.

Describing the search space concept is done best with a concrete example. We will use a USS example to illustrate all the search space aspects we have discussed up to this point. Consider, for example, a device in coverage conditions requiring the NPDCCH to be transmitted with up to 2 repetitions. R_{max} will in this case be set to 2. It is further assumed that the scheduling periodicity is configured to be eight times longer than the maximum repetition interval, i.e., G is set to 8. Finally, an offset α_{offset} of 1/8 is selected.

With these parameter settings, the search period is $T = R_{max}G = 16$ subframes. Figure 7.44 illustrates the search periods according to a reference case. For this reference case, the starting subframes are the ones satisfying (SFN $\times$ 10 + SN) mod $T = 0$, when the offset value is set to 0. According to the present example, the offset value is set to 1/8 of the search period, i.e., the starting subframe is shifted by two subframes.

Table 7.19 NPDCCH UE-specific search space candidates

| R_{max} | R | NCCE Indices of Monitored NPDCCH Candidates | |
		AL = 1	AL = 2
1	1	{0},{1}	{0,1}
2	1	{0},{1}	{0,1}
	2	–	{0,1}
4	1	–	{0,1}
	2	–	{0,1}
	4	–	{0,1}
≥8	$\dfrac{R_{max}}{8}$	–	{0,1}
	$\dfrac{R_{max}}{4}$	–	{0,1}
	$\dfrac{R_{max}}{2}$	–	{0,1}
	R_{max}	–	{0,1}

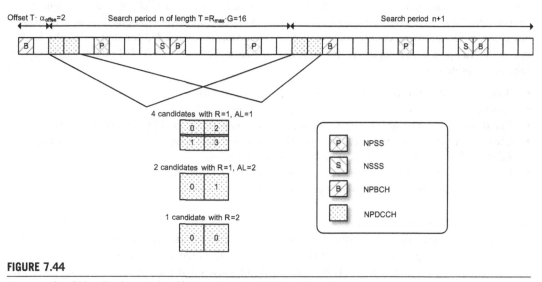

FIGURE 7.44

An example of User Equipment-specific search space (USS) configuration.

According to Table 7.19, with $R_{max} = 2$, the search space may have NPDCCH repetition factor of $R = 1$ or 2. Furthermore, as indicated in Table 7.19, for the case of $R = 1$, AL 1 may be used and thus both NCCE0 and NCCE1 (see Figure 7.19 in Section 7.2.4.5) are individually a search space candidate. For AL 2, NCCE0 and NCCE1 are used jointly as a search space candidate. All the search space candidates are illustrated in Figure 7.44, including the following set of seven candidates within a search period:

- Four candidates with $R = 1$ and $AL = 1$,
- Two candidates with $R = 1$ and $AL = 2$, and
- One candidate with $R = 2$.

It should be noted that the device needs to monitor a set of search space subframes that are not taken by NPBCH (subframe 0), NPSS (subframe 5), NSSS (subframe 9, in even-numbered SFN), and SI.

The search space candidates shown in Table 7.19 to a large extent also applies to Type-2 CSS, with the only exception that Type-2 CSS candidates are only based on $AL = 2$. Furthermore, Type-2 CSS and USS share the same set of values for G, $G \in \{1.5, 2, 4, 8, 16, 32, 48, 64\}$ [7]. Considering the maximum repetition factor of NPDCCH is 2048, the values of search period for Type-2 CSS and USS are $4 < T \leq 131072$.

7.3.2.2 Scheduling
In this section, we describe how scheduling for UL and DL transmissions works. When the network needs to schedule a device, it sends a DCI addressed to the device during one of the search space candidates that the device monitors. The C-RNTI, masking the DCI CRC, is used to identify the device. The NPDCCH carries a DCI that includes resource allocation (in both time and frequency domains), modulation and coding scheme (MCS), and information needed for supporting the HARQ operation. To allow low-complexity device implementation, Release 13 NB-IoT adopts the following scheduling principles:

- A device only needs to support one HARQ process in the DL.
- A device only needs to support one HARQ process in the UL.
- The device need not support simultaneous transmissions of UL and DL HARQ processes.
- Cross-subframe scheduling (i.e., DCI and the scheduled data transmission do not occur in the same subframe) with relaxed processing time requirements.
- Half-duplex operation at the device (i.e., no simultaneous transmit and receive at the device) allows time for the device to switch between transmission and reception modes.

7.3.2.2.1 Uplink Scheduling
DCI Format N0 is used for UL scheduling. Table 7.20 shows the information carried in DCI Format N0.

An UL scheduling example is illustrated in Figure 7.45 based on the same USS search space configuration example in Figure 7.44. We will highlight a few important aspects in this example and relate them to the scheduling information presented in Table 7.20.

For UL data transmissions, subframe scheduling with at least an 8 ms time gap between the last DCI subframe and the first scheduled NPUSCH subframe is required. This time gap allows the device

Table 7.20 DCI Format N0 used for scheduling NPUSCH Format 1		
Information	**Size [bits]**	**Possible Settings**
Flag for format N0/N1	1	DCI N0 or DCI N1
Subcarrier indication	6	Allocation based on subcarrier index
		3.75 kHz spacing: {0}, {1}, …, or {47}
		15 KHz spacing: 1-tone allocation: {0}, {1}, …, or {11} 3-tone allocation: {0, 1, 2}, {3, 4, 5}, {6, 7, 8}, {9, 10, 11} 6-tone allocation: {0, 1,…,5} or {6, 7,…,11} 12-tone allocation: {0, 1,…,11}
NPUSCH scheduling delay	2	8, 16, 32, or 64
DCI subframe repetition number	2	The R values in Table 7.19
Number of RUs	3	1, 2, 3, 4, 5, 6, 8, or 10
Number of NPUSCH repetition	3	1, 2, 4, 8, 16, 32, 64, or 128
MCS	4	0, 1,…, or 12, for indexing the row of the NPUSCH TBS table (see Section 7.2.5.2)
Redundancy version	1	Redundancy version 0 or 2
New data indicator (NDI)	1	NDI toggles for new TB or does not toggle for same TB

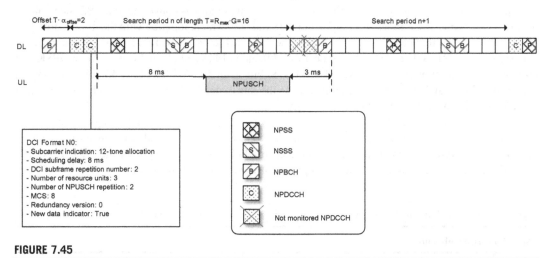

FIGURE 7.45

An uplink scheduling example.

to decode the DCI, switch from the reception mode to the transmission mode, and prepare the UL transmission. This time gap is referred to as the scheduling delay and is indicated in DCI. After the device completes its NPUSCH transmission, there is at least a 3-ms gap to allow the device to switch from transmission mode to reception mode and be ready for monitoring the next NPDCCH search space candidate. This means that according to this example the network cannot use the second search period to send the next DCI to the device because the device will skip both search space candidates since they are both within the 3-ms gap from the end of its NPUSCH transmission. The network scheduler needs to follow this timing relationship in determining when to send the DCI to the device.

DCI Format N0 provides the information about the starting subframe as well as the total number of subframes of the scheduled NPUSCH resources. As mentioned earlier, the time gap between the last NPDCCH subframe carrying the DCI and the first scheduled NPUSCH slot is indicated in the DCI. However, how does the device know which subframe is the last subframe carrying the DCI? This potential problem is resolved by including the information of NPDCCH subframe repetition number in the DCI. With this information, if the device is able to decode the DCI using any of the first available subframe in the search space, it knows that the DCI will be repeated in one more subframe, which is then the last subframe carrying the DCI. Generally speaking, with the information of NPDCCH subframe repetition number, if the device is able to decode the DCI using any of the subframes of a search space candidate, it can unambiguously determine the starting and the ending subframes of this specific search space candidate.

The total number of scheduled NPUSCH slots is determined by the number of RUs per repetition, the number of repetitions, and the length of an RU. The length of an RU is inferred from the number of subcarriers used for NPUSCH Format 1 (see Section 7.2.5.2). According to the example, 12 subcarriers are used, and one RU for 12-tone NPUSCH Format 1 is 1 ms. Thus, with 2 repetitions and 3 RUs per repetition the total scheduled duration is 6 ms as illustrated in Figure 7.45.

Modulation format is determined based on the MCS index, and the coding scheme is determined jointly based on the MCS index, number of RUs, and the redundancy version. According to the current example, the MCS index according to the DCI is 8. The MCS index is converted to a TBS index based on Table 7.21. Thus, MCS index 9 is mapped to TBS index 9, which, together with the information that each repetition uses three RUs, is used to determine TBS as 456 based on Table 7.10A. The number of data symbols per RU is determined based on Table 7.9 and is 144 symbols in this case. With QPSK and three RUs per repetition, there are overall 864 coded bits available. With TBS of 456 bits, code-word length of 864 bits, and redundancy version 0, using the rate matching framework of LTE, as detailed in [4], the code word can be generated accordingly.

Because only one HARQ process is supported in Rel-13, there is no need to signal the process number. From the device perspective, it only needs to know whether it needs to transmit the same or new TB. The HARQ acknowledgment is signaled implicitly using the NDI in the DCI. If NDI is toggled, the device treats it as an acknowledgment of the previous transmission.

Table 7.21 Relationship between MCS and TBS indexes for NPUSCH

Multitone Transmissions

I_{MCS}	0	1	2	3	4	5	6	7	8	9	10	11	12
I_{TBS}	0	1	2	3	4	5	6	7	8	9	10	11	12

Single-tone Transmissions

I_{MCS}	0	1	2	3	4	5	6	7	8	9	10		
I_{TBS}	0	2	1	3	4	5	6	7	8	9	10		

7.3.2.2.2 Downlink Scheduling

Scheduling of NPDSCH is signaled using DCI Format N1, which is shown in Table 7.22 including the parameter values of different information elements. Most of the general aspects of DL scheduling are similar to those used for UL scheduling, although the exact parameter values are different. For example, cross-subframe scheduling is also used for DL scheduling, but the minimum time gap between the last DCI subframe and the first scheduled NPDSCH subframe is 4 ms. Recall in the UL cross-subframe scheduling case, this gap is at least 8 ms. A smaller minimum gap in the DL case reflects that there is no need for the device to switch from receiving to transmitting between finishing receiving the DCI and starting NPDSCH reception.

An NPDSCH scheduling example is illustrated in Figure 7.46, again based on the same NPDCCH USS configuration example from Figure 7.44. First, the DCI indicates that there is no additional scheduling delay, and thus the scheduled NPDSCH starts after a minimum 4 ms gap following the last subframe carrying the DCI. The DCI also indicates that one subframe is used per repetition of NPDSCH and there are two repetitions. Based on this information, the device knows that there are two subframes scheduled for its NPDSCH reception. These two subframes are the first two available subframes from the scheduled starting point of NPDSCH. The available subframes are the ones not used by NPBCH, NPSS, NSSS, or SI and not indicated as invalid subframes. As illustrated, the first subframe after the scheduled NPDSCH starting point is available, but the next two subframes need to be skipped as they are used for NSSS and NPBCH. After the NPBCH subframe, the next subframe is available.

Table 7.22 DCI Format N1 for scheduling NPDSCH		
Information	**Size [bits]**	**Possible Settings**
Flag for format N0/N1	1	DCI N0 or DCI N1
NPDCCH order indication	1	Whether the DCI is used for NPDSCH scheduling or for NPDCCH order
Additional time offset for NPDSCH (in addition to a minimal 4-ms gap)	3	$R_{max} < 128$: 0, 4, 8, 12, 16, 32, 64, or 128 (ms) $R_{max} \geq 128$: 0, 16, 32, 64, 128, 256, 512, or 1024 (ms)
DCI subframe repetition number	2	The R values in Table 7.19
Number of NPDSCH subframes per repetition	3	1, 2, 3, 4, 5, 6, 8, or 10
Number of NPDSCH repetition	4	1, 2, 4, 8, 16, 32, 64, 128, 192, 256, 384, 512, 768, 1024, 1536, or 2048
MCS	4	0, 1,..., or 12, for indexing the row of the NPDSCH TBS table (see Table 7.5)
NDI	1	NDI toggles for new TB or does not toggle for same TB
HARQ-ACK resource	4	15 kHz subcarrier spacing: • Time offset value: 13, 15, 17, or 18 • Subcarrier index: 0, 1, 2, or 3 3.75 kHz subcarrier spacing: • Time offset value: 13 or 17 • Subcarrier index: 38, 39, 40, 41, 42, 43, 44, or 45

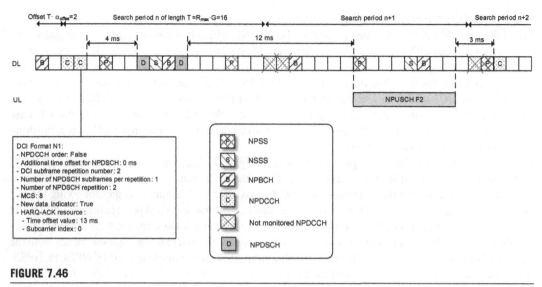

FIGURE 7.46

A DL scheduling example.

A major difference between DL and UL scheduling is that in the DL case, the scheduler also needs to schedule NPUSCH Format 2 resources for the signaling of HARQ feedback. This information is provided in the DCI in the format of a subcarrier index and a time offset. The time offset is defined between the ending subframe of scheduled NPDSCH and the starting slot of NPUSCH Format 2. NB-IoT requires such a time offset to be at least 13 ms, giving a gap of at least 12 ms between the end of NPDSCH and the start of NPUSCH Format 2. This gap is to allow sufficient NPDSCH decoding time at the device, time for switching from reception to transmission, and time for preparing the NPUSCH Format 2 transmission. As described in Section 7.2.5.2, NPUSCH Format 2 uses single-tone transmissions, either with 15 or with 3.75 kHz numerology. The RU for NPUSCH Format 2 is 2 ms for 15 kHz subcarrier numerology or 8 ms for 3.75 kHz numerology. Whether the device uses 15 or 3.75 kHz numerology is configured through RRC signaling. Furthermore, the NPUSCH Format 2 transmissions may be configured with multiple repetitions. This information, however, is not included in the DCI but signaled separately through higher-layer signaling [18]. The example in Figure 7.46 shows that the device is configured to use four repetitions for NPUSCH Format 2 (assuming 15 kHz subcarrier numerology).

After the device completes its NPUSCH Format 2 transmission, it is not required to monitor NPDCCH search space for 3 ms. This is to allow the device to switch from the transmission mode to getting ready for receiving the NPDCCH again. According to the example in Figure 7.46, the first subframe of search period $n+2$ is within the 3-ms gap and thus cannot be used for signaling a DCI to the device. The subframe immediately after the NPSS subframe is another NPDCCH search space candidate and is not within the 3-ms gap. Therefore, a DCI may be sent using this subframe.

7.3.2.3 Power Control

NB-IoT supports open-loop power control. The decision to not allow closed-loop power control was based on the following considerations:

- A data session for many IoT use cases is very short and thus not a good match for a closed-loop control mechanism, which takes time to converge.
- Closed-loop power control requires constant feedback and measurements, which are not desirable from device energy efficiency point of view.
- For devices in extreme coverage extension situations, the quality of channel quality measurements and reliability of power control command might be very poor.

Instead, NB-IoT uses open-loop power control based on a set of very simple rules. For NPUSCH, both Format 1 and Format 2, if the number of repetitions is greater than 2, the transmit power is the maximum configured device power, P_{max}. The maximum configured device power is set by the serving cell. Because NB-IoT maximum device power is 23 dBm, $P_{max} \leq 23$ dBm. If the number of NPUSCH repetitions is 1 or 2, the transmit power is determined by

$$P_{NPUSCH} = \max\{P_{max}, 10\log_{10}(M) + P_{target} + \alpha L\}(dBm), \qquad (7.19)$$

where P_{target} is the target received power level at the base station, L is the estimated path loss, α is a path loss adjustment factor, and M is a parameter related to the bandwidth of NPUSCH waveform. The bandwidth-related adjustment is used to relate the target received power level to target received SNR. The device uses the value of M according to its NPUSCH transmission configuration. The values of M for different NPUSCH configurations are shown in Table 7.23. The values of P_{max}, P_{target}, and α are provided by higher-layer configuration signaling.

The power control for NPRACH follows the same general principles. There may be multiple NPRACH configurations for supporting different coverage levels. The NPRACH preamble repetition levels may be different for these different NPRACH configurations. For NPRACH preambles not having the lowest repetition level, the maximum configured device power, P_{max}, is used in all transmissions. For NPRACH preambles having the lowest repetition level, the transmit power is determined based on the expression below.

$$P_{NPRACH} = \max\{P_{max}, P_{target} + L\}(dBm), \qquad (7.20)$$

Table 7.23 Bandwidth adjustment factors used in open-loop power control		
NPUSCH Configuration	**NPUSCH Bandwidth [kHz]**	**M**
Single-tone, 3.75 kHz subcarrier spacing	3.75	1/4
Single-tone, 15 kHz subcarrier spacing	15	1
3-tone, 15 kHz subcarrier spacing	45	3
6-tone, 15 kHz subcarrier spacing	90	6
12-tone, 15 kHz subcarrier spacing	180	12

where P_{target} is the target NPRACH received power level, which is indicated by the higher layers [18]. If the device does not get a response and has not used the maximum configured device power, it can increase its transmit power in its subsequent random access attempts until it reaches the maximum configured device power. This is referred to as power ramping.

7.3.2.4 Multicarrier Operation

To support a massive number of devices, NB-IoT also includes a multicarrier feature. In addition to the *anchor carrier*, which carries synchronization and broadcast channels, one or more *nonanchor carriers* can be provided. The notion of anchor and nonanchor carriers is explained in Section 7.2.2.1.

As a nonanchor carrier does not carry NPBCH, NPSS, NSSS, and SI, a device in idle mode camps on the anchor carrier, monitoring the paging messages on the anchor. When the device needs to switch from the idle to connected mode, the PRACH procedure also takes place on the anchor carrier. The network can use RRC configuration to point the device to a nonanchor carrier. Essential information about the nonanchor carrier will be provided to the device using dedicated signaling. During the remaining duration of the connected mode, USS monitoring and NPDSCH and NPUSCH activities all take place on the assigned nonanchor carrier. After the device completes the data session, it comes back to camp on the anchor carrier during the idle mode.

The nonanchor carriers can be allocated adapting to the traffic load of NB-IoT. Many IoT use cases generate highly delay-tolerant traffic. Such traffic can be delivered during off-peak hours in the network. The multicarrier feature allows the radio resources normally reserved for serving broadband or voice to be allocated for NB-IoT when the load of broadband and voice traffic is low. For example, during the middle of the night many of the LTE PRBs may be allocated as nonanchor NB-IoT carriers serving the IoT traffic. The network can, as an example, take advantage of the multicarrier feature to push firmware upgrades to a massive number of devices during the middle of the night. An example of replacing LTE PRBs with nonanchor NB-IoT carriers during off-peak hours of broadband and voice traffic is illustrated in Figure 7.47.

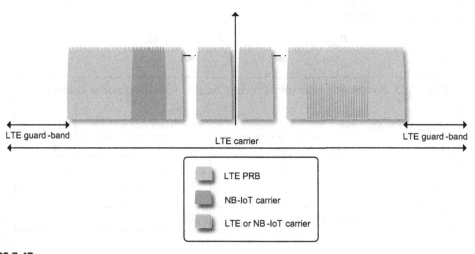

FIGURE 7.47

An example of multicarriers, replacing LTE PRBs with nonanchor NB-IoT carriers during off-peak hours of broadband and voice traffic.

7.4 **RELEASE 14 IMPROVEMENTS**

The previous chapters focused on describing the basic functionality specified for NB-IoT in 3GPP's Release 13. The work has continued in Release 14 with specification of the following features:

- Support for improved positioning accuracy.
- Increased data rates in UL as well as DL.
- Support for paging and random access on nonanchor carriers to support increased system capacity.
- Introduction of a new low device power class.
- Multicast to support, e.g., efficient distribution of firmware to a large number of devices.

In the next five sections the features implemented to support positioning, higher data rates, increased system capacity, introduction of a low device power class, and multicast support are introduced.

7.4.1 **SUPPORT FOR POSITIONING OF DEVICES**

To support positioning of devices is attractive as it opens new business opportunities such as people and goods tracking. Today, such services are associated with solutions based on GPS, but because low cost is highly important for IoT devices, it is desired to be able to support positioning using the NB-IoT cellular technology. This facilitates a low-cost device based on a single chip for both wireless connectivity and positioning services. Also, although GPS-based solutions work perfectly well outdoors, they may have limitations in determining the position of device indoors, where GPS coverage is limited.

The first solution designed for positioning of a device was *enhanced cell identity* (eCID), which is basically based on the estimated TA that determines the round-trip time between the device and the base station. It can be translated to the distance between the base station and device and allows improved positioning compared with only positioning a device based on the identity of the serving cell. Figure 3.42 illustrates the positioning accuracy of eCID and the impact from the synchronization accuracy of the device and the base station.

The second solution specified for NB-IoT is *Observed Time Difference of Arrival* (OTDOA). It is based on a device measuring the ToA on a set of DL *narrowband positioning reference signals* (NPRSs) transmitted from a set of time-synchronized base station surrounding the device. The device reports the narrowband positioning *Reference Signal Time Difference* (RSTD) to a positioning server. Each RSTD allows the positioning server to determine the position of the device to a hyperbola centered around the base stations transmitting the NPRS. If RSTDs between three or more base stations are reported, the positioning server will be able to determine multiple hyperbolas and fix the position of the device to the intersection of the hyperbolas. Figure 7.48 illustrates this, and the accuracy of the RSTD measurement is reflected by the width of the hyperbolas. High timing accuracy and many positioning base stations allowing multiple hyperbolas to be determined lead to a better positioning estimate.

Figure 7.49 depicts simulated NB-IoT OTDOA positioning accuracy in a perfectly time-synchronized network assuming two types of different radio propagation conditions, namely, *Typical Urban* (TU) [21] or *Extended Pedestrian A* [9]. As seen, the performance is heavily dependent on the radio environment. The time dispersion of the TU channel significantly impacts the ability of the device to determine the line of sight channel tap, which impacts the performance. The presented performance was evaluated for a device type using a high oversampling rate to be able to estimate the RSTD with high accuracy [19].

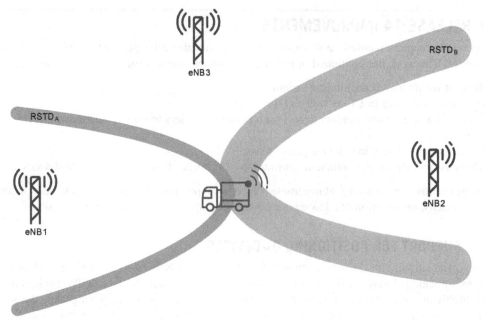

FIGURE 7.48

OTDOA positioning.

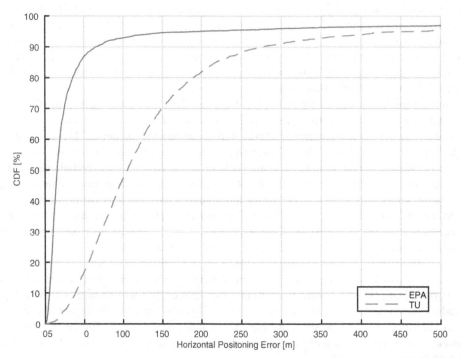

FIGURE 7.49

OTDOA positioning accuracy.

7.4.2 INCREASED DATA RATE

In Release 13 there was a strong focus on device complexity, while other factors such as latency and data rate were down prioritized. As a result, the NB-IoT Release 13 data rates (examined in Section 8.2 in detail) were modest. In the UL a 62.5 kbps and in the DL 25.5 kbps are achievable. Although NB-IoT is designed for ultra-low complexity, these numbers were seen as being on the lower end, especially for the DL expecting to support larger data transfers of firmware download to the devices.

Two aspects were identified as limiting the data rates, namely, the limited TBS and the time gaps associated with the NPDCCH scheduling of NPDSCH and NPUSCH. The TBS tables in Table 7.5 for the NPDSCH and Table 7.10A for the NPUSCH contain several unused code points. Table 7.6 also reveals that the there was room to increase the NPDSCH code rate. This observation also applies to the NPUSCH.

Tables 7.24 and 7.25 show the new TBS tables agreed in 3GPP Release 14. For both NPDSCH and NPUSCH, the maximum TBS was increased to 2536 bits. A new row in the TBS tables was also added to increase the code rate to improve the peak spectral efficiency for devices experiencing very good radio conditions. With this change, the DL data rate has been improved to 79 kbps and the UL data rate to 106 kbps. Because the DCI format N0 and N1 MCS information element indicating the TBS index contained spare code points in Release 13, the added support for the new block sizes had minimal impact on the overall system design. For the device, the increased TBS results in a slight increase in soft buffer memory requirements and computational complexity as more operations need to be carried out during the decoding of the data block.

The increase in TBS was the first step to increase the data rates. It did, however, not address the relatively high delay caused by the scheduling gaps illustrated in Figures 7.45 and 7.46 for the UL and DL, respectively. Therefore, in addition to increasing the TBS size, 3GPP Release 14 specified the support for two HARQ processes. The second HARQ process is intended to increase the throughput by reducing the relative overhead of the scheduling gaps. The minimum gap between the end of an

Table 7.24 3GPP Release 14 NPDSCH TBSs								
	Number of RUs, One RU = One NPDSCH Subframe							
NPDSCH TBS Index	1	2	3	4	5	6	8	10
0	16	32	56	88	120	152	208	256
1	24	56	88	144	176	208	256	344
2	32	72	144	176	208	256	328	424
3	40	104	176	208	256	328	440	568
4	56	120	208	256	328	408	552	680
5	72	144	224	328	424	504	680	872
6	88	176	256	392	504	600	808	1032
7	104	224	328	472	584	680	968	1224
8	120	256	392	536	680	808	1096	1352
9	136	296	456	616	776	936	1256	1544
10	144	328	504	680	872	1032	1384	1736
11	176	376	584	776	1000	1192	1608	2024
12	208	440	680	904	1128	1352	1800	2280
13	224	488	744	1128	1256	1544	2024	2536

Table 7.25 3GPP Release 14 NPUSCH Format 1 TBSs

TBS	Number of RUs							
	1	2	3	4	5	6	8	10
0	16	32	56	88	120	152	208	256
1	24	56	88	144	176	208	256	344
2	32	72	144	176	208	256	328	424
3	40	104	176	208	256	328	440	568
4	56	120	208	256	328	408	552	696
5	72	144	224	328	424	504	680	872
6	88	176	256	392	504	600	808	1000
7	104	224	328	472	584	712	1000	1224
8	120	256	392	536	680	808	1096	1384
9	136	296	456	616	776	936	1256	1544
10	144	328	504	680	872	1000	1384	1736
11	176	376	584	776	1000	1192	1608	2024
12	208	440	680	1000	1128	1352	1800	2280
13	224	488	744	1128	1256	1544	2024	2536

Table 7.26 Supported data rates for NB-IoT device category N2

	NPDSCH	NPUSCH [kbps]
Category N2	79	106
Category N2 with support for two HARQ processes	127	158.5

NPDCCH candidate and the first NPDSCH, or NPUSCH, has been, e.g., reduced to 2 ms. This further increases the achievable throughput to 127 kbps for the DL and 158.5 kbps for the UL. To support the new TBS, the two HARQ processes, and the thereto associated increase in device complexity, a second NB-IoT *device category* has been introduced (Table 7.26).

7.4.3 SYSTEM ACCESS ON NONANCHOR CARRIERS

The basic design of NB-IoT aimed at supporting at least 60,680 devices/km^2. In 3GPP Release 14, the target was increased to support 1,000,000 devices/km^2 as a consequence of a desire to make NB-IoT a 5G technology [22]. As explained in Section 4.6.1 these loads can be assumed to correspond to 6.8 and 112.2 access arrivals per second, respectively. To keep the collision rate of competing preambles low, a sufficiently high amount of radio resources need to be reserved for the NPRACH.

For *slotted* type of access schemes such as the NB-IoT NPRACH, a collision probability between competing preambles can be estimated as [20]:

$$\text{Prob(collision)} = 1 - e^{-\gamma/\text{L}} \tag{7.21}$$

where L is the total number of random access opportunities per second and γ is the average random-access intensity. If a collision probability of 1% is targeted for a random-access intensity of 6.8 access arrivals per second, the total number of random access opportunities provided by the system needs to be set at 677 per second. NB-IoT can at most provide 1200 access opportunities per second by configuring an NPRACH resource for CE level 0 spanning 48 subcarriers and reoccurring with a periodicity of 40 ms. It is clear that already this requires the NPRACH to consume a large portion of the available UL resources. If a collision rate of 10% is acceptable, it is sufficient to configure 65 access opportunities per second. But to support 112.2 access arrivals per second even 10% collision rate will require the configuration of 1065 access opportunities per second. This in combination with the required support of devices in extended coverage motivated the introduction of random access on the nonanchor carriers. To also support mobile terminated reachability for a very high number of users, also paging support has been added to the nonanchor carriers in 3GPP's Release 14.

7.4.4 SUPPORT OF A NEW DEVICE POWER CLASS

NB-IoT Release 13 supports two device power classes, namely, 20 and 23 dBm. 23 dBm is the most common power class in LTE and is also expected to gain most market traction in NB-IoT. The performance evaluation presented in Chapter 8 is therefore based on the 23 dBm device power class. The 20 dBm power class was specified with the target to simplify the integration of the PA on the chip to support a low complexity system on chip design.

The PA drain current expected for these two power classes may easily exceed 100 mA. This limits the ability of the so-called small *coin cell* batteries to support NB-IoT operation. To facilitate the use of coin cells to enable new uses cases such as *smart watches*, a 14 dBm power class is introduced in 3GPP Release 14. This significantly reduces the PA drain current and enables the use of simpler and more compact battery types.

The UL coverage is for obvious reasons negatively impacted by the reduction in device output power. The supported MCL for a 14 dBm device should therefore be expected to be relaxed from 164 to 155 dB. Despite the relaxation, this is still more than 10 dB higher than the MCL supported by GSM and LTE systems.

7.4.5 MULTICAST TRANSMISSION

To offer improved support of firmware or software updates, a group message delivery type of service known as *Single Cell Point to Multipoint* (SC-PTM) is introduced in NB-IoT Release 14. Similar to LTE-M SC-PTM, described in Section 5.4.2, in NB-IoT SC-PTM is intended to deliver single-cell transmission of multicast services. Two logical channels, *Single Cell Multicast Control Channel* (SC-MCCH) and *Single Cell Multicast Traffic Channel* (SC-MTCH), are defined for SC-PTM operations.

These logical channels are mapped on the NPDSCH. The NPDSCH is scheduled by the NPDCCH, which is transmitted in two new CSSs associated with the SC-MCCH and SC-MTCH. The *Type-1A*

CSS contains NPDCCH candidates, with the CRC scrambled by the *Single Cell RNTI*, that schedule the NPDSCH carrying the SC-MCCH. As SC-MCCH is similar to paging, i.e., it is broadcasted to a group of devices in a cell, the design of Type-1A CSS is based on Type-1 CSS, where the NPDCCH candidates can only start at the beginning of the search space.

Type-2A CSS contains NPDCCH candidates, with the CRC scrambled by the *Group RNTI*, that schedule the NPDSCH carrying the SC-MTCH. To maintain scheduling flexibility for the SC-MTCH, the design principle of Type-2 CSS is followed by Type-2A CSS, where several starting points are defined for the NPDCCH candidates.

The procedures of a configuring and transmitting SC-PTM services are illustrated in Figure 7.50. As shown in Figure 7.50, if a device is configured by a higher layer to receive SC-PTM service(s), on

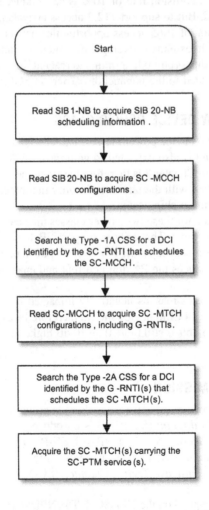

FIGURE 7.50

Procedure for acquiring SC-PTM services.

reading the SIB1-NB, the device identifies the scheduling information of *System Information Block Type 20-NB* (SIB20-NB). SIB20-NB contains the information required to acquire the SC-MCCH configurations associated with transmission of SC-PTM in the cell. In SC-MCCH, a device can further find the configuration information of the SC-MTCH that carriers the multicast service that a device is interested in. One SC-MCCH can be configured in a cell, and it may support up to 64 simultaneous SC-MTCH transmissions. Both anchor and nonanchor carriers can be used to carry SC-MCCH and SC-MTCH.

Retransmissions are not supported for the multicast packages, and therefore for each SC-PTM session, there is only a single transmission on the NPDSCH. Because of the high number of repetitions supported by the NPDCCH and NPDSCH an MCL of 164 dB can still be expected to be supported by NB-IoT SC-PTM services.

REFERENCES

[1] LoRa Alliance, LoRaWAN R1.0 Open Standard Released for the IoT, 2015. Available from: https://www.lora-alliance.org/kbdetail/Contenttype/ArticleDet/moduleId/583/Aid/23/PR/PR.

[2] Third Generation Partnership Project, Technical Report 45.820 v13.0.0, Cellular System Support for Ultra-Low Complexity and Low Throughput Internet of Things, 2016.

[3] 3GPP News, Standardization of NB-IOT Completed, 2016. Available from: http://www.3gpp.org/news-events/3gpp-news/1785-nb_iot_complete.

[4] Third Generation Partnership Project, Technical Specification 36.212 v13.3.0, Evolved Universal Terrestrial Radio Access (E-UTRA) and Evolved Universal Terrestrial Radio Access Network (E-UTRAN); Multiplexing and Channel Coding, 2016.

[5] Third Generation Partnership Project, Technical Specifications 24.301 v13.8.0, Technical Specification Group Core Network and Terminals; Non-access-Stratum (NAS) Protocol for Evolved Packet System (EPS), 2016.

[6] Third Generation Partnership Project, Technical Specifications 24.301 v13.5.0, Technical Specification Group Radio Access Network; Evolved Universal Terrestrial Radio Access Network (E-UTRAN); S1 Application Protocol (S1AP), 2017.

[7] Third Generation Partnership Project, Technical Specifications 36.331 v13.3.0, Evolved Universal Terrestrial Radio Access (E-UTRA) and Evolved Universal Terrestrial Radio Access Network (E-UTRAN); Radio Resource Control (RRC); Protocol Specification, 2016.

[8] C.E. Shannon, Communication in the presence of noise, Proceedings of the Institute of Radio Engineers 37 (1) (1949) 10−21.

[9] Third Generation Partnership Project, Technical Specifications 36.104 v13.3.0, Group Radio Access Network; E-UTRA, UTRA and GSM/EDGE; Multi-standard Radio (MSR) Base Station (BS) Radio Transmission and Reception (Release 13), 2016.

[10] E. Dahlman, S. Parkvall, J. Sköld, 4G: LTE/LTE-Advanced for Mobile Broadband, Academic Press, Oxford, 2011.

[11] Third Generation Partnership Project, Technical Specifications 36.101 v12.14.1, Evolved Universal Terrestrial Radio Access (E-UTRA); User Equipment (UE) Radio Transmission and Reception, 2017.

[12] Third Generation Partnership Project, Technical Specifications 36.211 v13.3.0, Evolved Universal Terrestrial Radio Access (E-UTRA) and Evolved Universal Terrestrial Radio Access Network (E-UTRAN); Physical Channels and Modulation, 2016.

[13] H. Bölcskei, A.J. Paulraj, Space−frequency coded broadband OFDM systems, in: Proc. IEEE Wireless Commun. Netw. Conf., Chicago, IL, USA, vol. 1, September 2000, pp. 1−6.

[14] Ericsson, R1-160094 NB-IoT − Design Considerations for Zadoff-Chu Sequences Based NB-PRACH, 3GPP TSG RAN1 Meeting #1 NB-IoT, 2016.

[15] M.P. Latter, L.P. Linde, Constant envelope filtering of complex spreading sequences, Electronics Letters 31 (17) (Aug. 1995) 1406−1407.

[16] Qualcomm, R1-161981 NB-PSS and NB-SSS Design, 3GPP TSG RAN1 Meeting#2 NB-IoT, 2016.

[17] Third Generation Partnership Project, Technical Specifications 36.304 v13.3.0, Evolved Universal Terrestrial Radio Access (E-UTRA) and Evolved Universal Terrestrial Radio Access Network (E-UTRAN); User Equipment (UE) Procedures in Idle Mode, 2016.

[18] Third Generation Partnership Project, Technical Specifications 36.213 v13.3.0, Evolved Universal Terrestrial Radio Access (E-UTRA) and Evolved Universal Terrestrial Radio Access Network (E-UTRAN); Physical Layer Procedures, 2016.

[19] Ericsson, R1-1608698 OTDOA Performance for NB-IoT, 3GPP TSG RAN1 Meeting #87, 2016.

[20] Third Generation Partnership Project, Technical Report 37.868 v11.0.0, Study on RAN Improvements for Machine-Type Communications, 2011.

[21] Third Generation Partnership Project, Technical Specifications 05.05 v8.20.0, Radio Access Network; Radio transmission and reception, 2005.

[22] Third Generation Partnership Project, Technical Report 38.913 v14.2.0, Study on Scenarios and Requirements for Next Generation Access Technologies, 2017.

NB-IoT PERFORMANCE

CHAPTER OUTLINE

Cellular Internet of Things. http://dx.doi.org/10.1016/B978-0-12-812458-1.00008-3

Abstract

This chapter presents the Narrowband Internet of Things (NB-IoT) performance in terms of coverage, data rate, latency, battery lifetime, and system capacity. All these performance aspects differ between the three operation modes of NB-IoT. Thus, in most cases the performance of each NB-IoT operation mode is individually presented. It shows that all the three operation modes of NB-IoT meet the Third Generation Partnership Project (3GPP) performance objectives agreed for Cellular IoT.

8.1 PERFORMANCE OBJECTIVES

NB-IoT share the same performance objectives as EC-GSM-IoT in terms of coverage, minimum data rate, service latency, device battery life, system capacity, and device complexity as presented, for example, in Sections 1.2.3 and 4.1. In addition, as mentioned in Section 7.1, NB-IoT also aims to achieve deployment flexibility, including a stand-alone mode for deployment using refarmed GSM spectrum as small as 400 kHz, facilitating one NB-IoT carrier plus a 100 kHz guard-band toward the surrounding GSM carriers; an in-band mode for deployment using an LTE Physical Resource Block (PRB); and a guard-band mode for deployment using the guard-band of an LTE carrier. In this chapter, we will present the NB-IoT performance for each of these three operation modes.

In addition to discussing the performance achieved at the targeted *Maximum Coupling Loss* (MCL) level of 164 dB, which represents 20 dB coverage enhancement (CE) compared to GSM/General Packet Radio Service (GPRS), this chapter discusses the following aspects:

- Best achievable performance for devices, which are in good coverage. The performance as such is then defined by the limitations set by the technology.
- Performance of devices in normal coverage, for example, within 144 dB coupling loss from the base station, which corresponds to the MCL of GSM/GPRS.
- Performance of devices with moderate requirements on extended coverage, i.e., 10 dB coverage enhancement compared to GSM/GPRS.

Performance of NB-IoT, like any other system, depends heavily on the device and base station implementations. In the performance evaluations presented here we adopt commonly used 3GPP assumptions such as those described and referenced to in Chapters 4 and 6. Additional NB-IoT specific assumptions will be described in this chapter.

As of today, the performance of NB-IoT based on the finalized 3GPP Release 13 specifications has not been thoroughly documented in the open literature. Although the results are abundant in 3GPP contributions and in the 3GPP Technical Report 45.820 *Cellular System Support for Ultra-low Complexity and Low Throughput Internet of Things* [1], most of these results were not evaluated exactly according to the Release 13 specifications. Therefore, the results presented in this chapter are to a large extent based on authors' own evaluations. For certain performance aspects where the performance results are available in the open literature, the results presented herein are well aligned with the archived results in the literature thus far, e.g., References [2–4].

8.2 **COVERAGE AND DATA RATE**

Like EC-GSM-IoT and LTE-M, all three operation modes of NB-IoT aim to achieve coverage, in terms of MCL, up to 164 dB and at this coverage level a data rate of at least 160 bps. In this chapter we start by describing the methodologies used to evaluate the coupling loss for each of the NB-IoT physical channels followed by actual performance evaluations and thereto corresponding results.

8.2.1 **EVALUATION ASSUMPTIONS**

8.2.1.1 *Requirements on Physical Channels and Signals*

For NB-IoT to meet the coverage requirement, all the physical channels must have *adequate performance* at 164 dB MCL. We will describe what is considered to be adequate performance for various physical channels and signals below.

8.2.1.1.1 Synchronization Signals

The synchronization signals, *Narrowband Primary Synchronization Signal* (NPSS) and *Narrowband Secondary Synchronization Signal* (NSSS), need to be detected with a 90% detection rate. A successful detection of NPSS and NSSS includes identifying the physical cell identity, achieving time synchronization to the downlink (DL) frame structure with ~ 2.5 μs accuracy, and achieving frequency synchronization within ~ 50 Hz. The 2.5 μs time synchronization and 50 Hz frequency synchronization accuracy is considered adequate as these residual errors will only give rise to small performance degradation during the device's subsequent idle and connected modes operation. Note that after the cell selection, the device may employ a timing tracker and *automatic frequency correction* to further refine its time and frequency references.

Because NPSS and NSSS are transmitted periodically, in principle, the device can extend its synchronization time to achieve better coverage. This is where the latency requirement presented in Section 8.4 becomes relevant. The required synchronization time for achieving a 90% detection rate must allow an overall service latency requirement of 10 seconds, to be met. Efficient synchronization does also improve the device battery life evaluated in Section 8.5.

8.2.1.1.2 Control and Broadcast Channels

The Narrowband Physical Broadcast Channel (NPBCH) carries the *Master Information Block* (MIB), which needs to be detected with 90% probability, i.e., support 10% *Block Error Rate* (BLER). Like in the case of the synchronization signals, a device can in principle compensate a poor receiver implementation by repeated attempts to decode the NPBCH until the MIB is received successfully. Also this would, however, negatively impact the overall performance, and efficient NPBCH acquisition improves both latency and battery life.

The Narrowband Physical Downlink Control Channel (NPDCCH) carries *Downlink Control Information* (DCI), which is a control message containing information for uplink (UL) and DL resource scheduling. In these evaluations, the NPDCCH is evaluated at a BLER target of 1% to secure robust operation even at the most extreme CE levels.

The Narrowband Physical Uplink Shared Channel (NPUSCH) Format 2 (F2) carries Hybrid Automatic Repeat Request Acknowledgments (HARQ-ACK) for *Narrowband Physical Downlink*

Shared Channel (NPDSCH) transmissions. The requirement for such signaling is just as the NPDCCH targeting a 1% BLER.

Also for the *Narrowband Physical Radom Access Channel* (NPRACH), a 1% miss detection rate is targeted. A higher miss detection rate target would facilitate higher capacity on the NPRACH, which just as the EC-GSM-IoT *Extended Coverage Random Access Channel* (EC-RACH) and LTE-M *Physical Random Access Channel* (PRACH), is a collision based channel. However, to secure a robust *timing advance* (TA), estimated on the NPRACH, from the very the start of the NB-IoT connection a low miss detection rate target has been chosen. A TA estimation accuracy in the range of 3 μs is considered adequate as this will secure good NPUSCH performance.

8.2.1.1.3 Traffic Channels
The NPDSCH and NPUSCH Format 1 (F1) carry DL and UL data for which the achievable data rate is a suitable criterion. For NB-IoT, a data rate of 160 bps is required at the MCL of 164 dB. In this chapter, the data rate is evaluated after the first HARQ transmission for which a BLER requirement of 10% is targeted. An erroneous transmission can be corrected through HARQ retransmissions. Thus 10% BLER for the initial transmission is more than adequate.

8.2.1.2 Radio Related Parameters
The radio related parameters used in the NB-IoT evaluations are summarized in Table 8.1. These assumptions were agreed in 3GPP [1], and most of them are identical to those used for the EC-GSM-IoT evaluations as summarized in Section 4.2.1.2.

On the device side, the oscillator accuracy assumed is 20 ppm, which reflects the expected accuracy of oscillators used in a low-cost device. For the 900 MHz band, such an oscillator accuracy gives rise to an initial frequency error up to 18 or -18 kHz. As mentioned in Section 7.2.2.1, for in-band and guard-band operations, there is also a frequency raster offset contributing to the initial frequency offset. The raster offset can be as large as ±7.5 kHz. Considering both device oscillator inaccuracy and a worst-case raster offset, the initial cell selection is thus evaluated with a frequency offset uniformly distributed within the interval of [7.5 − 18, 7.5 + 18] kHz, i.e., [−10.5, 25.5] kHz, or [−7.5 − 18, −7.5 + 18 kHz], i.e., [−25.5 ,10.5] kHz. Such a frequency offset not only introduces a time varying phase rotation in the baseband signal, but also results in a timing drift. For example, a frequency offset of 25.5 kHz (including the raster offset) at 900 MHz translates back to a 28.3 ppm error, which means for every 1 million samples, the timing reference at the device may drift by 28.3 samples relative to a correct timing reference. For initial cell selection evaluation, the effect of timing drift is modeled according to the exact initial frequency offset. One of the objectives of the initial cell selection is to reduce such an initial frequency offset to a low value so that the reception, or transmission, of other physical channels in the subsequent communication with the base station will not suffer serious degradation. However, considering that the oscillator used in a low-cost device may not hold on to a precise frequency, a frequency drift after initial frequency offset correction is also modeled. The assumed drift rate is 22.5 Hz/s. In addition to this, the noise figure (NF) assumed is 5 dB, and all the evaluations are done for devices equipped with one transmit antenna and one receive antenna using a maximum transmit power level of 23 dBm.

On the base station side, the NF assumed is 3 dB. Regarding antenna configuration and power level, the assumptions reflect different deployment scenarios for different operation modes. The stand-alone mode is intended for deployment in an existing GSM network. Most of the GSM base stations

Table 8.1 Simulation assumptions

Parameter	Value
Frequency band	900 MHz
Propagation condition	Typical Urban (see Reference [5])
Fading	Rayleigh, 1 Hz Doppler spread
Device initial oscillator inaccuracy	20 ppm (applied to initial cell selection)
Raster offset	Stand-alone: 0 Hz; in-band and guard-band: 7.5 kHz
Device frequency drift	22.5 Hz/s
Device NF	5 dB
Device antenna configuration	One transmit antenna and one receive antenna
Device power class	23 dBm
Base station NF	3 dB
Base station antenna configuration	Stand-alone: one transmit antenna and two receive antennas
	In-band and guard-band: two transmit antennas and two receive antennas
Base station power level	43 dBm (stand-alone), 35 dBm (in-band and guard-band) per 180 kHz
Number of NPDCCH/NPDSCH REs per subframe	Stand-alone 160; in-band: 104; guard-band: 152
Valid NB-IoT subframes	All subframes not carrying NPBCH, NPSS, and NSSS are assumed valid subframes

use one transmit antenna and 43 dBm transmit power and these assumptions are reused for NB-IoT stand-alone operation. The in-band and guard-band modes are intended for deployment in an LTE network. An LTE base station typically uses multiple transmit antennas with each transmit branch associated with a power amplifier of maximum power level of 43 dBm. 3GPP ended up using the assumption of two transmit antennas and thus a 46 dBm maximum base station total transmit power level when summed over the two transmit antennas. However, note that such a total transmit power level is over the entire LTE carrier bandwidth. For example, a total 46 dBm transmit power over 10 MHz LTE carrier bandwidth means that one PRB can only have 29 dBm if the power is evenly distributed over all the 50 PRBs within the LTE carrier. It is a common assumption during 3GPP standardization that PRB power boosting can be applied on an NB-IoT anchor carrier. For in-band and guard-band deployments within a 10 MHz LTE carrier, 6 dB power boosting has been assumed. This results in 35 dBm power for the anchor NB-IoT carrier. For LTE carrier bandwidth of 5, 15, and 20 MHz, power boosting of 3, 7.8, and 9 dB are assumed. This results in a 35 dBm transmit power level for an NB-IoT anchor carrier for all these LTE bandwidth variants. Note that power boosting is only assumed for the anchor carrier, and for a nonanchor carrier it is assumed that no power boosting is applied although this is a permitted configuration. Finally, for all modes of operations two base station receive antennas are assumed.

Furthermore, the amount of available resources for NB-IoT operation also depends on the operation mode. Obviously, for the in-band mode certain resource elements in the DL are taken by the legacy LTE channels, giving rise to fewer resources available to NB-IoT. This has impact on performance. In a subframe, there are 12 subcarriers and 14 *Orthogonal Frequency-Division Multiplexing* (OFDM) symbols, giving rise to a total of 168 resource elements per subframe. The stand-alone operation assumes one *Narrowband Reference Signal* (NRS) antenna port, which takes 8 resource elements per subframe, see Section 7.2.4.3. Thus, the number of resource elements available to NPDCCH or NPDSCH per subframe is 160. For the guard-band operation, 2 NRS antenna ports are assumed, which need 16 resource elements for NRS per subframe, leaving 152 resource elements available to NPDCCH or NPDSCH per subframe. For the in-band operation, we study a case where the first 3 OFDM symbols in every subframe are taken by LTE *Physical Downlink Control Channel* (PDCCH) and there are 2 *Cell-specific Reference Signal* (CRS) antenna ports. The total number of resource elements taken by LTE PDCCH and CRS per subframe is 48. In addition, the 2 NRS ports take 16 resource elements, leaving 104 resource elements available to NPDCCH or NPDSCH per subframe.

8.2.2 DOWNLINK COVERAGE PERFORMANCE

As discussed in Section 8.2.1, both the base station transmit power level and number of resource elements for certain DL channels depend on the operation mode. We will therefore present DL coverage performance for all the three different operation modes. The DL link budget for the 3 operation modes to achieve 164 dB MCL are presented in Table 8.2. As shown, the required signal-to-interference-plus-noise power ratio (SINR) is -4.6 dB for the stand-alone operation and -12.6 dB for the in-band and guard-band operations. We will therefore discuss the performance of all the DL physical channels at these SINRs accordingly. For completeness, the performance at the edge of the normal coverage level, i.e., 144 dB coupling loss, and 10 dB CE level, i.e., 154 dB coupling loss, will also be shown. The required DL SINR's for 154 and 144 dB coupling loss levels are 5.4 and 15.4 dB, respectively, for the stand-alone mode, and -2.6 and 7.4 dB, respectively, for both the in-band and guard-band modes.

Table 8.2 NB-IoT downlink link budget for achieving 164 dB MCL in different operation modes

#	Operation Mode	Stand-alone	In-band	Guard-band
1	Total base station Tx power [dBm]	43	46	46
2	base station Tx power per NB-IoT carrier [dBm]	43	35	35
3	Thermal noise [dBm/Hz]	-174	-174	-174
4	Receiver NF [dB]	5	5	5
5	Interference margin [dB]	0	0	0
6	Channel bandwidth [kHz]	180	180	180
7	Effective noise power [dBm] = (3) + (4) + (5) + $10 \log_{10}(6)$	-116.4	-116.4	-116.4
8	Required DL SINR [dB]	-4.6	-12.6	-12.6
9	Receiver sensitivity [dBm] = (7) + (8)	-121.0	-129.0	-129.0
10	Receiver processing gain	0	0	0
11	Coupling loss [dB] = (2) − (9) + (10)	164.0	164.0	164.0

8.2.2.1 Synchronization Signal

Initial synchronization performance in terms of required synchronization time is shown in Tables 8.3–8.5 for stand-alone, in-band, and guard-band operations, respectively. Note that the stand-alone operation has the best performance thanks to a higher transmitted power level. The in-band and guard-band operations have the same transmit power level and thus have similar performance. The slightly longer synchronization time in the in-band case is because of that LTE CRS punctures NPSS/NSSS, giving rise to performance degradation due to puncturing. However, as indicated by the results in Tables 8.4 and 8.5, the impact of NPSS/NSSS puncturing is small.

Table 8.3 NB-IoT initial synchronization performance for stand-alone operation

Coupling Loss	144 dB	154 dB	164 dB
Synchronization time [ms] for 50% devices	24	24	64
Synchronization time [ms] for 90% devices	84	104	264
Synchronization time [ms] for 99% devices	144	194	754
Average synchronization time [ms]	36	42	118

The 144 and 164 dB performance are based on Reference [2].

Table 8.4 NB-IoT initial synchronization performance for in-band operation

Coupling Loss	144 dB	154 dB	164 dB
Synchronization time [ms] for 50% devices	24	44	434
Synchronization time [ms] for 90% devices	84	124	1284
Synchronization time [ms] for 99% devices	154	294	2604
Average synchronization time [ms]	38	64	582

The 144 and 164 dB performance are based on Reference [2].

Table 8.5 NB-IoT initial synchronization performance for guard-band operation

Coupling Loss	144 dB	154 dB	164 dB)
Synchronization time [ms] for 50% devices	24	44	354
Synchronization time [ms] for 90% devices	84	124	1014
Synchronization time [ms] for 99% devices	154	294	2264
Average synchronization time [ms]	38	60	470

8.2.2.2 NPBCH

NPBCH is carrying the MIB-NB. All the three operation modes share the same resource mapping and thus the performance of MIB-NB acquisition only depends on the transmit power level and coverage. The NPBCH has a 640 ms *transmission time interval* (TTI) in which every subframe 0 carries a code subblock as illustrated in Section 7.2.4.4. When the device is in good coverage, receiving one NPBCH subframe may be sufficient to acquire MIB-NB. Recall that each code subblock is self-decodable and therefore the smallest MIB-NB acquisition time is only 10 ms. However, for devices in poor coverage, combining multiple NPBCH repetitions and multiple NPBCH code subblocks is necessary for the MIB-NB to be acquired. There are 64 NPBCH subframes in an NPBCH TTI. Thus, the device can jointly decode up to 64 NPBCH subframes. If the device still fails to acquire MIB-NB after jointly decoding 64 NPBCH subframes, the most straightforward approach is to reset the decoder memory and start a new decoding attempt in the next NPBCH TTI. With such an approach, the device simply tries again in the new TTI and sees if it can acquire MIB-NB. This is often referred to as the *keep-trying* algorithm. The results presented below are based on the *keep-trying* algorithm.

NB-IoT physical channel performance when channel estimation is required depends heavily on how the receiver acquires its channel estimate. In coverage limited scenarios, using NRS's in one subframe may not be sufficient to ensure adequate channel estimation accuracy. Cross-subframe channel estimation, where NRS's from multiple subframes are jointly used for channel estimation, is highly recommended. We will highlight the impact of channel estimation accuracy when the NPUSCH Format 1 performance is discussed (see Section 8.2.3.2). For now, it suffices to point out that for the NPBCH results presented below, cross-subframe channel estimation with NRS's collected within a 20 ms window is used.

Tables 8.6 and 8.7 show the MIB-NB acquisition time required for achieving a success rate higher than 90%. The average acquisition time is also shown. Because the in-band and guard-band operations share the same transmit power level, they also share the same MIB-NB acquisition performance.

Table 8.6 MIB-NB acquisition time for stand-alone operation

Coupling Loss	144 dB	154 dB	164 dB
Acquisition time [ms] required for achieving 90% success rate	10	20	170
Average acquisition time [ms]	10.3	16.2	83.9

Table 8.7 MIB-NB acquisition time for in-band and guard-band operation

Coupling Loss	144 dB	154 dB	164 dB
Acquisition time [ms] required for achieving 90% success rate	10	60	640
Average acquisition time [ms]	11.1	37.5	357.1

The keep-trying algorithm, although straightforward, is far from optimal. In MIB-NB, the information element that surely changes from one TTI to the next is the 6 bits used to indicate *system frame number* (SFN) and hyper SFN. This information changes in a predictable manner. If representing these bits as a decimal number n in one NPBCH TTI, the number in the next TTI is $n + 1$ mod 64. Exploiting this relationship, it is possible to jointly decode NPBCH subframes across TTI boundaries and significantly improve the performance of MIB-NB acquisition. A similar technique described in Reference [6] was demonstrated to significantly reduce the acquisition time of the LTE MIB. Besides the SFN number, the 1-bit *access barring* (AB) indicator may change more often than the other remaining information elements in MIB-NB. The MIB-NB decoder may need to hypothesize whether the AB indicator changes or not if it wants to jointly decode NPBCH subframes across TTI boundaries.

8.2.2.3 NPDCCH

The performance of NPDCCH is shown in Tables 8.8—8.10. As described in Section 7.2.4.5, NPDCCH may use a repetition factor as low as one and as high as 2048. The repetition factors required for achieving 1% BLER at DCI reception for the three coverage levels and the three operation modes are shown. Because each repetition of NPDCCH is mapped to one subframe, each repetition corresponds to 1 ms transmission time. However, not all subframes are available for NPDCCH. On the anchor carrier, NPDCCH transmissions need to skip subframes taken by NPBCH, NPSS, NSSS, and *System Information Block Type 1* (SIB1-NB). For example, for repetition factor 8, there are no 8 consecutive subframes available for NPDCCH. The shortest possible total TTI to fit in 8 NPDCCH subframes on the anchor carrier is actually 9 ms. This is illustrated in Figure 8.1.

Table 8.8 NPDCCH performance on the anchor carrier for stand-alone operation

Coupling Loss	144 dB	154 dB	164 dB
Repetition required for achieving 1% BLER	1	8	128
Total TTI required for achieving 1% BLER [ms]	1	9	182
Average time required for device to receive DCI correctly [ms]	1.0	1.6	18.0

Table 8.9 NPDCCH performance on the anchor carrier for in-band operation

Coupling Loss	144 dB	154 dB	164 dB
Repetition required for achieving 1% BLER	2	32	256
Total TTI required for achieving 1% BLER [ms]	2	44	364
Average time required for device to receive DCI correctly [ms]	1.1	5.4	78.9

Table 8.10 NPDCCH performance on the anchor carrier for guard-band operation

Coupling Loss	144 dB	154 dB	164 dB
Repetition required for achieving 1% BLER	2	16	256
Total TTI required for achieving 1% BLER [ms]	2	22	364
Average time required for device to receive DCI correctly [ms]	1.0	3.3	51.8

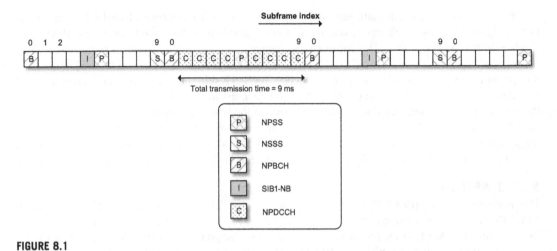

FIGURE 8.1

Illustration of NPDCCH with repetition factor 8 will have at least 9 ms total transmission time on the anchor carrier.

8.2.2.4 NPDSCH Channel

The performance of NPDSCH for different coverage levels and the three operation modes is shown in Tables 8.11–8.13. Here, we only show the performance for the maximum *transport block size* (TBS) of 680 bits as defined in 3GPP Release 13. In general, the entries on the same row of the TBS table (see Section 7.2.4.6) are expected to have approximately the same performance. Like NPDCCH, on the anchor carrier the subframes available for NPDSCH need to exclude those taken by NPBCH, NPSS, NSSS, and SIB1-NB. This results in the total transmission time longer than the total number of subframes needed for NPDSCH. We see that at 164 dB MCL, the data rates of NPDSCH measured over the total transmission interval of NPDSCH are 2.5, 0.47, and 0.62 kbps, for stand-alone, in-band, and guard-band operations, respectively. As described in Section 7.3.2.2, NPDSCH transmissions are scheduled by DCI transmitted in the NPDCCH and acknowledged by the HARQ-ACK bit carried in NPUSCH Format 2. There are also timing relationships between these different channels in that specific timing gaps that are defined between these physical channels. When all these factors are accounted for, the effective physical layer data rates are lower than if simply measured over the total NPDSCH transmission interval. We will give examples to illustrate how effective physical layer data rates are calculated in Section 8.3. In Tables 8.11–8.13, the effective physical layer data rates are presented based on anchor carrier configuration when the overheads such as NPSS, NSSS, NPBCH, and SIB1-NB are accounted for.

Table 8.11 NPDSCH performance on the anchor carrier for stand-alone operation

Coupling Loss	144 dB	154 dB	164 dB
TBS [bits]	680	680	680
Number of subframes per repetition	4	6	6
Number of repetitions	1	4	32
Number of subframes used for NPDSCH transmission	4	24	192
Total TTI required	4 ms	32 ms	272 ms
Data rate measured over NPDSCH TTI	170 kbps	21.3 kbps	2.5 kbps
Physical layer data rate (accounting for scheduling cycle)	19.1 kbps	8.7 kbps	1.0 kbps

Table 8.12 NPDSCH performance on the anchor carrier for in-band operation

Coupling Loss	144 dB	154 dB	164 dB
TBS [bits]	680	680	680
Number of subframes per repetition	10	8	8
Number of repetitions	1	16	128
Number of subframes used for NPDSCH transmission	10	128	1024
Total TTI required	12 ms	182 ms	1462 ms
Data rate measured over NPDSCH TTI	56.7 kbps	3.7 kbps	0.47 kbps
Physical layer data rate (accounting for scheduling cycle)	15.3 kbps	2.4 kbps	0.31 kbps

Table 8.13 NPDSCH performance on the anchor carrier for guard-band operation

Coupling Loss	144 dB	154 dB	164 dB
TBS [bits]	680	680	680
Number of subframes per repetition	8	5	6
Number of repetitions	1	16	128
Number of subframes used for NPDSCH transmission	8	80	768
Total TTI required	9 ms	112 ms	1096 ms
Data rate measured over NPDSCH TTI	75.6 kbps	6.1 kbps	0.62 kbps
Physical layer data rate (accounting for scheduling cycle)	15.3 kbps	3.8 kbps	0.37 kbps

8.2.3 UPLINK COVERAGE PERFORMANCE

Unlike the DL, the UL link budget does not depend on the operation mode. Thus, the results presented below, if the operation mode is not explicitly specified, apply to all three operation modes.

8.2.3.1 Narrowband Physical Random Access Channel

The NPRACH link budget for the three coverage levels is shown in Table 8.14. The link budget establishes the required SINR for NPRACH for each coverage level. Based on the required SINR, a suitable repetition factor is needed to achieve the required performance. A good description of NPRACH detection algorithm and its performance can be found in Reference [7]. Table 8.15 shows the repetition level needed for each coverage level, and the achieved performance. The results in Table 8.15 are based on Reference [7].

Table 8.14 NPRACH link budget				
#	Coupling Loss	144 dB	154 dB	164 dB
1	Total device Tx power [dBm]	23	23	23
2	Thermal noise [dBm/Hz]	−174	−174	−174
3	Base station receiver NF [dB]	3	3	3
4	Interference margin [dB]	0	0	0
5	Channel bandwidth [kHz]	3.75	3.75	3.75
6	Effective noise power [dBm] = (2) + (3) + (4) + 10 $\log_{10}$(5)	−135.3	−135.3	−135.3
7	Required UL SINR [dB]	14.3	4.3	−5.7
8	Receiver sensitivity [dBm] = (6) + (7)	−121.0	−131.0	−141.0
9	Receiver processing gain	0.0	0.0	0.0
10	Coupling loss [dB] = (1) − (8) + (9)	144.0	154.0	164.0

Table 8.15 NPRACH performance			
Coupling Loss	144 dB	154 dB	164 dB
Number of repetitions	2	8	32
Total TTI required [ms]	13	52	205
Detection rate	99.71%	99.76%	99.16%
False alarm rate	0/100,000	0/100,000	13/100,000
TA estimation error [us]	[−3, 3]	[−3, 3]	[−3, 3]

8.2.3.2 NPUSCH Format 1

We have mentioned that the performance of NB-IoT physical channels depend on the accuracy of the channel estimation. To achieve good channel estimation accuracy, cross-subframe channel estimation is highly recommended. An example are shown in Figure 8.2. It can be seen that there is a very substantial performance difference between single subframe based channel estimation and cross-subframe channel estimation. All the NPUSCH results presented in the remainder of this chapter are based on cross-subframe channel estimation with 8 ms estimation window, if not specified otherwise. For cases when the total NPUSCH TTI is lower than 8 ms, cross-subframe channel estimation is based on using all the NPUSCH subframes.

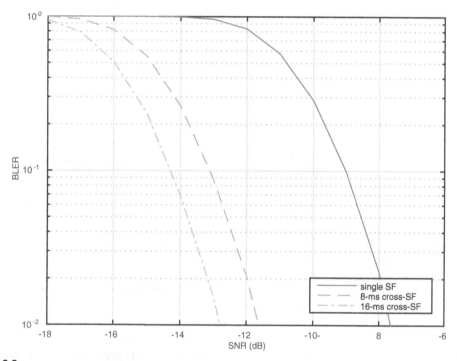

FIGURE 8.2

NPUSCH Format 1 performance. Single-tone transmission with 15 kHz numerology, TBS 1000, 80-ms transmission time per repetition, and 64 repetitions.

As described in Section 7.2.5.2, there are a few different transmission formats for NPUSCH Format 1. Figure 8.3 shows the performance of these different transmission formats. The data rates shown here are NPUSCH Format 1 data rates measured over the total NPUSCH transmission interval and do not account for the scheduling aspects. It can be seen that when coupling loss is high (e.g., greater than 150 dB), all these transmission formats achieve approximately the same performance. In such an operation regime, it is thus advantageous to use a more spectrally efficient signal-tone transmission. However, for devices in good coverage, multi-tone transmission can be used to allow these devices to transmit at a higher data rate. As shown, when the coupling loss is lower than 141 dB, the 12-tone format achieves the highest data rate.

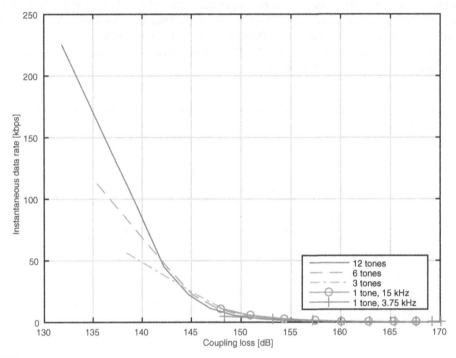

FIGURE 8.3

NPUSCH Format 1 performance, L1 data rate versus coupling loss for various transmission formats. (TBS = 1000, 8-ms cross-subframe channel estimation.)

NPUSCH Format 1 performance at 144, 154, and 164 dB coupling loss is summarized in Table 8.16. Here we choose the most suitable transmission format for each of these coupling loss levels. Like the performance of NPDSCH shown in Section 8.2.2.4, two types of data rates are shown in Table 8.16. First, data rates measured over total NPUSCH Format 1 transmission interval are shown. In addition, we also show data rates when the time it takes to signal the scheduling grant, via NPDCCH, and the associated timing relationship are accounted for. We refer to such rates as physical layer data rates. Note that because NPDCCH performance depends on the operation mode, UL physical layer data rates at different coverage levels also depend on the operation mode. To get the performance at 164 dB coupling loss as shown in Table 8.16, an improved channel estimation with 32-ms cross-subframe channel estimation is used. Using a higher degree of cross-subframe channel estimation requires a higher baseband complexity at the base station. Furthermore, performance improvement from more aggressive cross-subframe channel estimation will be more significant for devices of low mobility. According to the device coupling distribution in Reference [1], there is only a small percentage of devices requiring an extreme CE level. Thus, the base station can afford to use more advanced channel estimator to help these devices improve performance. Furthermore, devices that need extreme CE are expected to be located indoors or underground. These devices indeed will benefit from more aggressive cross-subframe channel estimation.

Table 8.16 NPUSCH Format 1 performance

Coupling Loss	144 dB	154 dB	164 dB
TBS [bits]	1000	1000	1000
Subcarrier spacing [kHz]	15	15	15
Number of tones	3	1	1
Number of resource units per repetition	8	10	10
Number of repetitions	1	4	32
Total TTI required [ms]	32	320	2560
Data rate measured over NPUSCH Format 1 TTI	28.1 kbps	2.8 kbps	371 bps
Physical layer data rate, stand-alone	18.8 kbps	2.6 kbps	343 bps
Physical layer data rate, in-band	18.7 kbps	2.4 kbps	320 bps
Physical layer data rate, guard-band	18.7 kbps	2.5 kbps	320 bps

8.2.3.3 NPUSCH Format 2

NPUSCH Format 2 is used to transmit HARQ-ACK for NPDSCH and its performance is summarized in Table 8.17. It can be seen that the target of 164 dB coupling loss can be achieved with 64 repetitions.

Table 8.17 NPUSCH Format 2 performance based on 1% BLER

Number of Repetitions	TTI [ms]	MCL [dB]
1	2	152.2
2	4	155.0
4	8	157.2
8	16	159.2
16	32	161.2
32	64	163.6
64	128	165.5
128	256	167.6

8.3 PEAK PHYSICAL LAYER DATA RATES

NB-IoT supports a range of data rates. For a device in good coverage, a higher data rate can be used, and as a result, reception and transmission can be completed in shorter time, which allows the device to return to the deep sleep mode sooner, thereby improving the battery lifetime. In this section, we discuss the performance of NB-IoT peak physical layer data rates. Similar to the discussions in Sections 8.2.2.4 and 8.2.3.2, two types of data rates are shown. First, data rates measured over total NPDSCH or NPUSCH Format 1 transmission interval are shown. We will refer to such data rates as instantaneous peak physical layer data rates. In addition, we also show peak data rates when the time it takes to receive or transmit other control signaling, for example, DCI or HARQ-ACK, and the associated timing relationship are accounted for. We refer to such rates simply as peak physical layer data rates.

The instantaneous peak physical layer data rates are purely determined based on the NPDSCH and NPUSCH configurations. For example, the maximum Release 13 NPDSCH TBS is 680 bits and can be mapped to as short a time duration of 3 subframes, i.e., 3 ms, in the stand-alone and guard-band modes. The peak L1 downlink data rate is thus 226.7 kbps in these operation modes. For the in-band mode, the shortest time duration for NPDSCH carrying a 680-bit transport block is 4 ms, giving an instantaneous peak physical layer downlink data rate of 170 kbps. NB-IoT instantaneous peak physical layer data rates are summarized in Table 8.18.

Table 8.18 Instantaneous peak physical layer data rates (Release 13 Category N1 devices)

	Stand-alone [kbps]	In-band [kbps]	Guard-band [kbps]
NPDSCH	226.7	170.0	226.7.0
NPUSCH multi-tone	250.0	250.0	250.0
NPUSCH single-tone (15 kHz)	21.8	21.8	21.8
NPUSCH single-tone (3.75 kHz)	5.5	5.5	5.5

The instantaneous peak physical layer data rate can be used as a performance metric for cross-technology comparison as this is often included in the data sheet.

The instantaneous peak physical layer data rates, however, do not account for the protocol aspects, and thus may not be a good indicator when it comes to e.g., latency. For these, data rate taking scheduling delays and time restrictions is a better indicator. For NB-IoT, NPDCCH search space configuration (see Section 7.3.2.1) and the timing relationship (see Section 7.3.2.2) affect this throughput, which we here just refer to as the physical layer data rate. For example, the instantaneous peak physical layer NPDSCH data rate of 226.7 kbps will become 25.5 and 24.3 kbps physical layer data rates on anchor and nonanchor carriers, respectively, when the limitation due to NPDCCH search space and timing relationship are accounted for. A nonanchor carrier example is illustrated in Figure 8.4. Here, the NPDCCH search space is configured to have $R_{max} = 2$ and $G = 4$, giving a search period of 8 ms. Recall that the device is not expected to monitor search space candidates after it receives a DCI and until 3 ms after it completes its HARQ feedback. For example, the search space candidate in the 24th subframe in Figure 8.4 will not be monitored by the device. The earliest opportunity that the base station can send the second DCI to the device is the 25th subframe. The timing relationship illustrated is according to the minimum gaps between the DCI and NPDSCH and between the NPDSCH and NPUSCH Format 2 allowed by the specifications. The gaps between the DCI's can be thought of as the time it takes to transmit each transport block. As shown, it takes 56 ms to complete two full scheduling cycles delivering two transport blocks of 680 bits, giving rise to a physical layer data rate of 24.3 kbps. To illustrate an example of peak physical layer data rate on an anchor carrier is more complicated as the presence of NPSS, NSSS, and NPBCH results in postponements of NPDCCH and NPDSCH. However, interestingly such postponements in some cases end up resulting in a slightly higher physical layer data rate. This is due to that a postponement of NPDCCH extends the time for certain search space candidates, which in some cases makes it possible to send a DCI to the device without waiting for an extra NPDCCH search period.

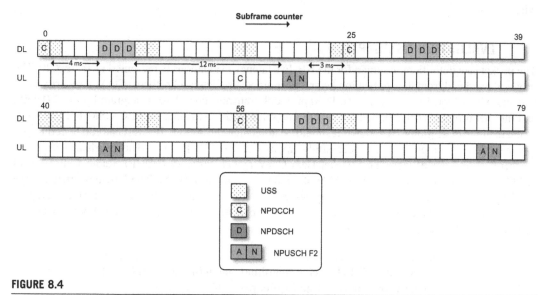

FIGURE 8.4

Example of peak downlink physical layer data rate on a nonanchor carrier (stand-alone and guard-band). NPDCCH search space configuration includes $R_{max} = 2$, $G = 4$.

Table 8.19 summarizes the peak physical layer data rates. The differences between the peak instantaneous physical layer data rates and peak physical layer throughputs data rates for the NPUSCH single-tone cases are relatively smaller. This is due to the actual NPUSCH transmissions take longer time for the single-tone cases, and therefore the timing relationship and NPDCCH search space constraint have smaller impact.

Table 8.19 Peak physical layer data rates

	Stand-alone [kbps]	In-band [kbps]	Guard-band [kbps]
NPDSCH	25.5	22.7	25.5
NPUSCH multi-tone	62.5	62.5	62.5
NPUSCH single-tone (15 kHz)	15.6	15.6	15.6
NPUSCH single-tone (3.75 kHz)	4.8	4.8	4.8

8.4 LATENCY

The concept of Exception Reporting was just as for EC-GSM-IoT introduced for NB-IoT in 3GPP Release 13. Also for NB-IoT it is required that a device can deliver an Exception Report to the network within 10 seconds.

8.4.1 EVALUATION ASSUMPTION

The packet size definitions assumed for the NB-IoT latency evaluations is presented in Table 8.20. The protocol overhead from the application layer Constrained Application Protocol (COAP), security layer Datagram Transport Layer Security (DTLS) protocol, transport layer User Datagram Protocol (UDP), and internet protocol (IP) assumed are exactly the same as those assumed for EC-GSM-IoT and LTE-M evaluations in Sections 4.4 and 6.4. It should be noted that the 40 byte IP overhead can be reduced if Robust Header Compression is successfully applied at the PDCP layer. The overheads from the protocols below the IP layer, i.e., Packet Data Convergence Protocol (PDCP), Radio Link Control (RLC), and Medium Access Control (MAC), are the same as for LTE-M: The 5 byte overhead at the PDCP layer is due to header and integrity protection; The overhead from RLC and MAC layers is eololib dependent on the message type and the message content. As a rule of thumb, 4 bytes overhead from RLC/MAC headers can be assumed.

Table 8.20 NB-IoT packet definitions including application, security, transport, IP, and radio protocol overheads

Type	UL Exception Report Size [bytes]
Application data	20
CoAP	4
DTLS	13
UDP	8
IP	40
PDCP	5
RLC	2
MAC	2
Total	94

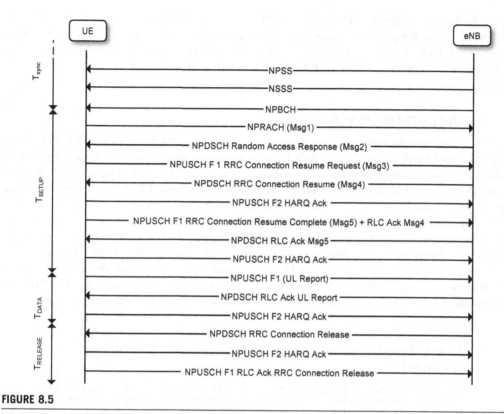

FIGURE 8.5

NB-IoT data transfer based on the RRC resume procedure.

When accessing a cell, the *Radio Resource Control (RRC) Connection Resume Request* message contains the information element *Establishment Cause*, which may signal that a device requests radio resources for transmitting an Exception report of high importance [8]. The process of transmitting such a report using the RRC Resume procedure is depicted in Figure 8.5. For simplicity, assignment messages conveyed over the NPDCCH are omitted from the figure. The depicted procedure is based on the NB-IoT 3GPP Release 13 specifications where RLC Acknowledge Mode is always used.

Four specific parts are identified in Figure 8.5, namely:

- T_{SYNC}: The time to acquire synchronization.
- T_{SETUP}: The time to perform the Random Access procedure and setup the connection.
- T_{DATA}: The time to transmit the data.
- $T_{RELEASE}$: The time to release the connection.

Clear differences are seen in figure 8.5 compared to the EC-GSM-IoT message flow depicted in Section 4.4.2 where, for example, the data transmission starts already in the second UL transmission. It can also be noted that the MIB-NB (carried in NPBCH) is read before accessing the system. This is needed to acquire frame synchronization, access barring information, and system information status.

The transmission times for the NPDSCH and the NPUSCH Format 1 were based on a selected modulation and coding scheme that could carry the message at the targeted coupling loss at a BLER of

10%. For the NPRACH, NPUSCH Format 2 and NPDCCH, a 1% BLER was assumed. For the NPBCH, NPSS, and NSSS the 90th percentile synchronization time was used in the evaluations. For the NPUSCH Format 1, a sub-PRB allocation was used in combination with cross subframe channel estimation to optimize the link budget at the 10% BLER target. The link level performance associated with these assumptions is presented in Section 8.2.

8.4.2 LATENCY PERFORMANCE

Table 8.21 summarizes the time to deliver the exception report described in Section 8.4.1. Results for stand-alone, guard-band, and in-band modes of operations are reported at coupling losses of 144, 154, and 164 dB. For in-band mode, the NB-IoT carrier is assumed to operate in a 10 MHz LTE carrier that limits the DL power and the number of available resource elements due to LTE CRS transmission. Also, in case of guard-band mode, the DL power is restricted by the LTE *power spectral density* and the emission requirements. In guard-band and stand-alone modes, the full set of resource elements are available for NB-IoT which improves DL performance compared to in-band. For stand-alone mode, the full base station power can be allocated to the NB-IoT carrier that improves the DL performance. Using a higher power will improve DL coverage and hence lower the latency needed to deliver the exception report. In good coverage conditions, the time to acquire synchronization, reading the MIB-NB and waiting for an access opportunity on the periodically occurring NPRACH dominates the latency.

Table 8.21 NB-IoT exception report latency [s] using the RRC resume procedure [10]

Coupling Loss [dB]	Stand-alone	Guard-band [s]	In-band [s]
144	0.3	0.3	0.3
154	0.7	0.9	1.1
164	5.1	8.0	8.3

8.5 BATTERY LIFE

Battery life is a key aspect of the NB-IoT technology and a 10-year target was assumed in the design. The next two sections describe battery life evaluations for NB-IoT and the expected performance.

8.5.1 EVALUATION ASSUMPTIONS

To verify the battery life objective, a simple traffic model was assumed, where the packet sizes presented in Table 8.22 were considered. These packets sizes contain overheads from application, internet, and transport protocols presented in Table 8.20. To understand the actually transferred data over the air interface, also additional overhead from PDCP, RLC, and MAC layers needs to be added to the values in Table 8.22.

Table 8.22 Packet sizes on top of the PDCP layer for evaluation of battery life [9]			
Message Type	UL Report		DL Application Acknowledgment
Size	200 bytes	50 bytes	65 bytes
Arrival rate	Once every 2 h or once every day		

For NB-IoT operation, it was assumed that the device operation can be associated with the power consumption levels shown in Table 8.23. The transmitter power consumption assumes a power amplifier with 45% efficiency, demanding 440 mW plus 60 mW power consumption from baseband and other circuitry. While transmitter and receiver power consumption depends on characteristics of the supported access technology the power levels when the device is in different levels of sleep, i.e., not actively transmitting or receiving, are more related to the hardware architecture and design. Due to this, the assumptions on power consumption in light and deep sleep are aligned for NB-IoT, EC-GSM-IoT, and LTE-M.

Table 8.23 NB-IoT power consumption [9]			
Tx, 23 dBm	Rx	Light Sleep	Deep Sleep
500 mW	80 mW	3 mW	0.015 mW

Figure 8.6 illustrates the UL and DL packet flow modeled in the battery life evaluations. Not illustrated are NPDCCH transmissions and a 1 second period of light sleep in-between the end of the UL report and the start of the DL data transmission. A 20 second period of light sleep before the device goes back to power save mode was also assumed in the model. This period was assumed and supported by the configured Active timer during which the device uses a Discontinuous Reception cycle that facilitates DL reachability. Just as in the latency evaluations a 10% BLER was targeted for the NPBCH, NPDSCH, and NPUSCH Format 1. A 1% BLER was assumed for the NPDCCH, NPUSCH Format 2, and NPRACH. For the synchronization time, the average NPSS and NSSS acquisition time was used and not the 90th percentile value used in the latency evaluations. The link level performance associated with these assumptions is presented in Section 8.2.

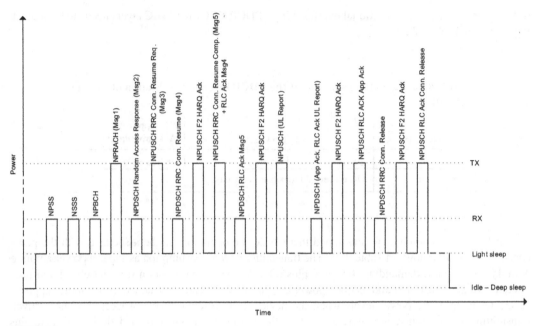

FIGURE 8.6

NB-IoT packet flow used in the evaluation of battery life.

8.5.2 BATTERY LIFE PERFORMANCE

The resulting battery life for the packet flow depicted in Figure 8.6 is presented in Table 8.24. Again, slightly different performance is seen for the different modes of system operation. UL performance is, however, not impacted by the system mode of operation. Due to this, and the fact that the UL data transmission dominates the power consumption, the differences seen are fairly small.

The general conclusion is that a 10-year battery life is feasible for a reporting interval of 24 h. It is also clear that the 2 hour reporting interval is a too aggressive target when the device is at 164 dB MCL. Just as in the case of EC-GSM-IoT, see Section 4.5, these evaluations were performed assuming ideal battery characteristics.

Table 8.24 NB-IoT stand-alone (S), guard-band (G), and in-band (I) battery life [10]											
	DL	UL	**Battery Life [Years]**								
Reporting Interval [Hours]	Packet Size [Bytes]	Packet Size [Bytes]	144 dB CL			154 dB CL			164 dB MCL		
			S	G	I	S	G	I	S	G	I
2	65	50	22.2	22.1	22.1	13	12.6	12.3	3.0	2.7	2.6
		200	20	20.0	20	7.9	7.8	7.7	1.4	1.3	1.3
24		50	36.2	36.1	36.1	33	32.8	32.6	19.3	18.4	18.0
		200	35.6	35.6	35.6	29	28.9	28.7	11.8	11.5	11.3

8.6 CAPACITY

During the initial feasibility study of NB-IoT in 3GPP Release 13, the objective was to design a system with the ability to handle a load of 60,680 devices/km^2 [1], as explained in Section 4.6. Later in Release 14 it was decided that NB-IoT should meet also the 5G requirement on connection density of 1,000,000 devices/km^2 [11]. This extreme load can be expected when people gather in a small area. Consider, for example, big events in popular sports that easily can attract hundred thousands of people to the location of the event.

At locations where this type of extreme load of the network is observed, it is likely that the cell grid is fairly dense, but, it cannot be ruled out that a macro deployment having a cell grid with a large intersite distance may carry a significant part of the load. Table 8.25 summarizes the 3GPP Release 13 and 14 requirements under the assumption that base stations are located on a hexagonal cell grid with an intersite distance of 1732 m resulting in a cell size of 0.87 km^2.

8.6.1 EVALUATION ASSUMPTIONS

To evaluate the NB-IoT system capacity, the traffic model described in Section 4.4.1.1 was used. It challenges mainly the UL with up to 80% of the users autonomously accessing the system to send reports. But the DL capacity is also tested, due to the network command message flow in the DL. While EC-GSM-IoT in Section 4.4 was evaluated exactly at the load of 60,680 users/km^2, NB-IoT was evaluated for a range of loads up to and beyond 100,000 devices per NB-IoT carrier.

Table 8.25 3GPP assumption on required system capacity [11]		
Assumed Scenario	**Devices/km^2**	**Devices/cell**
3GPP Release 13 assuming 40 devices per home in central London	60,680	52,547
3GPP Release 14 assuming large crowds in small areas e.g., at a sports event	1,000,000	865,970

Also the radio related modeling was to a large extent the same as described in Section 4.6.1. The channel model including outdoor to indoor loss and fast fading characteristics was reused. Also the macro deployment modeled as a hexagonal cell grid was assumed. In-band mode of operation with the NB-IoT carriers located on PRBs of a 10 MHz LTE carrier was assumed. The base station was assumed to use 46 dBm output power on the 50 PRBs or 29 dBm per PRB. The PRB carrying the anchor carrier was assumed to be power boosted by 6 dB leading to a total output power of 35 dBm for the anchor carrier.

Table 8.26 summarizes a set of the most relevant NB-IoT system simulation assumptions.

Table 8.26 System level simulation assumptions [12]

Parameter	Model
Cell structure	Hexagonal grid with 3 sectors per size
Cell intersite distance	1732 m
Frequency band	900 MHz
LTE system bandwidth	10 MHz
Frequency reuse	1
Base station transmit power	46 dBm
Power boosting	6 dB on the anchor carrier
	0 dB on nonanchor carriers
Base station antenna gain	18 dBi
Operation mode	In-band
Device transmit power	23 dBm
Device antenna gain	−4 dBi
Device mobility	0 km/h
Pathloss model	$120.9 + 37.6 \times \log_{10}(d)$, with d being the base station to device distance in km
Shadow fading standard deviation	8 dB
Shadow fading correlation distance	110 m
Anchor carrier overhead from mandatory downlink transmissions	NPSS, NSSS, NPBCH mapped to 25% of the downlink subframes.
Anchor carrier overhead from mandatory uplink transmissions	NPRACH mapped to 7% of the uplink resources.

8.6.2 CAPACITY PERFORMANCE

The NB-IoT system capacity was simulated assuming a network load up to 110,000 users per carrier. This was done both for anchor carriers and nonanchor carriers. Figure 8.7 shows how the DL NPDCCH and NPDSCH average resource utilization increases more or less linearly on the anchor carrier as a function of the user arrival intensity. Depicted is only the resource utilization of the resources available for NPDCCH and NPDSCH transmission. Resources used for transmission of NPSS, NSSS, and NPBCH are not taken into account. Also for the UL, a linear increase in average NPUSCH Format 1 and Format 2 resource utilization is observed. For the UL the resources allocated for NPRACH coupling loss 144, 154, and 164 dB are also included in the presented statistics.

The x-axis in Figure 8.7 is defined in terms of device arrival rate in the system. Thanks to the traffic pattern described in Table 4.14, a linear relation between this rate and the absolute load is possible to derive. Two important reference points are as follows:

- 6.8 arrivals per second correspond to the 3GPP targeted load of 60,680 users/km^2.
- 11.2 arrivals per second correspond to the 3GPP targeted load of 100,000 users/km^2.

It may be surprising to see that despite the UL heavy traffic model, it is DL capacity that is limiting performance. This can be explained by the NPSS, NSSS, and NPBCH transmissions that takes up 25% of the available DL subframes. Other contributing factors to the DL load are the NPDCCH scheduling mechanism and the relatively signaling intensive connection setup and release procedures. It should also be noticed that mandatory narrowband system information transmission, i.e., Narrowband System Information Block, and load related to paging was not modeled. Taking also these transmissions into account would further increase the stress on the anchor DL capacity.

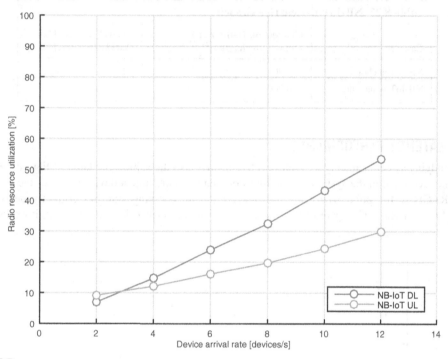

FIGURE 8.7

NB-IoT anchor carrier average utilization of resources available for data transmission.

In the 3GPP work on 5G, it was required that 99% of all users need to be served at the peak capacity of the system [11]. Although it may seem like a stringent requirement, it is a reasonable objective for a system that was designed to reach the last mile. For the anchor carrier a 1% *outage* was reached at 67,000 devices/km^2, corresponding to 7.5 user arrivals per second. Outage is here defined as the percentage of users not served by the system. So beyond this point the load increase, see in Figure 8.7, comes at the expense of a rapidly increasing outage. It should also be clear that a more relaxed outage requirement would be directly translated into a higher system capacity.

For the nonanchor carriers, a load up to 110,000 devices/km^2 was achieved at the point where 1% outage was observed. This significant capacity increase compared to the anchor is explained by the fact that a nonanchor has no downlink overhead in terms of synchronization and broadcast channels. As a result, the resource utilization recorded on the nonanchor was fairly even between UL and DL.

Table 8.27 summarizes the system capacity achievable on an anchor carrier and on a nonanchor carrier given the herein made assumptions. Given the high per carrier capacity, it is clear that NB-IoT has the potential to serve 1,000,000 devices/km^2 if 10 or more carriers are configured in a cell. It is important to note that the 3GPP Release 14 nonanchor support for random access as described in Section 7.4 was modeled in these simulations, which is necessary to facilitate the random access procedure at this high load.

Table 8.27 NB-IoT per carrier capacity		
Case	Connection Density at 1% Outage [devices/km^2]	Arrival Rate at 1% Outage [connections/s]
NB-IoT anchor	67,000	7.5
NB-IoT nonanchor	110,000	12.3

8.6.3 LATENCY PERFORMANCE

When studying system capacity it is relevant to consider also the achievable latency. Figure 8.8 shows the latency recorded in the system simulations counted from the time a device accesses the system on the NPRACH to the point where the upper layer in the receiving side has correctly decoded the received transmission. Compared to the latency analysis performed in Section 8.4.2, it excludes the

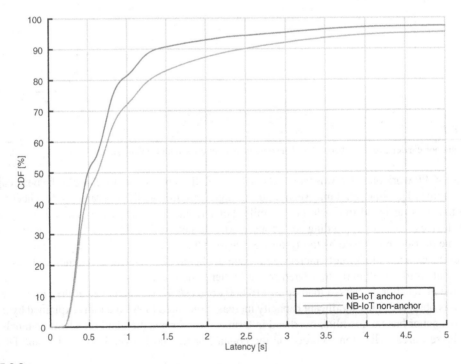

FIGURE 8.8

Service latency for NB-IoT anchor and nonanchor at the loads presented in Table 8.28.

time to synchronize to the system and acquire the MIB. The results also include aspects such as scheduling delays, whereas the results in Section 8.4 assumes an absolute priority of the exception report, resulting in no delays due to scheduling queues.

Table 8.28 captures a few samples from the graphs in Figure 8.8 including the 99th percentile not visible in the figure. It is clear that at these extreme loads, the last 10% of all devices needs to accept a considerable latency. This is mainly due to the high load that creates queues for the available resources. In addition, it should be remembered that latency is one of the qualities that have been traded off for high coverage and high capacity in the design of NB-IoT. In case a low latency is required, it can be indicated to the network that the message is an exception report, as described in Section 8.4, enabling the network to prioritize its transmission.

Table 8.28 Service latency at extreme load extracted from Figure 8.8		
Service Latency	NB-IoT Anchor [s]	NB-IoT Nonanchor [s]
At the 50th percentile	0.49	0.62
At the 90th percentile	1.4	2.5
At the 99th percentile	39	35

8.7 DEVICE COMPLEXITY

NB-IoT aims to offer competitive module price. Like EC-GSM-IoT, an NB-IoT module can be implemented, to a large extent, as a system on chip (SoC), and most of the descriptions in Section 4.7 regarding different functionalities on the SoC apply to NB-IoT as well. A summary of NB-IoT design parameters affecting the device complexity is tabulated in Table 8.29. Parameters both for the Release 13 NB-IoT device category Cat N1 and the Release 14 Cat N2 are included.

The number of soft channel bits in Release 13 is based on that a maximum NPDSCH TBS of 680 bits is used, which are attached with 24 cyclic redundancy check bits and then encoded by the LTE rate-1/3 *tail-biting convolutional code* (TBCC), giving rise to a maximum 2112 coded bits. Rate-matching, either through puncturing or repetition, is used to match the actual number of encoded bits available according to NPDSCH resource allocation. This results in a code rate lower or higher than the original code rate of 1/3. In case of a lower code rate than 1/3, there will be more encoded bits than 2112. However, rate-matching process can be handled at the device decoder to first combine the repeated bits. Thus, fundamentally, the decoder buffer can be dimensioned based on the rate-1/3 TBCC to support 2112 soft bits as indicated in Table 8.29.

Regarding baseband complexity, the most noteworthy operations are the *Fast Fourier Transform* (FFT) and decoding operations during connected mode, and the NPSS synchronization required during cell selection and reselection procedures.

Table 8.29 Overview of NB-IoT category N1 (Cat N1) and Cat N2 device complexity

Parameter	Value
Operation modes	FDD (Release 13 and 14)
Duplex modes	Half duplex
Half duplex operation	Type B
Device RX antennas	1
Power class	20, 23 dBm
Maximum bandwidth	180 kHz
Highest downlink modulation order	QPSK
Highest uplink modulation order	QPSK
Maximum number of supported DL spatial layers	1
Maximum DL TBS size	Cat N1: 680 bits
	Cat N2: 2536 bits
Number of HARQ processes	Cat N1: 1
	Cat N2: up to 2
Peak instantaneous DL physical layer data rate	Cat N1: 226.7 kbps
	Cat N2: 282.0 kbps
DL channel coding type	TBCC
Physical layer memory requirement	Cat N1: 2112 soft channel bits
	Cat N2: 7680 soft channel bits
Layer 2 memory requirement	4000 bytes

In NB-IoT, in principle only a 16-point FFT is needed. The complexity of N-point FFT is $6N\log_2 N$ real operations. There are 14 OFDM symbols per subframe, and therefore the complexity associated with FFT demodulation is ~ 5376 real-value operations per subframe.

The TBCC used is a 64-state code that can be decoded using 2 times trellis wrap-around [13,14]. To decode the largest TBS of 680 bits in Release 13, the device essentially needs to process $680 \times 2 \times 64 = 87{,}040$ state metrics. The operation involved with processing each state is calculating the two path metrics merging onto the state and select a surviving path. This takes 3 real-value operations. Thus, the complexity of decoding a TBS of 680 bits is ~ 260k operations. Consider the most computationally demanding case in Release 13, the device receives TBS 680 bits in 3 NPDSCH subframes and has 12 ms before signaling HARQ-ACK. The complexity of FFT and TBCC decoding is approximately $\frac{5376 \times 3 + 26{,}000}{15 \times 10^{-3}} \approx 18.5$ millions of operations per second (MOPS).

Regarding cell selection or reselection procedures, the complexity is dominated by NPSS detection, which requires the device to calculate a correlation value per sampling time interval. It was shown in Reference [15] that the complexity of NPSS detection is less than 30 MOPS. Note also that, the device does not need to simultaneously detect NPSS and perform other baseband tasks.

Considering the most computationally demanding baseband functions as discussed above, NB-IoT device can be implemented with baseband complexity less than 30 MOPS.

REFERENCES

[1] Third Generation Partnership Project, Technical Report 45.820 v13.0.0, Cellular System Support for Ultra-low Complexity and Low Throughput Internet of Things, 2016.

[2] A. Adhikary, X. Lin, Y.-P. Eric Wang, Performance evaluation of NB-IoT coverage, in: Proc. IEEE Veh. Technol. Conf., Montreal, Canada, September 2016, 2016.

[3] R. Ratasuk, N. Mangalvedhe, J. Kaikkonen, M. Robert, Data channel design and performance for LTE narrowband IoT, in: Proc. IEEE Veh. Technol. Conf., Montreal, Canada, September 2016, 2016.

[4] R. Ratasuk, B. Vejlgaard, N. Mangalvedhe, A. Ghosh, NB-IoT system for M2M communication, in: Proc. IEEE Wireless Commun. and Networking Conf., Doha, Qatar, April 2016, 2016.

[5] Third Generation Partnership Project, Technical Specifications 05.05 v8.20.0, Radio Access Network; Radio Transmission and Reception, 2005.

[6] Ericsson, R1-152190 PBCH repetition for MTC, 3GPP TSG RAN1 Meeting #80bis, 2015.

[7] X. Lin, A. Adhikary, Y.-P. Eric Wang, Random access preamble design and detection for 3GPP narrowband IoT systems, IEEE Wireless Communication Letters 5 (6) (December 2016) 640–643.

[8] Third Generation Partnership Project, Technical Specifications 36.331 v13.3.0, Evolved Universal Terrestrial Radio Access (E-UTRA) and Evolved Universal Terrestrial Radio Access Network (E-UTRAN); Radio Resource Control (RRC); Protocol Specification, 2016.

[9] Ericsson, R1-1701044 on mMTC, NB-IoT and eMTC Battery Life Evaluation, TSG RAN1 Meeting#1 NR, 2017.

[10] Ericsson, R1-1705189, Early Data Transmission for NB-IoT, 3GPP TSG RAN1 Meeting #88, 2017.

[11] Third Generation Partnership Project, Technical Report, 38.913, v14.1.0, Study on Scenarios and Requirements for Next Generation Access Technologies, 2016.

[12] Ericsson, R1-1703865 on 5G mMTC Requirement Fulfilment, NB-IoT and eMTC Connection Density, 3GPP TSG RAN1 Meeting #88, 2017.

[13] Y.-P. Wang, R. Ramesh, A. Hassan, H. Koorapaty, On MAP decoding for tail-biting convolutional codes, in: Proc. IEEE Information Theory Symposium, 1997.

[14] Ericsson, R1-073033 Complexity and performance improvement for convolutional coding, 3GPP TSG RAN1 Meeting #49bis, 2007.

[15] Qualcomm, R1-161981 NB-PSS and NB-SSS Design, 3GPP TSG RAN1 Meeting#2 NB-IoT, 2016.

THE COMPETITIVE INTERNET OF THINGS TECHNOLOGY LANDSCAPE

CHAPTER OUTLINE

Abstract

This chapter provides an overview over connectivity solutions for Internet of Things (IoT) devices. An overview of short-range radio and low power wide area technologies that operate in unlicensed spectrum is given. Also the characteristics of using unlicensed spectrum are described. The properties of unlicensed technologies are compared with the CIoT technologies that operate in licensed spectrum. Finally, the performance of the different CIoT technologies, Extended Coverage Global System for Mobile Communications Internet of Things (EC-GSM-IoT), Long-Term Evolution for Machine-Type Communications (LTE-M), and Narrowband Internet of Things (NB-IoT) is reviewed and compared with each other. Benefits of the different CIoT technologies are discussed.

Cellular Internet of Things. http://dx.doi.org/10.1016/B978-0-12-812458-1.00009-5

9.1 IoT CONNECTIVITY TECHNOLOGIES IN UNLICENSED SPECTRUM

9.1.1 UNLICENSED SPECTRUM

9.1.1.1 Unlicensed Spectrum Regulations

CIoT networks, such as EC-GSM-IoT, LTE-M, and NB-IoT operate in licensed spectrum. This means that mobile network operators have acquired long-term spectrum licenses from regulatory bodies in the country/region after, for example, an auction process. Such licenses provide an operator with exclusive spectrum usage right for a carrier frequency. Such spectrum licenses may also be combined with an obligation to build out a network and provide network coverage and communication services in a certain area within a certain time frame. This obligation in combination with the cost of the license motivates mobile network operators to invest upfront into a network infrastructure. This exclusive spectrum usage right provides the prospect of good financial returns on the investment obtained via communication services within the lifetime of the license. There are also other spectrum bands, which do not abide to the rules of licensed spectrum. In unlicensed or license-exempt spectrum any device is entitled to transmit as long as it fulfills the regulation without requiring any player from holding a license. These regulatory requirements have the objective to harmonize and ensure efficient use of the spectrum.

Unlicensed spectrum bands differ for different regions in the world. In the following an overview of the usage of unlicensed spectrum is provided for two bands, one at around 900 MHz and one at 2.4 GHz. These are of particular relevance due to their ability to cater for IoT services and popular wireless communication standards such as Wi-Fi and Bluetooth have been specified for these bands. The sub-GHz range around 900 MHz provides attractive propagation characteristics in terms of facilitating good coverage. The 2.4 GHz range is interesting because it is considered to be a global band, which is important for systems targeting a global footprint. While the 2.4 GHz band is globally harmonized, the sub-GHz range has regional variations. However, most regions have some unlicensed spectrum allocation even if they differ in their specifics. A more detailed description is here provided for the US unlicensed spectrum at 902–928 MHz and the European unlicensed spectrum at 863–870 MHz. In Europe, some differences in the allocations of the 863–870 MHz band exist on a per country basis. There has been significant market traction for IoT connectivity solutions operating in the unlicensed frequency domain in these two regions. In other regions, the unlicensed spectrum allocation in the sub-GHz range varies for different countries. For example, the allocations in Korea and Japan are overlapping with the US spectrum region, and China has an allocation that is below the European spectrum allocation, see e.g., Reference [1]. Radio technology standards that are addressing the unlicensed spectrum around 900 MHz, such as, IEEE 802.11ah, are typically designed in a way, in which they provide a common technology basis for different channelization options in this spectrum range; the detailed channelization is then adopted to the region where it is deployed, see e.g., Reference [1] for the channelization of IEEE 802.11ah. IEEE 802.11ah is the basis upon which Wi-Fi HaLow is built.

For the United States, the usage of unlicensed spectrum for communication devices is regulated by the Federal Communications Commission (FCC) and it is specified in Title 47 Code of Federal

Regulations Part 15 [2]. For Europe, the spectrum rules are specified by the European Conference of Postal and Telecommunications Administrations (CEPT), which is a coordination body of the telecommunication and postal organizations within Europe. As of today, 48 countries are members of CEPT [3]. The CEPT recommendation for usage of short-range devices in unlicensed spectrum is described in Reference [4]. This recommendation is the basis for European Telecom Standards Institute (ETSI) harmonized standards, which specify technical characteristics and measurement methods for devices that can be used by device implementers to validate their devices for conforming with the regulated rules. Such ETSI standards are as follows:

- ETSI standard EN 300 220 for short-range devices operating in 25 MHz–1 GHz [5,6],
- ETSI standard EN 300 440 for radio equipment to be used in the 1–40 GHz frequency range [7,8],
- ETSI standard EN 300 328 for data transmission equipment operating in the 2.4 GHz ISM band and using wide band modulation techniques. Direct-sequence spread spectrum (DSSS), frequency hopping spread spectrum (FHSS), and Orthogonal Frequency-Division Multiplexing (OFDM) are considered to be wide band modulation techniques [9].

Some of the more relevant spectrum usage rules for unlicensed spectrum at 863–870 MHz in Europe is given in Table 9.1 and for 902–928 MHz in the United States in Table 9.2. The tables present, e.g., the maximum allowed radiated power and requirements for interference mitigation. While the ETSI regulations in the band 863–870 MHz mandate the power in terms of *Effective Radiated Power* (ERP), i.e., the radiated power assuming a half-wave dipole antenna, the FCC sets

Table 9.1 European unlicensed spectrum at 863–870 MHz, for more details see Reference [4]				
Spectrum Band [MHz]	EIRP [dBm]	Mitigation Requirement	Bandwidth	Other
863–870	16.1	<0.1% duty cycle or LBT		FHSS
863–870	16.1	<0.1% duty cycle or LBT + AFA		DSSS and other non-FHSS wide band techniques
863–870	16.1	<0.1% duty cycle or LBT + AFA	≤100 kHz, for 1 or more channels; modulation bandwidth ≤300 kHz	
868–868.6	16.1	<0.1% duty cycle or LBT + AFA		
868.7–869.2 MHz	16.1	<0.1% duty cycle or LBT + AFA		
869.4–869.65	29.1	<10% duty cycle or LBT + AFA		
869.7–870	9.1; 16.1	For 4.8 dBm: No requirements; For 11.8 dBm: ≤1% duty cycle or LBT + AFA		

Table 9.2 US unlicensed spectrum at 902–928 MHz, for more details see Reference [2]

Spectrum Band [MHz]	EIRP [dBm]	Mitigation Requirement	Bandwidth [kHz]	Other
902–928	36	Dwell time per hopping channel: <0.4 s/20 s	≤250	Frequency hopping with ≥50 hopping channels
902–928	30	Dwell time per hopping channel: <0.4 s/10 s	200–500	Frequency hopping with ≥25 hopping channels
902–928	36		≥500	Digitally modulated

requirements in terms of conducted power in combination with allowed antenna gain. Here we have converted these requirements to *Equivalently Isotropically Radiated Power* (EIRP), i.e., the radiated power assuming an isotropic antenna, according to the following equation:

$$\text{EIRP} = \text{ERP} + 2.15 \text{ dB} \tag{9.1}$$

The 2.15 dB offset stems from the different reference antennas, i.e., dipole and isotropic, assumed for EIRP and ERP. It can be noticed that the permitted radiated power is higher for the unlicensed spectrum in the United States than in Europe. The European recommendations rely on duty cycle limitations, i.e., the percentage of time a transmitter may be active within a defined time span, while the US regulations defines maximum dwell times, i.e., the maximum continuous time by which a transmitting device may use a specific radio resource (e.g., a specific hopping channel). Regardless whether the limitations are on *duty cycle* or *dwell time*, these limits aim to avoid persistent interference. A short allowed dwell time can be understood to limit the coverage of a system, whereas a strict duty cycle requirement may limit coverage and system capacity. The duty cycle also limits the availability of downlink and uplink transmission opportunities which has a negative impact on service latency.

European recommendation allows to deviate from duty cycle limitations in case *listen-before-talk* (LBT) is used, i.e., when a transmitter first observes if a channel is not used by other devices before starting to transmit. Sometimes LBT needs to be combined with *adaptive frequency agility* (AFA), which is a method for avoiding transmission on already occupied channels (see e.g., Reference [1]).

The different rows in Tables 9.1 and 9.2 can be understood to cater for different types of applications and different types of equipment. For more details please see References [2,4].

The spectrum usage recommendation for unlicensed spectrum at 2400–2483.5 MHz in Europe is given in Table 9.3 and the rules for the United States in Table 9.4. Also in this band the maximum

Table 9.3 European unlicensed spectrum at 2400–2483.5 MHz, for more details see Reference [4]

Spectrum Band [MHz]	EIRP [dBm]	Mitigation Requirement	Other
2400–2483.5	10		
2400–2483.5	20	Spectrum sharing mechanism like LBT	For wide band data transmission and radio local area networks.
			For wide band modulations other than FHSS, the maximum EIRP-density is limited to 10 mW/MHz

Table 9.4 US unlicensed spectrum at 2400–2483.5 MHz, for more details see Reference [2]

Spectrum Band [MHz]	EIRP	Mitigation Requirement	Bandwidth	Other
2400–2483.5	30 dBm	Dwell time per hopping channel: <(0.4 s × X), X = number of hopping channels	≤250 kHz	Frequency hopping with ≥15 and <75 nonoverlapping hopping channels
2400–2483.5	36 dBm	Dwell time per hopping channel: <(0.4 s × X), X = number of hopping channels	200–500 kHz	Frequency hopping with ≥75 nonoverlapping hopping channels
2400–2483.5	36 dBm and max 8 dBm/3 kHz power spectral density (PSD)		≥500 kHz	Digitally modulated

allowed radiated power is higher in United States than in Europe. The European power limitation is given as EIRP. Higher radiated power is allowed for wide band transmission such as DSSS modulation, FHSS, and OFDM under the condition of using a mitigation method such as LBT.

9.1.1.2 Spectrum Coexistence in Unlicensed Spectrum

Different types of radio equipment and technologies can transmit on any frequency of the unlicensed spectrum. As depicted in Figure 9.1, the simultaneous transmissions of different devices interfere with each other. As a result, this interference may lead to that one or both of the transmissions depicted in

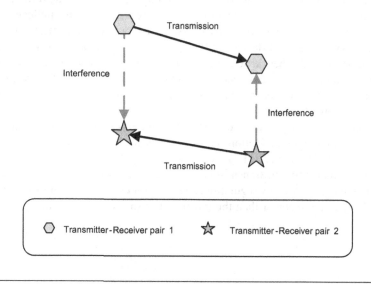

FIGURE 9.1

Transmission of two different unlicensed devices to their corresponding receivers.

the figure fail. Spectrum coexistence mechanisms are mechanisms that limit the interference that a transmitter may cause on other nearby devices.

The simultaneous usage of unlicensed spectrum can occur between homogenous or heterogeneous types of devices, which means devices that use the same or different wireless communication technologies. A wireless communication technology designed for unlicensed spectrum, often, has some mechanism, which specifies how the spectrum is shared among different devices of that communication technology, so that each device has good transmission opportunities, while at the same time minimizing the interference to other devices. The coordination scheme typically follows the spectrum regulation introduced earlier but may also have additional technology specific features. While such a coordination scheme can be applied within one wireless communication technology, it can typically not provide the same level of interference mitigation in-between different wireless communication technologies.

The spectrum regulation for unlicensed spectrum provides requirements on devices, which shall provide technology neutral coexistence, i.e., independent of a particular wireless communication technology being used. Spectrum regulation has thereby mainly two types of communication devices in mind: *adaptive devices* and *nonadaptive devices*.

Nonadaptive devices are considered to transmit in unlicensed spectrum, while staying unaware of the other types of devices. To limit the amount of interference that can be generated, the devices are limited as follows:

- In the radiated power they may use, see Tables 9.1–9.4, sometimes further limited by an allowed power spectral density,
- By a duty cycle or dwell time, see Tables 9.1 and 9.2, which limit what fraction of time a device may transmit on a channel.

By these means the total amount of interference a device may cause on others is restricted.

In contrast, an adaptive device is aware of other devices in its vicinity, which also make use of the same channel. As a result, it can adapt its transmission, reducing the interference to other devices. This provides a device with the opportunity to transmit for a longer amount of time if there are few devices in the surrounding. In contrast, if many devices in the vicinity want to use the unlicensed spectrum, the adaptive device reduces the frequency of its transmissions and provides less interference. The common way to provide this type of adaptive device is LBT, where a device before a transmission listens on the radio channel and observes if other devices are communicating, and after a *clear channel assessment* (CCA) it transmits for a limited time. In addition, technical standards consider a fair use of spectrum among devices, and thus consider that at a high utilization of the spectrum an adaptive device transmits less than in case of a low utilization.

The most common form of unlicensed spectrum usage is for short-range communication. One reason is that short-range communication provides a certain level of interference robustness. The amount of interference that a receiver is exposed to depends very much on the location of the interferer. Assuming that different transmitters in unlicensed spectrum use similar output powers, e.g., the maximum that is allowed by regulation, an interferer can be considered as a *strong interferer* if it is located closer to the receiver than the intended transmitter. Devices that jointly form a local

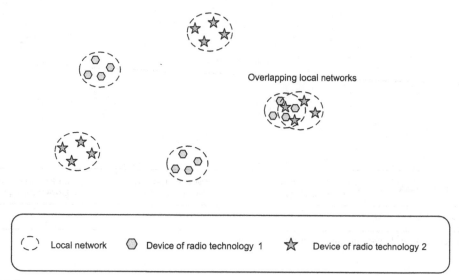

FIGURE 9.2

Coexistence among different groups of unlicensed devices. Intersystem interference is most severe if different groups are overlapping in space.

network typically have some coordination or coexistence functionality provided by the wireless communication standard they are using, which avoids or reduces interference within the local network. However, other unlicensed radio technologies are typically not part of this interference coordination. In Figure 9.2, it is depicted how different groups of devices use various unlicensed communication technologies. If these devices are separated in space, the interference is limited because it is typically significantly below the power levels of the communication within the group. However, if different unlicensed radio technologies operate at the same location, significant interference can occur.

Figure 9.3 shows the challenge of long-range communication in unlicensed spectrum. With long-range communication it becomes more likely that an interferer is located closer to the receiver than the intended transmitter. The example of the figure shows a long-range system that is designed to cover a large path loss of e.g., 150 dB for transmission over several kilometers. There may exist several other local unlicensed networks using the same spectrum in vicinity of the long-range receiver. If the long-range receiver is, e.g., placed on the roof of a building, there may be some local unlicensed networks used in the same or neighboring buildings, e.g., for home automation. Because these devices are significantly closer to the long-range receiver, they may cause interference at the location of the long-range receiver, which is significantly higher, by e.g., several 10's of dB, than the strongly attenuated signal of the long-range transmitter, which is coming from far away.

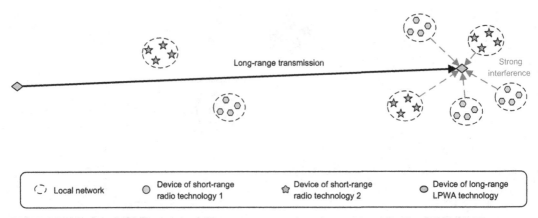

Long-range transmission

Strong interference

| ⟨⟩ Local network | ○ Device of short-range radio technology 1 | ⬠ Device of short-range radio technology 2 | ○ Device of long-range LPWA technology |

FIGURE 9.3

Coexistence between long-range and short-range devices.

Furthermore, if we assume that the devices in the local network are adaptive devices, which e.g., use LBT to avoid interfering with other devices, this operation is likely to fail to adapt to long-range transmitters that are far away because the long-range signal is so strongly attenuated that it is below a sensitivity threshold used for CCA. As long as unlicensed spectrum is barely used, such interference situations may be unlikely. If it is anticipated that unlicensed IoT use cases (and other use cases) will drive the deployment of various local area networks using unlicensed spectrum the inference in unlicensed spectrum will increasingly play a role; long-range unlicensed radio technologies are more exposed to this interference.

One aspect that is worth to mention for unlicensed spectrum, is about the maximum transmit power that may be emitted by the device. The maximum transmit power provided by spectrum regulation is typically given with respect to a certain reference configuration of emission. The reason is that different antenna configurations have different power emission patterns, which leads to that the antenna has different gains in different directions. Often maximum power levels are defined as either effective isotropically radiated power or as ERP, as shown above for Tables 9.1–9.4. For EIRP an ideal isotropic antenna is assumed, whereas for ERP a half-wave dipole antenna is assumed, which has a 2.15 dB antenna gained compared with an isotropic antenna in direction of the highest antenna gain as specified in Eq. (9.1). When the maximum transmit power is specified as EIRP or ERP, a transmitter with any antenna configuration should not emit power in any direction that would exceed the maximum power that could be emitted in this direction with an isotropic or dipole antenna, respectively. A practical result of this is that if a real antenna has an antenna gain of X dB with a maximum allowed radiated power of Y dBm EIRP, then the transmitter must limit its conducted power at the antenna port to Y-X dBm. One consequence is that downlink performance can be substantially limited compared with uplink, if a base station antenna with significant antenna gain is used. In uplink direction, the base station can use the antenna gain at the receiver side to improve the link budget. In the reverse downlink direction, the base station has to compensate the antenna gain by a reduced transmit power resulting in that the base station antenna gain does not improve the actual link budget, such as maximum path loss. This is one significant difference for a communication system operating in unlicensed spectrum compared with licensed spectrum, where the antenna gain could also be used in downlink direction and where higher maximum radiated powers are permitted in downlink.

Another cause for asymmetry of uplink and downlink in unlicensed spectrum can be found regarding capacity for nonadaptive devices that are limited by a duty-cycle. As an example, we assume a large number of N devices connected to a single base station, each device transmitting at a rate R_u and being limited by a maximum duty cycle of D_u. The maximum achievable uplink capacity is then limited by the maximum data rate and the duty cycle limitation for each device. Assuming an ideal situation of all transmissions being successful by neglecting all possible collisions and interference situations, the upper bound of maximum achievable uplink capacity C_u becomes in this case

$$C_u = N \cdot D_u \cdot R_u, \tag{9.2}$$

In downlink, the capacity of the base station is limited by its own maximum duty cycle. This duty cycle has to be used for the transmission to all N devices, in contrast to uplink where a duty cycle is valid per device. The upper bound of maximum achievable downlink capacity C_d under ideal error free transmissions becomes then

$$C_d = D_d \cdot R_d, \tag{9.3}$$

for a downlink data rate R_d and a maximum duty cycle of D_d. Assuming the same uplink and downlink parameters, the downlink capacity is thus by a factor of N smaller compared with the uplink capacity, which leads to a significantly lower achievable effective downlink data rate per device. To some extent this can be compensated by configuring downlink transmissions to certain subbands that allow higher duty cycles (increasing D_d), see e.g., Table 9.1.

9.1.2 RADIO TECHNOLOGIES FOR UNLICENSED SPECTRUM

Unlicensed spectrum enables immediate market access for any new radio technology with minimal regulatory requirements. As a result, a very large number of different Machine-to-Machine (M2M) connectivity solutions have been developed and brought to the market making use of this spectrum. Most of them are designed to satisfy a very particular application and communication needs. Examples are connectivity for remote-controlled lighting, baby monitors, electric appliances, etc. For many of those systems the entire communication stack has been designed for a single purpose. Even if it enables, in a wider sense, an environment with a wide range of connected devices and objects, it is based on M2M technology solution silos usually without end-to-end IP connectivity and instead via proprietary networking protocols as depicted on the left hand side in Figure 9.4. This is quite different from the vision of the IoT (depicted on the right hand side in Figure 9.4), which is based on a common IP-based connectivity framework for connecting devices and smart objects, which enables the IoT at full scale.

Different industrial alliances and standardization organizations, such as IPSO Alliance, IETF, are promoting an IP-based framework for connecting smart objects and devices, are defining corresponding components, and are developing according open standards. Figure 9.5 provides an overview of such a harmonized protocol stack, which has at its center the common IP connectivity protocols, such as IPv6, transmission control protocol (TCP), user datagram protocol (UDP), and the security

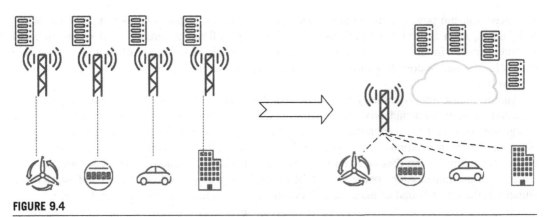

FIGURE 9.4

From M2M technology silos to the IoT.

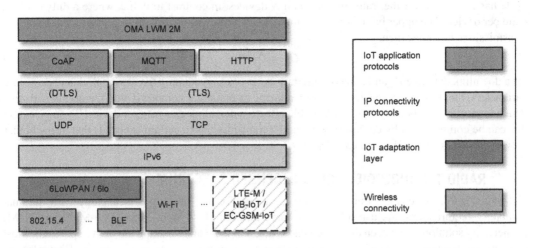

FIGURE 9.5

IoT protocol stack.

protocols transport layer security (TLS) and datagram transport layer security (DTLS). Several additional protocols have been developed to simplify the communication procedures for small and constrained devices. Besides the transaction protocol HTTP, which is already widely used for Internet services like WWW, a simplified IoT-focused version for constrained devices has been developed as the Constrained Application Protocol (CoAP). Also the Message Queue Telemetry Transport (MQTT) protocol is widely used as publish/subscribe messaging protocol for IoT services. Lightweight M2M, specified by the Open Mobile Alliance (OMA), is a device management protocol for IoT. With these IoT application and communication protocols, IP-based IoT services can be provided via a variety of communication technologies, which enable IP-based transmission. Some simple wireless communication technologies have difficulties to cope with the overhead provided by IP-based communication.

For this reason, the adaptation framework IPv6 over low power wireless personal area networks (6LoWPAN) has been specified in IETF to enable communication over very constrained wireless communication technologies; this work is continued in IETF under the label "6lo" in the working group IPv6 over Networks of Resource-constrained Nodes.

9.1.2.1 Short-Range Radio Solutions

In this section we provide an overview of the most promising unlicensed short-range radio communication technologies for the IoT, which are IEEE 802.15.4, Bluetooth Low Energy (BLE) and Wi-Fi HaLow. The choice to focus on these is made based on the following properties of those technologies: they address the communication requirements of IoT, they target an IP-based IoT solution according to Figure 9.5 and the right-hand side in Figure. 9.4, and they are based on open standards and are expected to reach a substantial economy of scale. We also address the capillary network architecture where any of these solutions may be used to provide the short-range connectivity within a larger network context.

9.1.2.1.1 IEEE 802.15.4

One of the early standards for 6LoWPAN is IEEE 802.15.4 [10−12]. It was standardized in 2003 at a time of intense research on wireless sensor networking technologies [13,14], and it is applicable for a wide range of IoT use cases, ranging from office automation, connected homes to industrial use cases. Several application-specific protocol stacks have been developed, which build on parts of the IEEE 802.15.4 standard (mostly the physical layer and to some extend the medium access control (MAC)) [15], including ZigBee, WirelessHART, ISA-100, and Thread.

IEEE 802.15.4 has been specified for the three frequency bands of 868 MHz (for Europe), 915 MHz (for United States), and 2.4 GHz (global) [16−18]. In the 2.4 GHz band, IEEE 802.15.4 has 16 channels available of 2 MHz bandwidth. It uses offset quadrature phase shift keying with DSSS and a spreading factor of 8. A gross data rate of 250 kbps is achievable, see References [16,18]. In 868 MHz one channel of 600 kHz is available, which uses differential binary phase shift keying modulation and DSSS with a spreading factor of 15. The achievable data rate at 868 MHz is 20 kbps [16,18]. In the 915 MHz 10 channels are available with a gross data rate of 40 kbps, see Reference [18].

IEEE 802.15.4 uses *carrier-sense multiple access with collision avoidance* (CSMA-CA) for access to the radio channel; this can be complemented with optional Automatic Repeat Request (ARQ) retransmissions. Typical coverage ranges are in the order of 10−20 m [11]. Two different topologies are supported: star topology and mesh (or peer-to-peer) topology. Two types of devices are defined: full-function devices that provide all MAC functionality and can act as network coordinator of the local network, and reduced function devices that can only communicate with a full-function device and are intended for very simple types of devices. The network can operate in beaconed mode, which allows a set of devices to synchronize to a superframe structure that is defined by the beacon transmitted by a local coordinating device. The channel access in this case is slotted CSMA-CA. In non-beacon mode, unslotted CSMA-CA is applied. In case of direct data transmission, a device transmits data directly to another device. In indirect data transmission, data is transferred to a device, e.g., from a network coordinator. When beacon transmission is active, the network coordinator can indicate the availability of data in the beacon; the device can then request the pending data from the network coordinator. In nonbeaconed transmission, the network coordinator buffers data and it is up to the device to contact the network coordinator for pending data.

In Reference [18] the performance of IEEE 802.15.4 has been evaluated in an experiment with an ideal link and devices placed at 1 m distance. It has been found that for a configuration at 2.4 GHz with a theoretical gross data rate of 250 kbps, a net data rate of 153 kbps was measured for direct transmission (from the device), and a net data rate of 66 kbps for indirect transmission (towards the device). Furthermore, it has been shown that the effective data rate and delivery ratio decrease with an increasing number of devices.

A major step for broader relevance of IEEE 802.15.4 for the IoT has been to address end-to-end IP-based communication. To this end the IETF working group 6LoWPAN has been chartered in 2005 and it has developed IETF standards for header compression and data fragmentation. The maximum physical layer payload size of 802.15.4 is limited to 127 bytes, which is further reduced by various protocol headers and optional security overhead and can leave as little as 81 bytes available for application data within an IEEE 802.15.4 frame. IETF has developed standards that provide header compression and IP packet fragmentation that enable the transmission of IPv6 over 802.15.4 networks [17,19−22]. In addition, the RPL routing protocol has been developed to enable IP mesh routing over IEEE 802.15.4 [17,22,23]. In 2014 the Thread group was formed with the objective to harmonize the usage of IEEE 802.15.4 together with 6LoWPAN for home automation.

IEEE 802.15.4 has been extended in IEEE 802.15.4g to address smart utility networks with an objective to improve coverage and support higher data rates [24−26]. To this end multiple new physical layers have been defined which can be used from a common MAC layer. Physical layer implementations are multirate frequency shift keying (MR-FSK), multirate orthogonal frequency-division multiplexing (MR-OFDM), and multirate offset quadrature phase shift keying (MR-OQPSK). MR-FSK has a benefit of good transmit power efficiency because of constant signal envelop, MR-OFDM enables higher data rates for frequency selective fading channels, and MR-OQPSK has the benefit of the original IEEE 802.15.4 modulation with cost-effective and easy design. A physical layer agnostic management protocol is based on a common signaling mode to allow a network configuration with interference coordination among multiple IEEE 802.15.4 transmitters [26].

9.1.2.1.2 BLE

Bluetooth has been developed as a technology for wireless short-range connectivity [27,28] and has established itself as a leading technology for personal area networking. With the release of the Bluetooth core specification 4.0 [29] in 2010 a novel transmission mode called Bluetooth Low Energy (BLE) was introduced, which considerably reduces power consumption compared with Bluetooth classic. BLE has been a significant first step to expand the Bluetooth ecosystem towards IoT.

BLE uses the 2.4 GHz ISM band. The spectrum is divided into 40 channels, with 2 MHz channel spacing, of which 37 are data channels and 3 are used as *advertising channels*. Frequency hopping is applied to mitigate the impact of interference. The modulation is based on *Gaussian Frequency Shift Keying* and a data rate of up to 1 Mbps can be achieved over-the-air. A master-slave architecture has been adopted to assign asymmetric roles to devices; peripheral devices perform only a minimum amount of functions to enable ultra-low power consumption, while central devices perform coordination functions. BLE has short connection setup and data transfer times so that applications can transfer authenticated data within a few milliseconds. BLE allows connection-oriented or connectionless communication. It supports fragmentation and reassembly of large data packets into small radio frames, which are then transmitted over the radio interface. This enables BLE to support data services with large packets (e.g., IP packets).

An analysis of BLE for building automation use cases has been performed in References [16,30−32]. With a single-hop deployment, the range for BLE in an indoor deployment setup is in the order of 10 m, and around five BLE gateways are needed to provide coverage in a 1000 m^2 office floor [32].

In 2014 the Bluetooth Special Interest Group (BT SIG), the standardization forum for Bluetooth, published the Internet Protocol Support Profile [33], which enables IP connectivity for BLE devices. Further, IETF has standardized a standard for end-to-end IPv6 connectivity over BLE [34], including header compression. This enables that end-to-end IP-based IoT services can be provided via BLE systems [35].

A further evolution of BLE has occurred recently with the launch of Bluetooth 5, the Bluetooth core specification 5.0 [36,37]. It comprises quadrupling of the communication range at low data rates (i.e., 125 kbps) and the doubling of the peak data rates (to 2 Mbps). The BT SIG has also announced that the development of an extension of BLE for mesh networking is ongoing, which would further increase the range of BLE, see e.g., Reference [38].

9.1.2.1.3 Wi-Fi

Wi-Fi based on IEEE 802.11 is one of the most used unlicensed radio access technologies and its focus is on providing high data rate services to a range of mainly consumer electronics devices. Very long battery life has not been in focus. Also the scalability of IEEE 802.11 has mainly addressed being able to provide a high total throughput for a number of connected devices in an area; from the early Wi-Fi version IEEE 802.11b to the version IEEE 802.11ac the theoretically achievable physical layer peak data rates have increased from 11 Mbps to 6.9 Gbps [39]. The typical operation of IEEE 802.11 is in the 2.4 and 5 GHz unlicensed spectrum.

IEEE 802.11ah is an amendment to the IEEE 802.11 standard that is focused on IoT applications. The Wi-Fi Alliance has chosen Wi-Fi HaLow as the marketing term to be used for the IEEE 802.11ah amendment. IEEE 802.11ah has some design targets that significantly differ from the high-data rate focused IEEE 802.11 variants. First, IEEE 802.11ah addresses the unlicensed spectrum below 1 GHz, which is in the range 902−928 MHz in the United States and 863−868 MHz in Europe; other regions also have unlicensed spectrum regions somewhere in the range 750−928 MHz [40]. Differences of the sub-1-GHz spectrum versus higher spectrum bands are as follows:

- The propagation conditions sub-1-GHz facilitate longer range. For a wide-area usage and spread of IoT devices, transmission range is a key property to provide sufficient coverage to IoT services with limited amount of access points. At the same time, the IoT devices are expected to transmit only limited amounts of data. For mobile broadband Wi-Fi usage, where devices are expected to transmit a lot of data, the extended range would mean that the channel is blocked for a longer time and the channel access time per device would reduce.
- There is less unlicensed spectrum available than at higher spectrum bands. This also means that the total capacity of data that can be provided within an area is lower than at higher spectrum bands. For mobile broadband focused Wi-Fi usage, this is a disadvantage because one focus is to provide high capacity in combination with high per user data rates. For IoT-focused Wi-Fi this is less of a problem, as the total amount of data transmitted even by a very large group of IoT devices is expected to remain modest.

The IEEE 802.11ah physical layer design is derived from the IEEE 802.11ac [1]. To address the lower spectrum bands, with less available bandwidth, and to enable robust long range transmission, the bandwidth of the IEEE 802.11ah has been scaled down by a factor of 10 compared to 802.11ac. That means that IEEE 802.11ah supports different carrier bandwidths of 2−16 MHz in comparison with the 20−160 MHz carriers of 802.11ac. In addition, an extra robust carrier configuration with 1 MHz bandwidth has been defined. Reference [1] describes a 24.5 dB link budget gain of IEEE 802.11ah at 900 MHz compared with 802.11n at 2.4 GHz. The gains stem from reduced path loss at low frequency (8.5 dB), reduced noise bandwidth due to narrower carriers (10 dB), further reduced noise bandwidth and repetition coding gains of the new robust 1 MHz carrier configuration (6 dB). The achievable data rates with IEEE 802.11ah are between 150 kbps and 347 Mbps. Several MAC features have been introduced to reduce power consumption for a client device and support more devices being connected to the same access point. IEEE 802.11 applies LBT in form of CSMA/CA. A larger number of connected devices lead to increased collision probabilities, which can be accentuated with the increased effect of hidden nodes with outdoor deployments [1]. To reduce the collision probability, the *restricted access window* (RAW) has been introduced. It divides the contention period into up to 64 RAW slots. Devices are allocated to particular RAW slots; and the number of devices, which are contending simultaneously for channel access, can be reduced to those devices being allocated to the same RAW slot. Device battery consumption can be significantly reduced, by enabling communication in uplink and downlink direction in new *bidirectional transmission opportunities*, where reverse link traffic can follow closely on forward link traffic. This enables long sleep cycles for devices. In addition, with a new *target wake time* the device and an access point can agree on certain fixed time periods, when data that the access point receives for a device shall be forwarded to the device. This reduces the amount of activity of a device to be able to receive data. Furthermore, the maximum idle period for a device has been extended in IEEE 802.11ah so that devices can be configured with sleep periods of up to around five years, and such devices only need to connect once every maximum idle period to the access point to avoid being automatically disassociated from the access point. IEEE 802.11ah also introduces new frame formats, which reduce the overhead of control information added in messages. This is significant for IoT traffic because the data payloads are often very small (e.g., a few bytes for a meter reading) and control info can quickly introduce significant overhead. For data transmission a short MAC frame format is added, and for control messages a *null data packet* has been introduced.

A more extensive description and evaluation of Wi-Fi IEEE 802.11ah can be found in References [39−43]

9.1.2.1.4 Capillary Networks

Short-range radio technologies provide the ability to build out connectivity efficiently to devices within a specific local area. Typically, these local—or capillary—networks need to be connected to the edge of a wide area communication infrastructure so that they have the ability, for example, to reach service functions that are hosted somewhere on the Internet or in a service cloud.

A capillary network needs a backhaul connection, which can be well provided by a cellular network. Their ubiquitous coverage allows backhaul connectivity to be provided practically anywhere, simply and, more significantly, without additional installation of network equipment. Furthermore, a capillary network might be on the move, as is the case for monitoring goods in transit, and therefore cellular networks are a natural solution. To connect a capillary network through a cellular network, a gateway is used between the cellular network and the capillary network, which acts just like any other cellular device towards the cellular network.

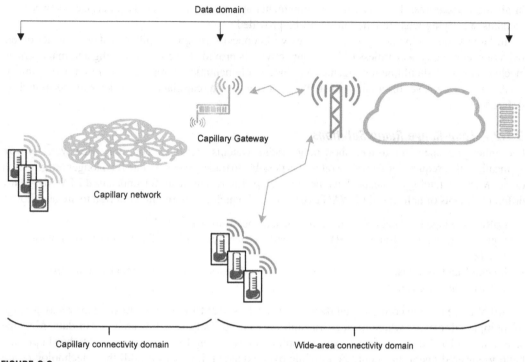

FIGURE 9.6

Capillary networks.

Figure 9.6 illustrates an architecture, which comprises three domains: the capillary connectivity domain, the wide-area connectivity domain, and the data domain. The capillary connectivity domain spans the nodes that provide connectivity in the capillary network, and the wide-area connectivity domain spans the nodes of the cellular network. The data domain spans the nodes that provide data processing functionality for a desired service. These nodes are primarily the connected devices themselves as they generate and use service data through an intermediate node, such as a capillary gateway. The capillary gateway would also be included in the data domain if it provides data processing functionality (for example, if it acts as a CoAP mirror server).

All three domains are separate from a security perspective, and end-to-end security can be provided by linking security relationships in the different domains to one another.

The ownership roles and business scenarios for each domain may differ from case to case. For example, to monitor the in-building sensors of a real estate company, a cellular operator might operate a wide-area network and own and manage the capillary network that provides connectivity to the sensors. The same operator may also own and manage the services provided by the data domain and, if so, would be in control of all three domains.

Alternatively, the real estate company might own the capillary network, and partner with an operator for connectivity and provision of the data domain. Or the real estate company might own and manage both the capillary network and the data domain with the operator providing connectivity only.

In all these scenarios, different service agreements are needed to cover the interfaces between the domains specifying what functionality will be provided.

In large-scale deployments, some devices will connect through a capillary gateway, while others will connect directly. Regardless of how connectivity is provided, the bootstrapping and management mechanisms used should be homogenic to reduce implementation complexity and improve usability.

A more extensive discussion of IoT connectivity via capillary networks can be found in Reference [44].

9.1.2.2 Long-Range Radio Solutions

For unlicensed spectrum usage, short-range radio systems are most common. However, for IoT applications that require very low data rates, it is possible to trade lower data rate for a longer transmission range. Many technology concepts have been developed in recent years for unlicensed LPWAN. Many different variants of unlicensed LPWAN exist; some of which are more often referred to are as follows:

- LoRa: developed for deployment in unlicensed spectrum below 1 GHz,
- Sigfox Ultra-Narrow Band (UNB): developed for deployment in unlicensed spectrum below 1 GHz,
- Ingenu Random Phase Multiple Access (RPMA): developed for deployment in unlicensed spectrum at 2.4 GHz.

All of those have in common that they target wireless M2M/IoT communication over a long range of multiple kilometers, where devices transmit only infrequently very low amounts of data. Message sizes are small and there is often a focus on uplink transmission. Devices are desired to be simple and battery-powered operation should be possible over extended time periods. All these technologies are proprietary and not standardized in standards developing organizations.

In the following we provide a briefly overview of the LoRa, Sigfox, and RPMA technologies.

9.1.2.2.1 LoRa

LoRa is a network technology designed to provide long-range connectivity to battery operated devices; it is specified within an industry alliance. The LoRa Alliance claims to provide a Maximum Coupling Loss (MCL) of 155 dB in the European 867–869 MHz band, and 154 dB in the United States 902–928 MHz band [45]. LoRa has the target to provide secure bidirectional communication. LoRa operates in the sub-GHz unlicensed frequency bands. The physical layer is based on Chirp spread-spectrum modulation technology, and it can use one or more channels. The channel bandwidth is mainly 125 kHz for European spectrum bands, and 125 or 500 kHz for US spectrum bands. Different data rates are supported and are reported to lie in the range of 300 bps–50 kbps. The selection of data rate is a trade-off between transmission duration, i.e., the time during which the message is transmitted over the air and range. LoRa does not deploy LBT but instead uses the duty cycle restrictions required by regulation and the maximum dwell time (i.e., the maximum time that a device may continuously occupy a channel). The system architecture comprises LoRa end devices, LoRa gateways, and a network server. LoRa gateways correspond to base stations in a cellular network. Communication is between the end device and the network server. The communication between the network server and the gateway is based on IP communication; the communication between the end device and a gateway is based on LoRa specific protocols without IP. IP communication is terminated at the LoRa gateway. When a device transmits in uplink, the message can be received by one or more gateways.

For bidirectional communication a downlink transmission opportunity is provided after an uplink transmission. If downlink data arrives for a device in between uplink transmission, the data needs to be buffered in the network and can only be transmitted during the devices downlink receive window, which follows on an uplink transmission by the device.

For more information on LoRa see Reference [46].

9.1.2.2.2 Sigfox

Sigfox is a proprietary technology and of which no specifications are publicly available. Some indicative properties of the so-called UNB communication scheme of Sigfox is provided in Third Generation Partnership Project (3GPP) contribution [47] and technical reports of an industry specification group in ETSI [48,49].

UNB is targeted for operation in sub-GHz unlicensed spectrum bands. The channel bandwidths are 600 Hz in the United States and 100 Hz elsewhere [49]. Data rates are limited in the order of some few hundreds of bits per second. The maximum payload size for uplink data is 12 bytes. UNB does not use LBT but applies duty cycle limitations per transmitter. The channel access scheme is based on ALOHA, which starts to deteriorate at higher loads when the channel utilization exceeds around 15% [47].

The Sigfox network architecture comprises devices, which communicate with Sigfox servers. The radio communication is between the devices and Sigfox access points or base stations. Devices can transmit at any time without prior synchronization to the network. Typically, messages are transmitted on three different uplink channels, which are randomly selected. The base stations observe the entire system bandwidth to detect and decode uplink data. Messages can be received by different base stations, which provide selection diversity.

Downlink transmission is "piggybacked" onto uplink transmissions. After an uplink transmission a device maintains an open receiver window to receive downlink data for a certain time. The server sends buffered data to the device after receiving an uplink message. If the server has received uplink data via multiple base stations, it selects one of the base stations for downlink transmission.

9.1.2.2.3 RPMA

RPMA is a proprietary LPWAN technology provided by the company Ingenu. In contrast to Sigfox and LoRa it operates in the unlicensed 2.4 GHz band, which is globally available. No technical specification of the technology is publicly available. RPMA uses DSSS modulation and applies a pseudo-random time of arrival that helps separating users that are multiplexed on the same radio resource [50]. Transmission consists of two slots, a downlink slot and an uplink slot. Variable packet sizes can be transmitted within a slot. RPMA uses closed loop power control and a large range of spreading factors can be selected. The transmission is adapted to the estimated link performance. It is claimed that RPMA supports an extreme maximum path loss of around 170 dB.

9.2 BENEFITS OF CIoT

In the previous section we provided an overview of interesting unlicensed wireless connectivity solutions for IoT. In this section we discuss how CIoT solutions differ from unlicensed connectivity solutions and what benefits they can provide. For a further discussion on IoT connectivity options, see also Reference [51].

One of the differentiators of CIoT connectivity is that it decouples connectivity provisioning from the IoT service realization. CIoT is built on the high-level paradigm that an independent operator provides suitable IoT connectivity essentially everywhere where an IoT service shall be realized. This means that when a new IoT services is established, no dedicated effort needs to be put in installing, managing, and operating an IoT connectivity solution. Instead the connectivity is realized via the network of an operator. This is different from unlicensed IoT connectivity solutions. In this case, an installation of a connectivity infrastructure is needed to provide connectivity at the location where the IoT service is to be realized. This comprises installing of base stations or access points, establishing backhaul connectivity, providing *authentication, authorization, and accounting* (AAA) infrastructure, maintaining and updating the connectivity network with security updates, etc. Furthermore, the connectivity needs to be monitored and managed throughout the lifetime of the IoT service. There is the potential that the total cost of ownership for providing and managing a connectivity infrastructure for a wide range of IoT services is lower than the cost of ownership of separate connectivity solutions per IoT service. This is in particular the case when the IoT service and the participating devices are spread over larger areas and are not confined to limited deployments. From the unlicensed technologies, Sigfox provides an operator model for Sigfox-based end-to-end connectivity, where operators build up a dedicated Sigfox infrastructure and connectivity can be purchased by end users.

One major benefit of CIoT solutions is that they provide a reliable long-term and future proof solution. CIoT is based on global standards with very large industry support by a large number of vendors, network, and service providers. The technology outlook is independent from the outlook of few individual market players; this is in contrast to proprietary technologies, which come with high risk concerning their long-term support. CIoT solutions are embedded into cellular communication networks, which are, and will be, an essential infrastructure for a society. Deployment plans are made over decades and systems are built to be highly reliable according to standards with high availability. CIoT systems are built for a global market and allow roaming over multiple operator networks. CIoT networks have full support for mobility of devices, which can also be handled over larger areas because of the wide area coverage and high availability. The rollout of CIoT capabilities, as well as future updates, takes mainly place as software updates to the installed network infrastructure.

One extremely important benefit of CIoT connectivity is that it provides reliable and predictable service performance also for future operation. CIoT uses dedicated spectrum. Radio resources are managed, interference is coordinated, and full quality of service is supported. Long-term guarantees are challenging to provide for any solution based on unlicensed spectrum. Both mobile broadband services and IoT services are predicted to continue to grow. In particular for IoT devices, an extremely strong growth is predicted to hundreds of billions of communicating devices within a decade. Many of those mobile broadband and IoT services will be using unlicensed spectrum, which means that a significant increase in utilization of unlicensed spectrum can be expected. This will in particular provide challenges to long-range unlicensed technologies as described in Section 9.1.1.2.

CIoT also follows the continuous evolution of cellular network technologies, where new capabilities and features are continuously added to the networks. This evolution is designed for backward compatible operation so that devices that cannot be upgraded to new functionality can continue to operate long-term according to the original capabilities, while new services and devices can simultaneously benefit from newer features.

A disadvantage of CIoT is the cost of licensed spectrum resources. This is a cost, which solutions for unlicensed bands do not need to bear. Another potential disadvantage of CIoT for an IoT service provider can be if the CIoT coverage is insufficient for a specific IoT use case. In this case, extra connectivity and corresponding network build-out may be needed to cover the entire IoT service area. If the additional coverage build-out is needed in few confined areas, such a build-out may be simpler and more flexibly arranged with a dedicated deployment rather than when an operator needs to be involved. One property that unlicensed long-range radio technologies have benefitted from is their fast time to market. Any proprietary technology has a timing benefit over standardized solutions that require harmonization and agreements within an entire industry segment. In case of unlicensed LPWAN versus CIoT, a benefit in time to market of unlicensed LPWAN has existed during the last few years, while the CIoT standards were being developed. As the CIoT standards are now finalized and products become widely available, this benefit of unlicensed LPWAN is disappearing. Instead the benefit shifts toward CIoT deployments, which can reach wide coverage quicker and at lower cost due to the reuse of the installed cellular communication network infrastructure.

9.3 CHOICE OF CIoT TECHNOLOGY
9.3.1 COMPARISON OF CIoT TECHNOLOGIES

The different CIoT technologies EC-GSM-IoT, NB-IoT, and LTE-M have been extensively analyzed in Chapters 3−8. Here we summarize and compare the performance and characteristics. For NB-IoT we consider in this summary only in-band and stand-alone deployment options for simplicity. The performance of guardband mode of operation is to a large extent similar to the in-band performance. The full NB-IoT performance analysis including guardband operation can be found in Chapter 8.

9.3.1.1 Coverage and Data Rate
The data rate in uplink and downlink for all CIoT technologies are summarized in Figures 9.7 and 9.8 for different coupling losses. All of those technologies have introduced extended coverage features,

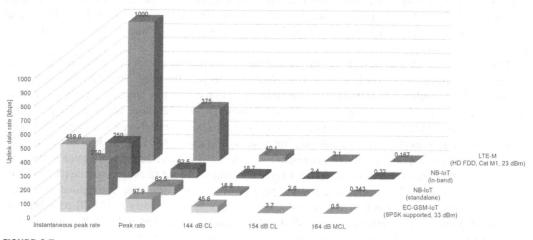

FIGURE 9.7

Coverage and physical layer data rate for uplink.

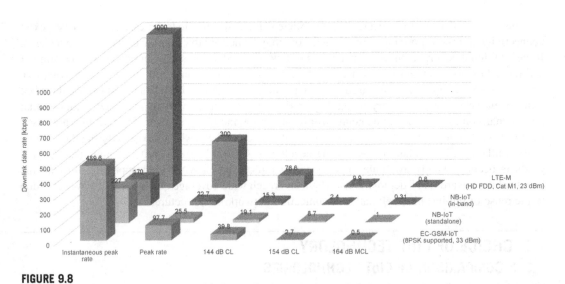

FIGURE 9.8

Coverage and physical layer data rate for downlink.

which enable an operation at a coupling loss of up to 164 dB. This is a significant extension of coverage range compared to what can be found in Global System for Mobile Communications (GSM), UMTS, or Long-Term Evolution (LTE) networks today. For EC-GSM-IoT the 164 dB coupling loss is based on a device with an output power of 33 dBm, as it is common in GSM networks. However, this means that a 10 dB higher device output power is needed for full extended range in EC-GSM-IoT compared to the device output power for NB-IoT and LTE-M for achieving the same uplink coverage. When looking more into details of the extended coverage results in Chapters 4, 6, and 8, it is seen that NB-IoT can operate at a lower control channel block error rate than EC-GSM-IoT and LTE-M at 164 dB MCL, making it more robust at extreme coverage. It can be noted that LTE-M and EC-GSM-IoT can apply frequency hopping, which provides some additional coverage robustness due to added frequency diversity.

Figures 9.7 and 9.8 also provide the *physical layer data rates* values for the different CIoT technologies. The *instantaneous peak physical layer data rate* specifies the achievable data rate of the data channels only. The other data rate values in the tables refer to the effective physical layer data rates for the transmission of a single message, where also the latencies for scheduling and control signaling is taken into account in the transmission time of the message. In this comparison it is assumed that half-duplex operation is used for all technologies but it should be noted that LTE-M devices can also be implemented with support for full-duplex operation which will achieve higher data rates (with peak rates close to the instantaneous peak physical layer data rates). These rates are provided for devices with different coupling loss to the base station: *peak physical layer data rate* corresponds to device with an ideal error free connection to a base station. Physical layer data rates at 144 dB coupling loss corresponds to the normal cell edge of the GSM or LTE radio cell, and 154 and 164 dB correspond to 10 and 20 dB of coverage extension compared to the cell edge of GSM.

What can be seen is that LTE-M can achieve significantly higher data rates in uplink and downlink compared to NB-IoT or EC-GSM-IoT. This is, in particular, the case for devices, which are within normal coverage of the radio cell. When devices are located in extended coverage areas, the uplink is limited by device output power, and all CIoT technologies make us of repetitions to achieve the required link quality. In extreme coverage situations like at 164 dB coupling loss, the achievable data rates for different technologies become quite similar when using the same output power. EC-GSM-IoT has at the 164 dB MCL a higher data rate than the other technologies due to the 10 dB higher output power of the device. Within the same LTE carrier, LTE-M has in general higher data rates than in-band NB-IoT.

All three technologies fulfill the 3GPP requirement on achieving 160 bps at the MCL of 164 dB.

9.3.1.2 Latency

The latency of the CIoT technologies has been evaluated with respect to an *exception report*, which is an infrequent high-important IoT message contained in an 85 byte IP packet, which is being transmitted from a device over the CIoT network. All technologies, LTE-M, NB-IoT, and EC-GSM-IoT, fulfill the 3GPP latency target of 10 s first defined in Release 13, as depicted in Figure 9.9. When a device is within normal coverage, LTE-M can achieve somewhat lower latencies due to the higher data rates provided by LTE-M. In extended coverage, EC-GSM-IoT can provide the lowest latency due to the higher device output power, which can provide higher data rates. Stand-alone NB-IoT has a lower latency compared with in-band NB-IoT due to the higher power used for downlink channels.

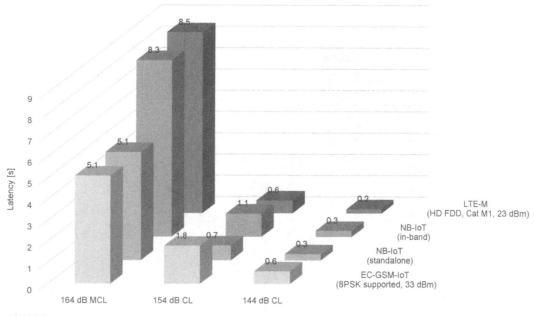

FIGURE 9.9

Latency for exception report.

9.3.1.3 Battery Lifetime

The battery lifetime has been analyzed for all CIoT technologies, assuming two AA batteries with a joint capacity of 5 Wh. A power amplifier efficiency of 45%—50% has been assumed for all three IoT technologies.

Overall, all CIoT technologies apply mechanisms to save battery lifetime for infrequent transmission of messages, as it is common for many IoT services. The main principles are that devices only become active for the transfer of data, and otherwise are put into a battery-saving sleep state. Efficient procedures have been defined, which minimize the signaling overhead associated with the data transfer. This is particularly important for small messages because any signaling overhead can then account for a significant part of the energy consumption.

For a daily report of a 200 byte message, the battery lifetimes for the different CIoT technologies is depicted in Figure 9.10. The results for different message sizes and periodicities of IoT data transfers are summarized in Table 9.5. Overall, all technologies enable battery lifetimes of 10 years, and for some cases even significantly longer. The biggest challenge for long battery lifetime is when a device is located in a very bad coverage position. In the extended coverage mode, very low data rates are used and many repetitions are applied for the data transfer. In this situation a device requires an extended effort for data transmissions, which reduces the opportunity for resting in a battery-saving sleep state. Accordingly, the battery lifetime is significantly reduced at the MCL of 164 dB for all CIoT technologies. With such a large coupling loss, a battery lifetime of 10 years can only be achieved, if data transfer events of a device occur rarely, like once per day. For more frequent data transfer events, like one message every 2 h, battery lifetimes of 1—3 years are achievable at a MCL of 164 dB.

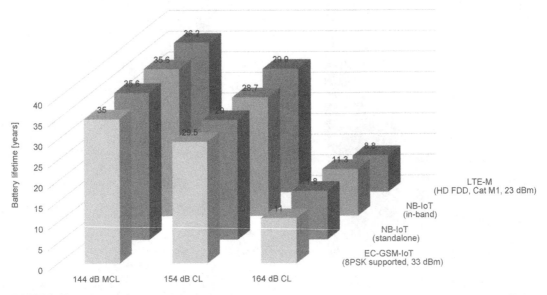

FIGURE 9.10

Battery lifetime for a device with a daily report of a 200 byte message.

Table 9.5 Battery lifetime

	Reporting Interval	DL Packet Size	UL Packet Size	Battery Life (Years)					
				144 dB CL		**154 dB** CL		**164 dB** MCL	
LTE-M HD-FDD CAT M1 23 dBm	2 h	65 bytes	50 bytes	23.7		13.9		2	
			200 bytes	22.3		8.7		0.9	
	24 h		50 bytes	36.5		33.4		15.5	
			200 bytes	36.2		29.9		8.8	
	Stand-alone (S), In-band (I)			S	I	S	I	S	I
NB-IoT 23 dBm	2 h	65 bytes	50 bytes	22.2	22.1	13	12.3	3.0	2.6
			200 bytes	20	20	7.9	7.7	1.4	1.3
	24 h		50 bytes	36.2	36.1	33	32.6	19.3	18.0
			200 bytes	35.6	35.6	29	28.7	11.8	11.3
EC-GSM-IoT 8PSK supported 33 dBm	2 h	65 bytes	50 bytes	22.6		13.7		2.8	
			200 bytes	18.4		8.5		1.2	
	24 h		50 bytes	36.0		33.2		18.8	
			200 bytes	35.0		29.5		11.0	

All three technologies fulfill, or indicate a potential of fulfilling, the 3GPP requirement on achieving 10 years battery life at the MCL of 164 dB.

9.3.1.4 Device Complexity

All CIoT technologies have introduced similar features to reduce device complexity, and thereby enabling low-cost CIoT devices. The following design objectives have been pursued for low device complexity:

- The frequency bandwidth used by the device for transmitting and receiving has been limited to avoid the high costs of wide-band front ends. For LTE-M the bandwidth that needs to be supported by a device is 1.4 MHz, which is significantly less than the maximum LTE channel bandwidth of 20 MHz. For NB-IoT and EC-GSM-IoT, the bandwidth that needs to be supported by the device is 200 kHz.
- The peak data rate has been limited to reduce processing and memory requirements for a device. For LTE-M the peak rate has been reduced to 1 Mbps; for NB-IoT the peak rate has been limited to below 300 kbps and for EC-GSM-IoT it is below 500 kbps.
- All CIoT technologies are specified so that CIoT device are not required to use more than one antenna.

- All CIoT technologies have been specified to support half-duplex operation in frequency-division duplex (FDD) bands. This avoids the needs for a device to integrate one or more costly duplex filters. LTE-M devices can be implemented with support for half-duplex frequency-division duplex (HD-FDD), full-duplex frequency-division duplex (FD-FDD) or time-division duplex (TDD) operation.
- All CIoT technologies have defined User Equipment (UE) categories with lower power classes. This enables a device to use cheaper power amplifiers. It can become an option to implement the power amplifier on the modem chip, and thereby avoiding the costs of a separate component. For EC-GSM-IoT a new 23 dBm device power class has been introduced, in addition to the 33 dBm device power class that is typically used for GSM. For LTE-M two device power classes are defined with 23 and 20 dBm output power. NB-IoT supports 23, 20, and 14 dBm.

The features above enable to reduce the device costs for CIoT devices. However, it must be noted, that the device cost is not entirely depending on the communication standard. The cost of the device depends also on what peripherals are added to the device, such as power supply, CPU, or the real-time clock.

In the end, the costs of the device depend on the market success and the market volume of the devices. A large economy of scale will help to reduce the production costs.

In summary, all CIoT options have introduced low device cost features, and for all technologies low complexity and low cost devices can be expected to appear on the market.

9.3.1.5 CIoT Capacity

The capacity of EC-GSM, LTE-M, and NB-IoT has been analyzed in Sections 4.6, 6.6, and 8.6, respectively. The traffic model for the capacity analysis is based on autonomous device reports sent by devices; the traffic assumptions are described in detail in Section 4.6.1.1 and on average a device is transmitting an autonomous report every ~ 128.5 min. The initial capacity requirement for CIoT in Release 13 has been to be able to serve 60,680 devices/km^2, which corresponds to 40 devices per household in a city like London. Assuming an intersite distance of 1732 m, the size of a radio cell is 0.87 km^2, and the number of devices that need to be supported per radio cell becomes 52,547. With the above traffic model this corresponds to ~ 6.8 message arrivals per second per cell. This traffic load can be provided by EC-GSM-IoT at an uplink radio resource utilization of 27%, where the percentage of failed access attempts is below 0.1%. A similar analysis has been performed for LTE-M and NB-IoT. This can be seen in Figure 9.11; for a detailed analysis of the performance depicted see Sections 4.6.2, 6.6, and 8.6.2. The arrival rate of messages has been increased up to the level where the failed access attempts remained below 1%; the results are listed in Table 9.6. For LTE-M, the system bandwidth spans a number of nonoverlapping LTE-M narrowbands of size 1.08 MHz (6 PRBs). The smallest

Table 9.6 LTE-M and NB-IoT per carrier capacity		
Case	Connection Density at 1% Outage (devices/km^2)	Arrival Rate at 1% Outage (connections/s)
NB-IoT anchor	67,000	7.5
NB-IoT nonanchor	110,000	12.3
LTE-M narrowband	361,000	40.3

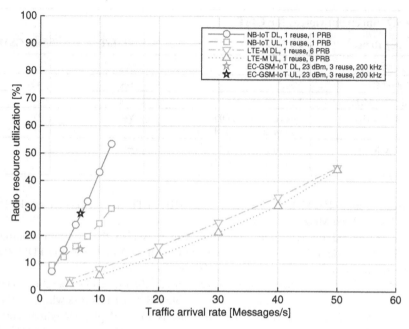

FIGURE 9.11

Capacity of NB-IoT anchor, LTE-M, and EC-GSM-IoT.

available LTE system bandwidth (1.4 MHz) contains a single LTE-M narrowband. An LTE-M narrowband can support up to 40.3 message arrivals per second at an outage probability of 1%, which corresponds to a 361,000 devices/km^2 or 314,070 devices/cell. For NB-IoT the 1% outage limit supports up to 7.5 arrivals per second per cell on an anchor carrier and 12.3 arrivals per second per cell on a nonanchor carrier. This corresponds to 67,000 devices/km^2 or 58,290 devices/cell as capacity limit for an NB-IoT anchor carrier; on a nonanchor carrier 110,000 devices/km^2 or 95,700 devices/cell can be supported (Table 9.6). The radio resource utilization depending on the traffic load for all CIoT technologies is shown in Figure 9.11.

For 5G, a capacity requirement has been defined to be able to serve 1,000,000 devices/km^2, as summarized in Table 10.1 in Section 10.1. With LTE-M this capacity can be provided with three LTE-M narrowbands, i.e., deployed within a 5 MHz LTE carrier. For NB-IoT one anchor carrier and nine nonanchor carriers are needed to be able to serve 1,000,000 devices/km^2; this corresponds to 10 × 180 kHz of spectrum.

9.3.1.6 CIoT Deployments

CIoT standards have been specified for deployments in different spectrum ranges as shown in Table 9.7. All CIoT technologies, LTE-M, NB-IoT, and EC-GSM-IoT, can be deployed in the cellular bands just below 1 GHz and those below and around 2 GHz. In addition, LTE-M and NB-IoT

Table 9.7 Spectrum ranges for CIoT

Frequency Range	LTE-M	NB-IoT	EC-GSM-IoT
Around 450 MHz	Yes	Yes	–
700–1000 MHz	Yes	Yes	Yes
Around 1500 MHz	Yes	Yes	–
1700–2100 MHz	Yes (incl. one TDD option at ∼1900 MHz)	Yes	Yes
2500–2700 MHz	Yes (incl. one TDD option at ∼2500 MHz)	–	–

can be deployed around 450 MHz and around 1500 MHz. LTE-M can further be deployed in the range around 2500–2700 MHz.

NB-IoT is configured for HD-FDD operation, which is also the common configuration for EC-GSM-IoT. LTE-M devices can be implemented with HD-FDD or FD-FDD support. For LTE-M, TDD bands around 1.9 GHz and 2.5 GHz are also specified.

All CIoT standards can in principle be deployed for stand-alone CIoT operation. The minimum spectrum required for such a deployment is listed in Table 9.8. For LTE-M, a whole LTE carrier of at least the minimum system bandwidth of 1.4 MHz needs to be deployed, which corresponds to 2 × 1.4 MHz for FDD (because of paired spectrum) or one 1.4 MHz carrier for TDD. EC-GSM-IoT could be deployed in 2 × 600 kHz of FDD spectrum, assuming three GSM carriers being used with a reuse factor of 3, see Section 4.8 for performance evaluation of this scenario. For NB-IoT a minimum of 2 × 200 kHz are needed for a stand-alone FDD deployment. However, it is most likely that CIoT systems are embedded in a mobile broadband system, which means that an LTE or GSM carrier is used for both mobile broadband and IoT traffic, and the radio resources are dynamically shared between the two types of traffic. Still for NB-IoT and EC-GSM-IoT a stand-alone operation is imaginable, e.g., to enable a migration of GSM spectrum resources to a future cellular technology, such as LTE or 5G, while maintaining a minimum allocation to continue serving existing IoT customers. For NB-IoT it can also be considered that it can be deployed in any spectrum that remains available for an operator, for example, when an allocated band cannot be fully exploited with the carrier bandwidths that are defined for LTE. Then an NB-IoT carrier can be configured adjacent to the LTE carrier within the band of an operator. The full list of bands that are specified for CIoT for 3GPP Releases 13 and 14 are defined in References [52–54], and are listed in Tables 9.9 and 9.10.

Table 9.8 Minimum spectrum allocation for deploying a CIoT network

	Stand-Alone Deployment	Embedded in a Mobile Broadband Network
EC-GSM-IoT	FDD only: 2 × 600 kHz	Usage of a very small fraction of spectrum of the mobile broadband network
NB-IoT	FDD only: 2 × 200 kHz	
LTE-M	FDD: 2 × 1.4 MHz	
	TDD: 1 × 1.4 MHz	

Table 9.9 Spectrum bands defined for LTE-M and NB-IoT in Release 13 and 14

E-UTRA Operating Band	Uplink Spectrum (MHz)	Downlink Spectrum (MHz)	LTE-M	NB-IoT
31	452.5–457.5	462.5–467.5	HD- or FD-FDD Release 13	HD-FDD Release 14
12	699–716	729–746	HD- or FD-FDD Release 13	HD-FDD Release 13
28	703–748	758–803	HD- or FD-FDD Release 13	HD-FDD Release 13
17	704–716	734–746	–	HD-FDD Release 13
13	777–787	746–756	HD- or FD-FDD Release 13	HD-FDD Release 13
27	807–824	852–869	HD- or FD-FDD Release 13	–
26	814–849	859–894	HD- or FD-FDD Release 13	HD-FDD Release 13
18	815–830	860–875	HD- or FD-FDD Release 13	HD-FDD Release 13
5	824–849	869–894	HD- or FD-FDD Release 13	HD-FDD Release 13
19	830–845	875–890	HD- or FD-FDD Release 13	HD-FDD Release 13
20	832–862	791–821	HD- or FD-FDD Release 13	HD-FDD Release 13
8	880–915	925–960	HD- or FD-FDD Release 13	HD-FDD Release 13
11	1427.9–1447.9	1475.9–1495.9	HD- or FD-FDD Release 13	HD-FDD Release 14
21	1447.9–1462.9	1495.9–1510.9	HD- or FD-FDD Release 13	–
3	1710–1785	1805–1880	HD- or FD-FDD Release 13	HD-FDD Release 13
2	1850–1910	1930–1990	HD- or FD-FDD Release 13	HD-FDD Release 13
25	1850–1915	1930–1995	–	HD-FDD Release 14
39	1880–1920	1880–1920	TDD	–
70	1695–1710	1995–2020	–	HD-FDD Release 14
4	1710–1755	2110–2155	HD- or FD-FDD Release 13	–
66	1710–1780	2110–2200	–	HD-FDD Release 13
1	1920–1980	2110–2170	HD- or FD-FDD Release 13	HD-FDD Release 13
41	2496–2690	2496–2690	TDD	–
7	2500–2570	2620–2690	HD- or FD-FDD Release 13	–

Table 9.10 Spectrum bands defined for EC-GSM-IoT

GSM Band	Uplink Spectrum (MHz)	Downlink Spectrum (MHz)	Duplex Mode
GSM band 850	824–849	869–894	HD-FDD
Extended GSM 900 band	880–915	925–960	HD-FDD
digital cellular system (DCS) 1800 band	1710–1785	1805–1880	HD-FDD
personal communications service (PCS) 1900 band	1850–1910	1930–1990	HD-FDD

9.3.2 WHICH CIoT TECHNOLOGY TO SELECT

The choice of a CIoT solution is a decision that needs to be taken by different market players. On one hand, it is the mobile network operator that has to decide which CIoT technology to add to its existing network. On the other hand, it is the IoT device manufacturer and service provider, which select for which IoT connectivity options they develop their IoT service. It can be expected that different options of solutions will coexist.

9.3.2.1 The Mobile Network Operator's Perspective

For a mobile network operator, the decision about which CIoT technology to deploy and operate has multiple facets. There are in particular two sides that need to be considered:

- Long-term mobile network strategy and existing assets,
- IoT market segment strategy.

A typical mobile network operator has one or more cellular networks deployed. Increasingly, different radio technologies are provided via a single multiradio technology network. For example, the same base station can be used for GSM, UMTS/HSPA, or LTE transmission. But there are also deployments where the networks for 2G, 3G, and 4G are rather independent in their deployment and operation.

In addition, a mobile network operator has a spectrum license from typically a national regulator, which gives rights to operate a network in the assigned spectrum. Spectrum licenses are typical long-lasting, like e.g., 20 years; this is motivated by providing network operators with an economic safety. A return on investment for an extremely high network installation cost of a new technology can be planned over a long period. On expiry of a spectrum license, a spectrum licensing contest, like a spectrum auction, is initiated by the regulator for providing a new spectrum license. In general, any network build-out roadmap by an operator is a long-term decision and needs to consider at least the following elements:

- How long are the existing spectrum licenses valid, what technologies are allowed to be operated in the spectrum, and when is a new spectrum reallocation process planned by the regulator?
- What is the status of the network build-out for different radio technologies, in particular GSM and LTE, and what is the operator's market share?
- What is the network build-out of competing operators and what is their market share?

- What is the strategic intent of an operator concerning IoT?
- What services are planned to be provided, and on what roles does the operator intend to address the IoT market (e.g., as connectivity provider or also as service provider/enabler)?
• What is the market maturity for IoT services?
• What IoT segment would the operator like to address?

It shall be noted that the above questions are raised from a perspective of the operation of an operator network in a specific country. However, several operators are active in multiple countries and even on multiple continents. Even if the decision is largely made per country, an operator may want to harmonize decisions over multiple regions in which it operates networks.

When looking at the CIoT technology options, the following characteristics can be identified, which will influence an operator's decision.

As a baseline, we assume that there is a very large incentive by an operator to reuse existing mobile network infrastructure for deploying any of the CIoT technologies. Naturally, EC-GSM-IoT can be easily deployed based on a GSM infrastructure and by using GSM spectrum. The GSM network resources and the GSM spectrum would be shared between GSM usage and EC-GSM-IoT usage. LTE-M and NB-IoT can be deployed based on an LTE infrastructure and by using the LTE spectrum; LTE network and spectrum resources would be shared between LTE, LTE-M, and NB-IoT usage. In most network configurations, it can be expected that the deployment of EC-GSM-IoT, LTE-M, and NB-IoT can be realized as a software update to the deployed GSM or LTE networks. This implies that the introduction of the CIoT into the market can be realized by operators rather quickly and at a low total cost of ownership.

For IoT services it is expected that many services expect a long service lifetime of, e.g., a decade. This expectation should be addressed with a CIoT network. As a result, the decision of the CIoT technology is also coupled to the operator's long-term strategy for mobile networks focusing on telephony and mobile broadband services. If an operator intends to transition GSM deployments to, e.g., LTE or 5G in the coming future, an introduction of EC-GSM-IoT seems a questionable choice, as any long-term EC-GSM-IoT users would require to maintain the GSM infrastructure operational for a long time.

GSM is today still the cellular network technology, which covers the largest part of the globe. According to Reference [55] in 2016 around 90% of the world population was within coverage of a GSM network, and this is expected to increase to ~95% by 2022. In contrast, in 2016 only around 40% of the population were within the coverage of an LTE network, which is expected to increase to more than 80% by 2022. However, there are very large regional variations in the spread of cellular technologies. To give some example from Reference [55], in the Middle East and in Africa only 5% of mobile subscriptions in 2016 were LTE subscriptions (which also enable to use WCDMA/HSPA and GSM); this number is expected to increase to around 30% by 2022. Approximately 20% of the subscriptions are GSM-only and a majority of subscriptions is for WCDMA/HSPA (which also enables to use GSM but not LTE). In regions like Middle East and Africa, GSM is predicted to continue being an important technology for many years to come, even if in other regions the mid- and long-term role of GSM is less certain. Looking at North America, already in 2016 65% of mobile subscriptions were LTE subscriptions. For 2022 very few non-LTE-capable[1] subscriptions are foreseen. Such differences will result in different preferences of CIoT solutions for different operators in different regions.

[1]As LTE-capable subscription in 2022 we count here the predicted LTE subscriptions as well as 5G subscriptions, which are assumed to be fully compatible for operation in LTE networks.

While the reuse of existing network infrastructure and spectrum is an important aspect for an operator, a specific benefit of NB-IoT shall be pointed out in its spectrum flexibility. It is generally expected that existing GSM/LTE spectrum deployments are extended to also include EC-GSM-IoT/LTE-M traffic, so the IoT traffic will be on the same spectrum that is already deployed for telephony and mobile broadband services. For NB-IoT, the narrow system bandwidth of NB-IoT makes it suitable to be deployed also in spectrum that is not used for mobile broadband services today. Examples exist, where operators have spectrum allocations that do not fit with exact carrier bandwidths provided by LTE. As a result, a remainder of the spectrum allocation remains unused. NB-IoT provides the flexibility to make use of even small portions of idle spectrum resources that an operator may have. Such portions of spectrum resources can be even created by an operator, e.g., by emptying individual GSM carriers from GSM operation and reuse them instead for NB-IoT usage.

A further aspect of spectrum being available to an operator can influence the choice of CIoT technology. EC-GSM-IoT and NB-IoT have been specified for use in FDD bands. LTE-M, in contrast, can be used for both TDD and FDD bands and provides more opportunities to operators with significant TDD spectrum allocations.

Finally, an operator may align a CIoT plan with its future deployment plans of a 5G *New Radio* (NR) technology. Many early 5G NR deployments are planned for higher frequency bands; but some operators may also consider an NR deployment in low frequency bands that may be used for CIoT. In the ongoing 5G standardization activities within 3GPP, it has already been identified, that a backward compatible operation of NR with NB-IoT and LTE-M should be envisioned. For example, NB-IoT should be able to be operated in-band within an NR carrier in a similar way as if it is deployed in-band within an LTE carrier. With the same approach, LTE-M operation could be enabled in-band within an NR carrier. Two scenarios should be supported. In the first, NB-IoT is deployed on some carrier (e.g., some reassigned GSM carriers), and the entire GSM band is then at a later time refarmed to NR. The NB-IoT operation for existing IoT services should be able to continue within the NR carrier after the refarming. Similarly, in the second scenario, an LTE carrier should be possible to be refarmed to NR, while continuing the operation of LTE-M services on that carrier. Similar close coexistence of EC-GSM-IoT and NR on a common carrier will not be possible.

The considerations presented so far for choosing a CIoT technology by a mobile network operator have been based on what spectrum and what radio access technology the operator use or plan to use in future. The driving force in this regard is to reuse existing or planned mobile networks to achieve a low capital expenditure and operational expenditure for the deployment and operation of CIoT connectivity. Another major component in an operator assessment of CIoT is the IoT service strategy of the operator. Does the operator target a specific IoT market segment? And if so, what are the service requirements in this segment and what connectivity requirements does it imply? In this case the operator decision is largely based on how well a CIoT technology fulfills the service requirements, as discussed in Section 9.3.1.

9.3.2.2 The IoT Service Provider's Perspective

An IoT service provider targets with its offering a set of particular IoT services. For example, a focus may be on smart city applications or precision agriculture. The targeted IoT service implies a certain location where the service will be realized, i.e., where IoT devices will be located. For smart city services, this will be in urban areas, for precision agriculture this will be primarily in rural areas. The IoT service characteristics determine what kind of traffic profile needs to be supported for realizing the service. For smart city this may be regular monitoring of available parking spaces or notifications when waste containers have reached a certain fill level. For precision agriculture, it can be the monitoring of humidity and fertilization on fields or in green houses or the tracking of cattle. Other IoT service characteristics besides the traffic profile can be the maximum time that a device must operate on a battery.

Based on an analysis of the targeted IoT service, the connectivity requirement of the service becomes clear:

- What data rates need to be supported by the communication?
- Do devices need to run on battery for extended time periods?
- What device density is expected?
- Where are the devices located?
- Are devices in particularly hard to reach locations (e.g., in enclosures underground)?
- Are devices mobile over larger areas, possibly even across national borders?

Based on this review a service provider can determine:

- Which CIoT technologies provide sufficient performance for the targeted service, see Section 9.3.1.
- At what locations network coverage is needed.

It can be expected that coverage of multiple CIoT technologies is provided by one or more network operators at various locations. EC-GSM-IoT will be provided as technology in markets where the LTE network build-out is not so advanced. In other markets with significant LTE deployments, both LTE-M and NB-IoT will be found. An IoT service provider has to select a network operator that provides coverage and connectivity via a suitable CIoT technology at the targeted deployment area at a fair price.

Since IoT devices may be deployed and operated over long time spans, flexibility in reselecting a network provider is desirable. Embedded Subscriber Identity Modules (SIM) that enable to remotely reprovision devices and reselect network providers will play an increasing role for CIoT devices, see Reference [56].

REFERENCES

[1] M. Park, IEEE 802.11ah: sub-1-GHz license-exempt operation for the internet of things, Computer Communications 58 (March 2015) 53–69.
[2] Federal Communications Commission, Title 47 of the Code of Federal Regulations, Part 15 on Radio Frequency Devices, 2017. Available from: https://www.fcc.gov/general/rules-regulations-title-47.
[3] European Conference of Postal and Telecommunications Administrations (CEPT). Available from: http://www.cept.org/cept/.

[4] CEPT Electronic Communications Committee (ECC), ERC Recommendation 70-03, Relating to the Use of Short-Range Devices (SRD), Edition of October 2016, 2016.

[5] European Telecom Standards Institute, EN 300 220-1 V3.1.1, Short Range Devices (SRD) Operating in the Frequency Range 25 MHz to 1 000 MHz; Part 1: Technical Characteristics and Methods of Measurement, February 2017.

[6] European Telecom Standards Institute, EN 300 220-2 V2.4.1 Electromagnetic Compatibility and Radio Spectrum Matters (ERM); Short Range Devices (SRD); Radio Equipment to Be Used in the 25 MHz to 1 000 MHz Frequency Range with Power Levels Ranging up to 500 mW; Part 2: Harmonized EN Covering Essential Requirements under Article 3.2 of the R&TTE Directive, May 2012.

[7] European Telecom Standards Institute, EN 300 440-1 V1.6.1, Electromagnetic Compatibility and Radio Spectrum Matters (ERM); Short Range Devices; Radio Equipment to Be Used in the 1 GHz to 40 GHz Frequency Range; Part 1: Technical Characteristics and Test Methods, August 2010.

[8] European Telecom Standards Institute, EN 300 440 V2.1.1, Short Range Devices (SRD); Radio Equipment to Be Used in the 1 GHz to 40 GHz Frequency Range; Harmonised Standard Covering the Essential Requirements of Article 3.2 of Directive 2014/53/EU, January 2017.

[9] European Telecom Standards Institute, EN 300 328 V2.1.1, Wideband Transmission Systems; Data Transmission Equipment Operating in the 2,4 GHz ISM Band and Using Wide Band Modulation Techniques; Harmonised Standard Covering the Essential Requirements of Article 3.2 of Directive 2014/53/EU, November 2011.

[10] J.A. Gutierrez, M. Naeve, E. Callaway, M. Bourgeois, V. Milter, B. Heile, IEEE 802.15.4: a developing standard for low-power low-cost wireless personal area networks, IEEE Network 15 (5) (September/October 2001) 12–19.

[11] E. Callaway, P. Gorday, L. Hester, J.A. Gutierrez, M. Naeve, B. Heile, V. Bahl, Home networking with IEEE 802.15.4: a developing standard for low-rate wireless personal area networks, IEEE Communications Magazine (August 2002).

[12] J. Zheng, M.J. Lee, Will IEEE 802.15.4 make ubiquitous networking a reality?: a discussion on a potential low power, low bit rate standard, IEEE Communications Magazine (June 2004).

[13] I.F. Akyildiz, W. Su, Y. Sankarasubramaniam, E. Cayirci, Wireless sensor networks: a survey, Computer Networks 38 (2002) 393–422.

[14] I.F. Akyildiz, W. Su, Y. Sankarasubramaniam, E. Cayirci, A survey on sensor networks, IEEE Communications Magazine (August 2002).

[15] A. Willig, Recent and emerging topics in wireless industrial communications: a selection, IEEE Transactions on Industrial Informatics 4 (2) (May 2008) 102–124.

[16] N. Langhammer, R. Kays, Performance evaluation of wireless home automation networks in indoor scenarios, IEEE Transactions on Smart Grid 3 (4) (December 2012).

[17] M.R. Palattella, N. Accettura, X. Vilajosana, T. Watteyne, L.A. Grieco, G. Boggia, M. Dohler, Standardized protocol stack for the internet of (important) things, IEEE Communications Surveys and Tutorials 15 (3) (2013).

[18] J.-S. Lee, Performance evaluation of IEEE 802.15.4 for low-rate wireless personal area networks, IEEE Transactions on Consumer Electronics 52 (3) (August 2006).

[19] Internet Engineering Task Force (IETF), Request for Comments 4944, Transmission of IPv6 Packets over IEEE 802.15.4 Networks, September 2007.

[20] J.W. Hui, D.E. Culler, IPv6 in low-power wireless networks, Proceedings of the IEEE 98 (11) (November 2010).

[21] J.W. Hui, D.E. Culler, Extending IP to low-power, wireless personal area networks, IEEE Internet Computing (July/August 2008).

[22] I. Ishaq, D. Carels, G.K. Teklemariam, J. Hoebeke, F. Van den Abeele, E. De Poorter, I. Moerman, P. Demeester, IETF standardization in the field of the internet of things (IoT): a survey, Journal of Sensor and Actuator Networks 2 (2) (April 2013) 235–287.

[23] Internet Engineering Task Force (IETF), Request for Comments 6550, RPL: IPv6 Routing Protocol for Low-Power and Lossy Networks, March 2012.

[24] K.-H. Chang, B. Mason, The IEEE 802.15.4g standard for smart metering utility networks, in: Proceedings IEEE International Conference on Smart Grid Communications, Tainan City, Taiwan, 5–8 November 2012, 2012.

[25] C.-S. Sum, F. Kojima, H. Harada, Coexistence of homogeneous and heterogeneous systems for IEEE 802.15.4g smart utility networks, in: Proc. IEEE International Symposium on Dynamic Spectrum Access Networks (DySPAN), Aachen, Germany, 3–6 May 2011, 2011.

[26] C.-S. Sum, H. Harada, F. Kojima, L. Lu, An interference management protocol for multiple physical layers in IEEE 802.15.4g smart utility networks, IEEE Communications Magazine 51 (4) (April 2013) 84–91.

[27] J. Haartsen, Bluetooth—the universal radio interface for ad hoc, wireless connectivity, Ericsson Review 75 (3) (1998) 110–117. Available from: http://ericssonhistory.com/Global/Ericsson%20review/Ericsson%20Review.%201998.%20V.75/Ericsson_Review_Vol_75_1998_3.pdf.

[28] J.C. Haartsen, The bluetooth radio system, IEEE Personal Communications 7 (1) (August 2002) 28–36.

[29] Bluetooth Special Interest Group, Bluetooth Core Specification v4.0, June 2010.

[30] L.F. Del Carpio, P. Di Marco, P. Skillermark, R. Chirikov, K. Lagergren, P. Amin, Comparison of 802.11ah and BLE for a home automation use case, in: Proc. IEEE 27th Annual International Symposium on Personal, Indoor, and Mobile Radio Communications (PIMRC), 4–8 September 2016, 2016.

[31] L.F. Del Carpio, P. Di Marco, P. Skillermark, R. Chirikov, K. Lagergren, Comparison of 802.11ah and BLE in a Home Automation Use Case, to appear in Springer International Journal of Wireless Information Networks.

[32] P. Di Marco, R. Chirikov, P. Amin, F. Militano, Coverage analysis of bluetooth low energy and IEEE 802.11ah for office scenario, in: Proc. IEEE Int. Symposium on Personal, Indoor and Mobile Radio Communications, Workshop M2M Communication, August 2015, 2015.

[33] Bluetooth Special Interest Group, Internet Protocol Support Profile (IPSP), December 2014.

[34] Internet Engineering Task Force (IETF), Request for Comments 7668, IPv6 over BLUETOOTH Low Energy, October 2015.

[35] Bluetooth Special Interest Group, Internet Gateways, Bluetooth White Paper, February 2016.

[36] Bluetooth Special Interest Group, Bluetooth Core Specification v5.0, December 2016.

[37] P. Di Marco, P. Skillermark, A. Larmo, P. Arvidson, R. Chirikov, Performance Evaluation of the Data Transfer Modes in Bluetooth 5, to appear in IEEE Communications Standards Magazine special issue on IoT standards.

[38] C. Gomez, S. Darroudi, T. Savolainen, IPv6 Mesh over Bluetooth(R) Low Energy Using IPSP, 2016. Available from: https://tools.ietf.org/pdf/draft-gomez-6lo-blemesh-02.pdf.

[39] W. Sun, O. Lee, Y. Shin, S. Kim, C. Yang, H. Kim, S. Choi, Wi-fi could Be much more, IEEE Communications Magazine (November 2014).

[40] W. Sun, M. Choi, S. Choi, IEEE 802.11ah: a long range 802.11 WLAN at sub 1 GHz, Journal of ICT Standardization 1 (1) (July 2013) 83–108.

[41] T. Adame, A. Bel, B. Bellalta, J. Barcelo, M. Oliver, IEEE 802.11ah: the WiFi approach for M2M communications, IEEE Wireless Communications (December 2014).

[42] E. Khorov, A. Lyakhov, A. Krotov, A. Guschin, A survey on IEEE 802.11ah: an enabling networking technology for smart cities, Computer Communications 58 (March 1, 2015) 53–69.

[43] B. Badihi, L.F. Del Carpio, P. Amin, A. Larmo, M. Lopez, D. Denteneer, Performance Evaluation of IEEE 802.11ah Actuators, in: Proceedings 83rd IEEE Vehicular Technology Conference, Nanjing, China, 15−18 May, 2016.

[44] J. Sachs, N. Beijar, P. Elmdahl, J. Melen, F. Militano, P. Salmela, Capillary networks − a smart way to get things connected, Ericsson Review (September 2014).

[45] LoRA Alliance Technical Marketing Workgroup 1.0, White Paper, LoRaWAN, What Is It?, A Technical Overview of LoRa® and LoRaWAN, November 2015.

[46] LoRa Alliance, LoRaWAN™ Specification, V1.0, January 2015.

[47] Sigfox Wireless, GPC150052 Coperative Ultra Narrow Band Technology for Cellular IoT, 3GPP TSG GERAN1-GERAN2 Ad'hoc Meeting, Sophia Antipolis, France, 2−5 February 2015, 2015.

[48] ETSI Industry Specification Group on Low Throughput Networks (ISG LTN), GS LTN 002 V1.1.1, Low Throughput Networks (LTN); Functional Architecture, September 2014.

[49] ETSI Industry Specification Group on Low Throughput Networks (ISG LTN), GS LTN 003 V1.1.1, Low Throughput Networks (LTN); Protocols and Interfaces, September 2014.

[50] Ingenu, How RPMA Works, January 2016. Available from: https://www.youtube.com/watch?v=4beoZapuBXw.

[51] S. Andreevy, O. Galinina, A. Pyattaev, M. Gerasimenko, T. Tirronen, J. Torsner, J. Sachs, M. Dohler, Y. Koucheryavy, Understanding the IoT connectivity landscape: a contemporary M2M radio technology roadmap, IEEE Communications Magazine 53 (9) (September 2015) 32−40.

[52] Third Generation Partnership Project, Technical Specifications 36.101, v13.6.1, Evolved Universal Terrestrial Radio Access (E-UTRA); User Equipment (UE) Radio Transmission and Reception (Release 13), January 2017.

[53] Third Generation Partnership Project, Technical Specifications 36.101, v14.2.1, Evolved Universal Terrestrial Radio Access (E-UTRA); User Equipment (UE) Radio Transmission and Reception (Release 14), January 2017.

[54] Third Generation Partnership Project, Technical Specifications 45.005, v14.0.0, Technical Specification Group Radio Access Network; GSM/EDGE Radio Transmission and Reception (Release 14), March 2017.

[55] Ericsson, Ericsson Mobility Report, November 2016. Available from: https://www.ericsson.com/assets/local/mobility-report/documents/2016/ericsson-mobility-reportnovember-2016.pdf.

[56] GSMA, Remote SIM Provisioning for Machine to Machine, 2017. Available from: http://www.gsma.com/connectedliving/embedded-sim/.

5G AND THE INTERNET OF THINGS

10

CHAPTER OUTLINE

Abstract

This chapter describes the evolution of mobile networks toward 5G. Beyond an evolution of the mobile broadband experience, 5G will provide optimized connectivity for MTC and IoT services. An overview of the standardization activities toward massive Machine-Type Communications (mMTC), as well as critical Machine-Type Communications (cMTC), with Ultra Reliable and Low Latency Communications (URLLC) is presented.

10.1 5G VISION AND REQUIREMENTS

Mobile networks have been evolving continuously since their introduction around the 1980s. Approximately every 10 years a larger technology shift has been introduced towards a new generation of mobile network, as depicted in Figure 10.1. The first two mobile network generations were originally focused on mobile telephony, first based on analog transmission and later based on digital transmission. The 3G and 4G mobile network generations have introduced and established mobile broadband connectivity to the consumers. The introduction of 5G is anticipated around 2020; 5G broadens the use cases significantly beyond mobile broadband and consumer-focused services.

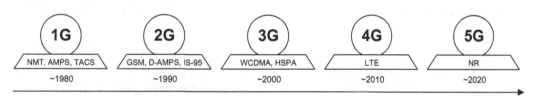

FIGURE 10.1

Mobile network evolution toward 5G.

The International Telecommunications Union (ITU) has defined a framework for the development of 5G in Reference [1]. At ITU, 5G is addressed under the term *International Mobile Telecommunications (IMT) for 2020 and beyond*, or *IMT-2020*. Reference [1] describes the targeted usage scenarios for IMT-2020 and the required capabilities that need to be provided by an IMT-2020 system. This is based on an assessment of traffic growth in mobile networks until today, as well as societal trends and new technology developments. The targeted usage scenarios for IMT-2020 are significantly broader than for earlier mobile network generations. Before 5G, the focus of mobile communications has been on human-centric communications, from telephony to mobile broadband services, providing multimedia and data services to human users. An evolution for enhancing mobile broadband usage is envisioned, in particular to cater for the strong mobile broadband traffic growth, as well as new service types such as 3D video or augmented and virtual reality. This usage scenario is denoted as enhanced mobile broadband (eMBB). In addition, new machine-centric communication services are defined as a specific usage scenario for 5G, also denoted as MTC. Two areas of machine-centric communications are identified, as depicted in Figure 10.2. Massive MTC (mMTC) corresponds to communication of often simple and many sensor devices, which transmit small amounts of data that are not delay sensitive. cMTC in contrast addresses machine-centric communication that has stringent requirements on latency, data rates, reliability, and availability. Example use cases are industrial manufacturing and production, fully automated intelligent transport systems and self-driving vehicles, automation in

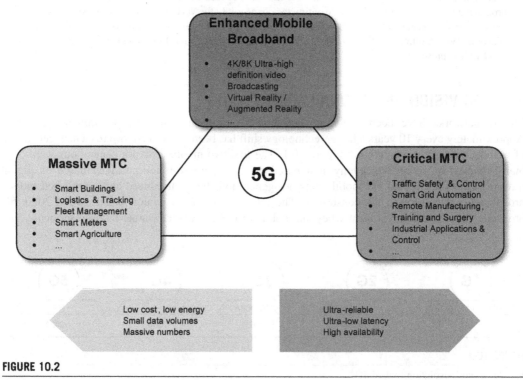

FIGURE 10.2

5G usage scenarios.

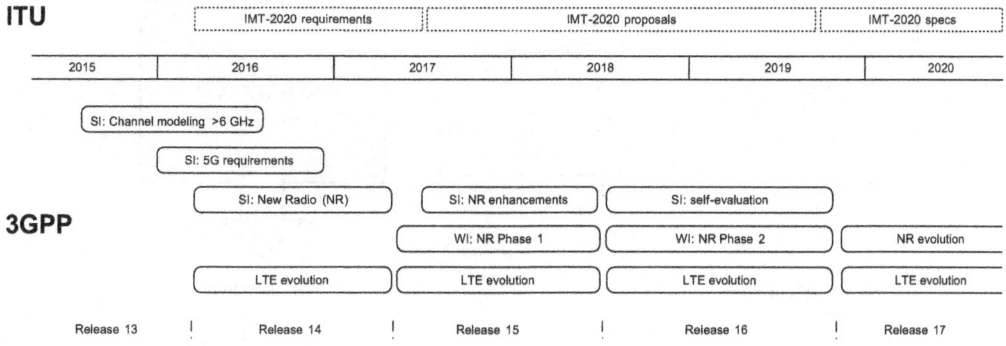

FIGURE 10.3

5G timeline at ITU and 3GPP.

future energy networks, etc. cMTC is also referred to as ultra reliable and low latency communications (URLLC).

For the development of 5G, ITU has set a timeline [2] for 5G development as depicted in Figure 10.3. In a first period until mid-2017, technical requirements on IMT-2020 are specified. This is followed by a period of developing 5G solutions, which are submitted as IMT-2020 proposals to ITU, where they are evaluated according to the IMT-2020 requirements. From the end of 2019, ITU-approved IMT-2020 specifications shall exist that enable a full market deployment of 5G systems. Draft IMT-2020 requirements have been developed by ITU in Reference [3] and are being prepared for approval.

Third Generation Partnership Project (3GPP) has adopted a time plan that matches the ITU time plan (see Figure 10.3). At the end of 2015, 3GPP has started studies to develop channel models for wireless networks operating above 6 GHz and to define 5G requirements. In Release 14 a study item for a new 5G *New Radio* (NR) air interface was concluded and a Release 15 work item has been defined for NR specification. Release 15 will provide a first phase of an NR specification, which will be extended with a second phase in Release 16 that builds on studies on NR enhancements in Release 15. NR Phase 2 specifications will be accompanied with a self-evaluation that will be jointly submitted with the 5G proposal to ITU.

In parallel to the NR technology track, 3GPP will continue the further evolution of LTE toward addressing 5G requirements. Initially, NR is expected to be primarily deployed in new spectrum that will be assigned to 5G, either in the range above 6 GHz or in new bands below 6 GHz (see Figure 10.4). In the longer run, NR will also migrate to carriers currently used by earlier mobile network standards. With a Long-Term Evolution (LTE), 5G capabilities can be introduced into carriers on which LTE is already operational. Accordingly, an LTE evolution toward 5G capabilities needs to be compatible with simultaneous usage of the same carrier for LTE communication with pre-5G devices. Tight interworking with LTE is an important requirement for NR, to allow continuous communication for 5G devices already during early NR deployments, when NR coverage is still limited.

The performance requirements set on 5G for MTC in 3GPP and ITU are listed in Table 10.1. In general, the requirements defined in 3GPP are more stringent than those defined in ITU. It can be seen that the requirements for mMTC correspond largely to the requirements of cellular IoT. The focus is on extended coverage for low data rates, long device battery lifetime, per message latency, and scalability to many devices. In addition, there are requirements on URLLC, where the focus is on very low latency and high reliability [4].

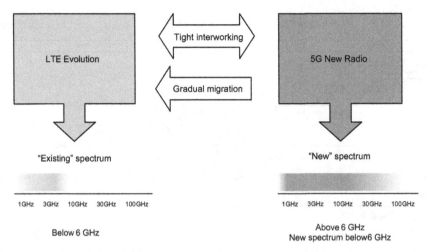

FIGURE 10.4

5G based on LTE and New Radio.

10.2 5G FOR IoT CONNECTIVITY

5G NR is defined as radio technology with a scalable OFDM (Orthogonal Frequency-Division Multiplexing) radio interface [4]. It is designed to be configurable for frequency bands ranging from sub-GHz up to millimeter wave spectrum at up to 100 GHz. Frame structure and radio procedures should be configurable to achieve very low latency. An ultra-lean design is envisioned not only to enable network energy saving but also to provide flexibility to introduce new features at a later release.

10.2.1 URLLC

For URLLC a key feature is the ability to transmit data with high reliability within a guaranteed latency bound. This combination of high reliability at low latency puts high requirements on a 5G system. One difficulty is that the time domain cannot be exploited to achieve high reliability, e.g., by continuous retransmissions. Instead, a radio interface must be designed and configured to provide a sufficient reliability level with either a single transmission or a very limited number of retransmissions.

Several design components enable URLLC services (see, e.g., Reference [6]). Very robust link adaptation is needed to provide high reliability. Furthermore, it is important to exploit diversity as much as possible. This is achieved in the spatial domain, by using multiple antennas, or in the frequency domain. For achieving low latency, appropriate data frame structures must be used. An OFDM system with scalable numerology enables to configure a larger subcarrier spacing so that symbol duration and slot durations decrease. Furthermore, a frame design with early reference symbols and control information is beneficial. With such a configuration, a receiver can start decoding received packets very early after initial estimation of the channel. For increased robustness during mobility, dual connectivity can be used when data is transmitted over multiple frequency layers.

Table 10.1 5G requirements for MTC			
Frequency range	**Primary 5G usage scenario**	**3GPP [5]**	**ITU [3]**
Coverage	mMTC	164 dB maximum coupling loss at a rate of 160 bps	–
Connection density	mMTC	1,000,000 device/km^2	1,000,000 devices/km^2
UE battery life	mMTC	10 years battery lifetime (15 years desirable) for a 5 Wh battery for a device sending daily 200 bytes uplink data followed by 20 bytes downlink data at a maximum coupling loss of 164 dB	–
Latency for infrequent small packets	mMTC	10 s for a 20 byte application packet in uplink at 164 dB maximum coupling loss and starting from the device being in the most "battery efficient" state	–
User plane latency	URLLC (and eMBB)	0.5 ms on average for URLLC	1 ms for URLLC
Reliability	URLLC, including enhanced Vehicle to Everything communication (eV2X)	URLLC: 32 bytes delivered over the radio within 1 ms with a $1-10^{-5}$ success probability	32 byte message delivered over the radio within 1 ms with a $1-10^{-5}$ success probability
		eV2X: 300 bytes delivered via a base station or via the sidelink within $3-10$ ms with a $1-10^{-5}$ success probability	
Control plane latency	URLLC (and eMBB)	10 ms	20 ms (10 ms encouraged)
Mobility interruption time	URLLC (and eMBB)	0 ms	0 ms

10.2.2 mMTC

5G design for mMTC is a continuation of the cellular IoT standards LTE Machine-Type Communications (LTE-M) and Narrowband Internet of Things (NB-IoT). In fact, it has been shown that NB-IoT and LTE-M already largely fulfill the mMTC 5G requirements [7,8]. For this purpose, it was decided in 3GPP not to specify any NR mMTC solution in 3GPP Release 15. Instead, it was decided to use the evolution of LTE-M and NB-IoT as the baseline for addressing the 5G mMTC requirements. Some improvements for LTE-M and NB-IoT have been identified for Release 15, which include an early data transmission in the random access procedure to reduce the latency and increase the scalability of cellular IoT. For LTE-M, resource allocation of less than a full Physical Resource Block has been identified as an important improvement. This increases the uplink capacity of LTE-M, in particular, when many devices are only accessible in extended coverage mode.

To evolve the service offering, it is desirable to migrate NR into spectrum bands used for existing mobile network technologies. MTC devices can have very long service lifecycle times and operate for many years. To ease the migration of bands toward NR, it is desirable that NR can coexist well with both NB-IoT and LTE-M on a common carrier. Ideally, NB-IoT and LTE-M can operate in-band within an NR carrier in a similar way as NB-IoT can be deployed in-band within an LTE carrier.

REFERENCES

[1] Radio Communication Sector of the International Telecommunications Union, ITU-R, Recommendation ITU-R M.2083-0, IMT Vision – Framework and Overall Objectives of the Future Development of IMT for 2020 and Beyond, September 2015. Available from: http://www.itu.int/rec/R-REC-M.2083-0-201509-I/en.

[2] International Telecommunication Union (ITU), ITU towards 'IMT for 2020 and beyond', 2017. Available from: http://www.itu.int/en/ITU-R/study-groups/rsg5/rwp5d/imt-2020/Pages/default.aspx.

[3] Radio Communication Sector of the International Telecommunications Union, ITU-R, Working Party 5D Document 5/40-E, Draft New Report ITU-R M.[IMT-2020.TECH PERF REQ], Minimum Requirements Related to Technical Performance for IMT-2020 Radio Interface(s), February 22, 2017. Available from: https://www.itu.int/md/R15-SG05-C-0040/en.

[4] E. Dahlman, G. Mildh, S. Parkvall, J. Peisa, J. Sachs, Y. Selén, J. Sköld, 5G wireless access: requirements and realization, IEEE Communications Magazine 52 (12) (December 2014) 42–47.

[5] Third Generation Partnership Project, Technical Report 38.913, v14.2.0, Technical Specification Group Radio Access Network; Study on Scenarios and Requirements for Next Generation Access Technologies; (Release 14), March 2017.

[6] J. Sachs, P. Popovski, A. Höglund, D. Gozalvez-Serrano, P. Fertl, Machine-type communications, in: A. Osseiran, J. Monserrat, P. Marsch (Eds.), 5G Mobile and Wireless Communications Technology, Cambridge University Press, 2016.

[7] Ericsson, R1-170511, On eMTC 5G mMTC requirement fulfilment, in: 3GPP TSG RAN#75 Meeting, Dubrovnik, Croatia, 6–9 March, 2017, 2017.

[8] Ericsson, R1-170512, On NB-IoT 5G mMTC requirement fulfilment, in: 3GPP TSG RAN#75 Meeting, Dubrovnik, Croatia, 6–9 March, 2017, 2017.

Index

Printed in the United States
By Bookmasters